ADVANCES IN ENZYMOLOGY
AND RELATED AREAS OF MOLECULAR BIOLOGY

Volume 58

LIST OF CONTRIBUTORS

RONALD BRESLOW, Department of Chemistry, Columbia University, New York, New York 10027.

IRWIN FRIDOVICH, Department of Biochemistry, Duke University Medical Center, Durham, North Carolina 27710.

HARTMUT KÜHN, Institute of Biochemistry, Humboldt University, DDR-1040 Berlin, GDR.

CHUN-YEN LAI, Central Research, Hoffmann-La Roche, Inc., Nutley, New Jersey 07110.

JONATHAN S. NISHIMURA, Department of Biochemistry, University of Texas Health Science Center, San Antonio, Texas 78284.

J. MICHAEL POSTON, Laboratory of Biochemistry, National Heart, Lung, and Blood Institute, National Institutes of Health, Bethesda, Maryland 20205.

SAMUEL M. RAPOPORT, Institute of Biochemistry, Humboldt University, DDR-1040 Berlin, GDR.

TANKRED SCHEWE, Institute of Biochemistry, Humboldt University, DDR-1040 Berlin, GDR.

ADVANCES IN ENZYMOLOGY

AND RELATED AREAS OF MOLECULAR BIOLOGY

Founded by F. F. NORD

Edited by ALTON MEISTER

CORNELL UNIVERSITY MEDICAL COLLEGE
NEW YORK, NEW YORK

VOLUME 58

1986

AN INTERSCIENCE® PUBLICATION

JOHN WILEY & SONS
New York · Chichester · Brisbane · Toronto · Singapore

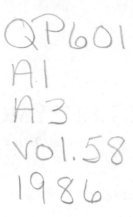

QP601
A1
A3
vol.58
1986

Library of Congress Catalog Card Number: 41-9213
ISBN: 0-471-88013-2

Printed in the United States of America

10 9 8 7 6 5 4 3 2 1

CONTENTS

ADVANCES IN ENZYMOLOGY
AND RELATED AREAS OF MOLECULAR BIOLOGY

Volume 58

ARTIFICIAL ENZYMES AND ENZYME MODELS

By RONALD BRESLOW *Department of Chemistry, Columbia University, New York, New York 10027*

CONTENTS

I. Introduction

There are two general motives behind the study of enzyme models and mimics of enzyme action. One purpose is to furnish information that will help in the understanding of natural biochemical catalysis. We generally "explain" things by showing that they are related to simpler phenomena, which we believe we understand. Thus explanations of enzyme mechanisms almost always refer to simple chemical analogs in which a proposed effect has been demonstrated. Some work on enzyme models has the purpose of supplying such analogs. For instance, most of the current work aimed at understanding how coenzyme B-12 catalyzes biochemical reactions is actually chemical work on analogs, trying to find the sort of chemistry needed to explain these remarkable transformations. The other general purpose of enzyme model and mimic studies is to try to extend the chemistry performed in living organisms and invent new chemical reactions

1

with Nature as our guide and teacher. When chemistry is discovered that can be used to explain the reactions catalyzed by coenzyme B-12, chemists will try to make small artificial catalysts using this same chemistry in order to perform the otherwise unprecedented rearrangements carried out in biochemical systems.

The contrast between these two purposes has to do with the direction of information flow. In the models used to help explain enzymes, chemical information is fed to biochemistry; in the attempt to construct enzyme mimics, information about biochemical processes is used to develop new chemical reactions. This is based on the general perception that synthetic chemistry would be greatly improved if it could achieve some of the high velocity and great selectivity characteristic of biochemical processes. However, the mimicking of enzymes runs the gamut from systems closely designed to imitate the detailed chemistry of a particular enzyme all the way to other systems that are designed only to incorporate some of the general principles (formation of an enzyme–substrate complex, geometric control of the reaction within the resulting complex) that are characteristic of all enzyme systems.

For the past 28 years we have been active in both of these areas. In this chapter we shall review our contributions (1,2) only occasionally referring to the work of others to put our studies in context. Of course such a personal review does not give an objective picture of the field.

II. Models Aimed at Clarifying Enzyme Mechanisms

A. THIAMINE

Thiamine pyrophosphate is the coenzyme for a large number of biochemical reactions, which all have the feature that they formally involve an acyl anion. Thus they are related to the well-known benzoin condensation, an organic reaction that also formally involves an acyl anion. It seemed likely that thiamine pyrophosphate played, in the biochemical system, the same sort of role that cyanide ion plays as a catalyst for the benzoin condensation, but it was not at all clear how this occurred.

After one false start (3) we demonstrated (4) that thiamine forms an anion (Fig. 1) that is very similar to cyanide ion in its chemistry and catalytic potential. This was done by studying the catalysis of

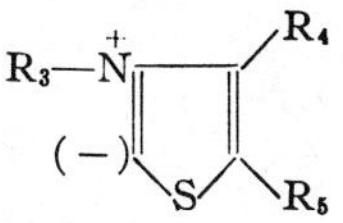

Figure 1. The thiamine anion that plays the key role in thiamine pyrophosphate catalyzed reactions.

the benzoin condensation by thiamine and other thiazolium salts. Because of the formal mechanistic analogy, it seemed clear (5) that this simple model system would be related to the biochemical reaction.

The mechanism elucidated in these model studies indeed proved to be the mechanism used enzymatically. Furthermore, the model studies permitted us to analyze the detailed relationship between the chemical structure of thiamine pyrophosphate and its biochemical function (6–9). In these model studies it was possible to show the importance of the electron-withdrawing character of the pyrimidine ring, and the importance of the aromatic character of the thiazolium ring. We also were able to show (10) that acetylthiamine (Fig. 2) was an active acetylating agent, as expected. Work by others added to this picture, and furnished the evidence that the biochemical reactions utilize the same mechanism. Later in this chapter we shall describe an artificial enzyme that incorporates thiamine.

B. METALLOENZYMES

We were apparently the first (11) to suggest that coenzyme B-12 reacts with substrates by a free radical hydrogen abstraction path. We examined a simple model system (12) in which a carbon radical removes a hydrogen atom from across a ring, and the resulting rearranged radical then couples with B-12, the cobalt radical species formed when coenzyme B-12 originally dissociates (Fig. 3). This model system furnishes chemical precedent for some of the steps

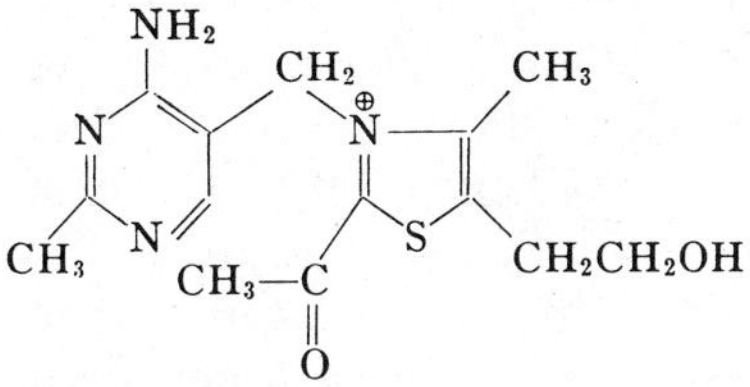

Figure 2. 2-Acetylthiamine, an active acetylating agent.

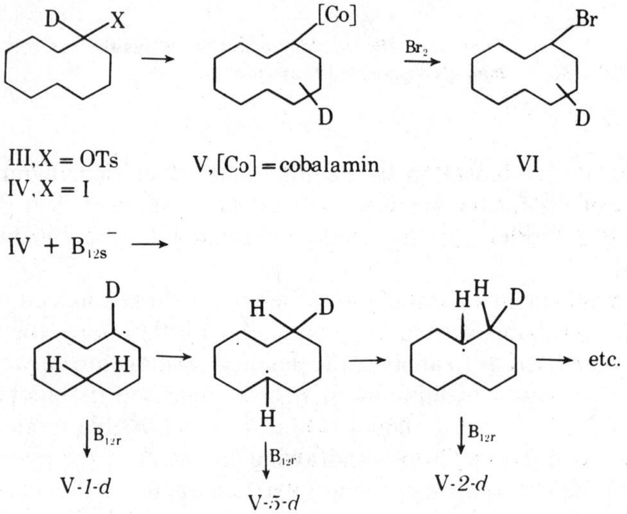

III, X = OTs   V, [Co] = cobalamin   VI
IV, X = I

IV + B$_{12s}^-$ ⟶

Figure 3.   A model system for part of the mechanism by which coenzyme B-12 reacts with substrates.

needed, but much more chemistry will be required before the biochemistry of coenzyme B-12 can be put into a fully understood chemical context.

Alkaline phosphatase is a zinc enzyme that catalyzes hydrolysis of phosphate monoesters by initial phosphorylation of an enzyme serine. Since some of the most interesting simple chemistry of phosphate monoesters involves reactions through metaphosphate anion, it seemed desirable to develop a diagnostic test for such a mechanism. In simple chemical model studies (13) we showed that the replacement of one of the oxygen atoms by a sulfur in p-nitrophenyl phosphate led to an increased rate of chemical reaction by the metaphosphate mechanism, but that a related replacement of oxygen by sulfur in a phosphate triester led to slower reaction by the addition–elimination mechanism. Thus this kinetic element effect, which was used for the first time as such a diagnostic, let us distinguish between these two mechanisms. We found that the enzyme alkaline phosphatase itself hydrolyzed the thiophosphate ester more slowly than it did the normal oxygen compound. This was consistent with the idea that the enzyme used an addition–elimination mechanism, not a metaphosphate mechanism, but of course it was not firm proof. The enzyme might well discriminate against the sulfur compound for reasons of size and shape, reasons that were not available to a simple chemical model. However, subsequent work with this enzyme seems to have confirmed our conclusion that it does not in fact use the metaphosphate mechanism in hydrolyzing phosphate monoesters.

One of the most widely studied zinc enzymes is carboxypeptidase A. Up until recently it was believed that the enzyme used a bound zinc, a glutamate carboxyl, and a tyrosine hydroxyl to catalyze cleavage of peptide substrates. For ester substrates it was clear that the tyrosine was not required, and recent genetic evidence suggests that the particular tyrosine identified, tyrosine 248, is not required for peptide cleavage either. However, with or without this phenol the principal mechanistic question has concerned the roles of the other two groups.

The glutamate carboxyl could be acting either as a nucleophile or as a base. In the first case it would attack the scissile carbonyl of the substrate amide, leading to an anhydride intermediate in which the substrate acyl group is attached to the glutamate. This anhydride would then cleave in a subsequent step. If instead the glutamate acts as a general base, it would deliver a water molecule to the scissile carbonyl, leading directly to the hydrolyzed products without an acyl-enzyme intermediate.

Although studies from our laboratory on the enzyme itself indicate strongly that the general base mechanism (Fig. 4) is correct (14–16), we have done some model studies related to the anhydride mechanism. In one study (17), we showed that a phenol would indeed catalyze the intramolecular attack of a carboxylate on an amide, producing an anhydride (Fig. 5). Thus this mechanism has good chemical precedent, although it is apparently excluded for the enzyme by our O-18 studies.

In another model system (18), we examined the cleavage of an anhydride by a neighboring zinc ion (Fig. 6). If the enzyme had used the anhydride mechanism, it would have had to use the zinc to catalyze the subsequent hydrolysis of this intermediate. We found that indeed a zinc could cleave the anhydride rapidly, in a model system. The mechanistic studies showed that the zinc was delivering a bound

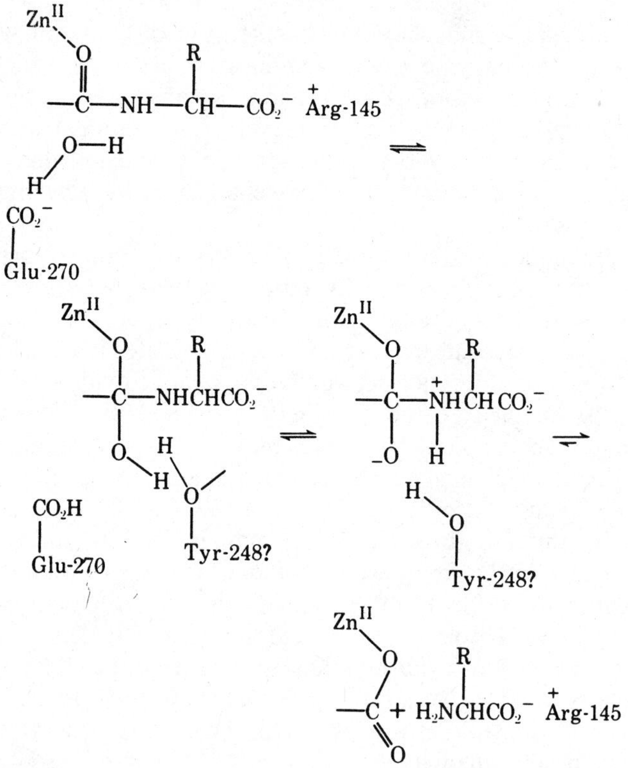

Figure 4. The general base mechanism for peptide hydrolysis by carboxypeptidase A.

water to the anhydride, in an attack by a zinc-hydroxide species. This mechanism, in the model system, shows a preference for delivering methanol rather than water (16). The great contrast between this finding and the inability of carboxypeptidase to permit any detectable substitution of methanol for water in its catalyzed reactions is one of the arguments that the enzyme does not use such an anhydride mechanism (16).

We had earlier studied some metal-catalyzed hydrolysis and hydration reactions to determine how big the accelerations could be, and to develop tools to establish the mechanisms of these processes.

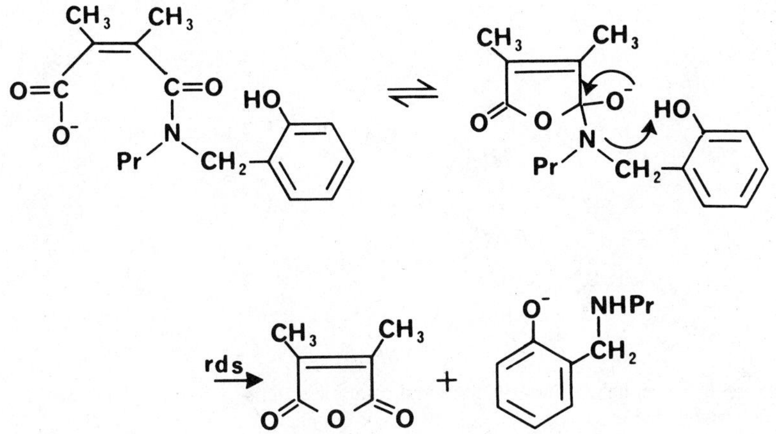

Figure 5. A model for the anhydride mechanism of carboxypeptidase, with phenol participation.

The metal-catalyzed hydration of 2-cyanophenanthroline (19) and of 2-cyanopyridine (20) both showed enormous accelerations as compared to reactions without the metals. Two catalytic mechanisms were possible: in one the metal acted to deliver a hydroxide from its coordination sphere (as in the zinc-catalyzed anhydride cleavage just discussed), and in the other mechanism the hydroxide attacked externally with the metal acting as a Lewis acid. We were able to get some evidence for the first mechanism from the fact that a co-

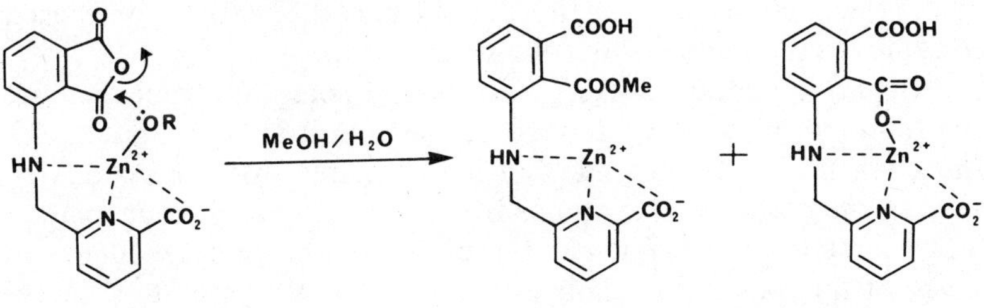

Figure 6. A model for the zinc-catalyed cleavage of a possible anhydride intermediate.

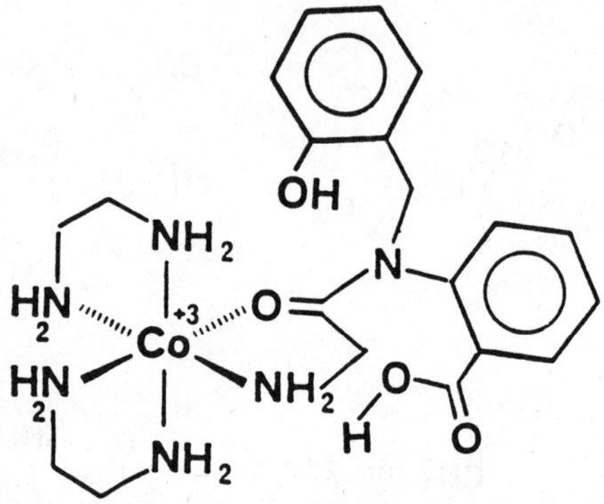

Figure 7. A model for the substrate and catalytic groups in carboxypeptidase. Related models have also been studied that lack the phenol.

ordinated hydroxyl group of a bound alcohol was also delivered by metal to the cyano group (20).

Of course the most direct model for carboxypeptidase action would incorporate a substrate, a metal binding group, and a carboxylate in the same species. A phenol might also be present or not, depending on whether a tyrosine is indeed to be invoked for the enzyme. Our first model study (21) of this sort simply showed that there was a modest cooperative catalysis of ester cleavage when a substrate contained both a bound nickel or zinc ion and a carboxylate group. More recently (22), we have constructed a model system (Fig. 7) that incorporates a peptide substrate, coordinated to the inert Co^{3+} ion, in which there is also a neighboring carboxylate and a phenol. In this latter case we find cooperative catalysis of the cleavage of the peptide by the phenol and the metal. The Co^{3+} ion is used to remove ambiguities about the mechanism.

In carboxypeptidase itself there is still some uncertainty about the function of the zinc. It seems clear that it is coordinated with the oxygen of the substrate, but it is possible that the water that attacks the carbonyl also comes from the zinc coordination sphere. The geometry is difficult, and there is as yet no direct evidence in favor of this possibility. However, it is clearly related to some of the hydrolysis mechanisms mentioned earlier and to a very effective zinc catalyzed cleavage in a model system described by Groves and

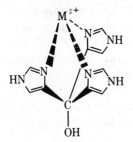

Figure 8. A model for the zinc-binding site of carbonic anhydrase. Many related ligands have also been synthesized.

Chambers (23). Such ambiguities are not present in model systems utilizing substitution inert metals such as Co^{3+}

In the principal enzymes that have been investigated, the ligands that bind zinc consist of at least some imidazoles. In carboxypeptidase A there are two imidazoles and a carboxylate, while in carbonic anhydrase the zinc is bound with three imidazoles. This led us to investigate the synthesis and study of ligand systems in which three imidazoles, or two imidazoles and a carboxylate, were mounted so as to be able to bind zinc ions. We developed synthetic methods (24,25) to prepare such ligands (Fig. 8) but found that they do not bind zinc as well as the enzymes do. The difficulty is undoubtedly a reflection of the flexibility in simple molecules. Flexibility is inhibited in enzymes by subsidiary interactions and hydrogen bonds to imidazoles, which help to immobilize them and aim them in the correct direction. Brown et al. have constructed several related ligands, based in part on these synthetic methods, and shown that in at least a few cases the resulting zinc complexes have some carbonic anhydrase activity (26). It is still a challenge to put together good models for the zinc binding sites in enzymes, and to demonstrate appropriate catalytic activities of the resulting complexes.

III.   Artificial Enzymes and Enzyme Mimics

It is one of the characteristics of enzyme-catalyzed reactions that strong and selective complexing of the enzyme with a substrate contributes greatly to the rate and selectivity of the process. Of course, for maximum velocity it is important that the activated complex be bound even more strongly than the substrate.

There are several aspects of enzyme chemistry that should be imitated in good artificial enzymes or enzyme mimics. Perhaps most striking is the *velocity* of enzyme-catalyzed reactions. The ability of carboxypeptidase A, for instance, to hydrolyze an amide at neutral pH and room temperature in a few milliseconds is very far away from the capability of current laboratory or industrial chemistry. The message from nature that such velocities are accessible is a powerful stimulus to chemists to learn how to achieve them in synthetic systems.

At least as interesting as the velocity of enzyme-catalyzed reactions is their *selectivity*. The selectivity of enzymes can be divided into four categories, all of which are attractive to imitate. First of all, an enzyme is selective for its *substrate*. Enzymes can operate in a mixture of substances, binding and acting on only particular components of the mixture. This kind of selectivity, which is not necessary in chemical reactions in which a single pure substrate could be used, is certainly important if we are ever to imitate complex systems that resemble living cells. The second kind of selectivity is with respect to the *chemical reaction* performed. In the most trivial sense one might want to perform an oxidation and not a hydrolysis, and this is already under pretty good control in normal chemical reactions as well as in biochemical ones. However, when one considers the many different reactions that various enzymes can perform on the amino acid serine with the coenzyme pyridoxal phosphate, the significance of such chemical selectivity becomes apparent. One enzyme may perform a transamination, another one may eliminate formaldehyde to form glycine, while another one may convert the serine to pyruvic acid and ammonia. In cases in which several chemical reactions are possible, the ability of enzymes to direct the chemistry is very much to be imitated as a way to improve yields and remove side reactions.

The third kind of selectivity is *regioselectivity*, sometimes called *positional selectivity*. Enzymes are able to direct chemical reactions into particular positions of a molecule even when those positions are not the most reactive in ordinary chemical terms. For instance, an enzyme may be able to oxidize a quite unreactive methyl group and leave untouched a much more sensitive double bond in a molecule. This kind of selectivity is very much worth imitating, since it is contrary to the normal situation in chemical synthetic reactions.

We have devoted a considerable effort to imitating such regioselectivity, as will be described later.

The final kind of selectivity attractive to chemists is *stereoselectivity*. This refers to the ability of an enzyme to perform chemical reactions on one side or the other of a double bond, for instance, or even more strikingly to achieve the selective synthesis of a D-amino acid or a L-amino acid without production of the other isomer. Chemists have devoted considerable energy to trying to develop stereoselective reactions with enzymelike selectivities. One approach to this is to try to use the actual style by which enzymes achieve this control.

It is generally understood that the most important aspect of both the velocity and the selectivity characteristics of enzymes results from the fact that they form complexes with substrates. Of course the most important thing an enzyme does is to bind the activated complex for a chemical reaction. Indeed, since binding is an interaction between two species that results in lowered energy, catalysis is by definition the binding of the activated complex. However, much of the geometric control and positioning characteristic of the transition state is normally also seen to at least some extent in the binding of the substrate. If the substrate were not bound in the early stages of a reaction it would generally be hopelessly improbable for the activated complex to bind in the precise geometry required for chemical reaction. Thus substrate binding, whose strength normally will increase as the reaction progresses to the activated complex, is a necessary component of any enzyme mimic. In addition, it is desirable that this binding result in good proximity of catalytic groups to the important parts of the substrate. There can also be rate contributions as well from the change of the medium when a substrate is extracted from a solution and put into the interior of a catalyst molecule.

In any enzyme mimic, then, the important first choice is the nature of the binding interaction to be used in holding the catalyst and substrate together. We started investigating this area by using metal coordination as a way to achieve this binding. Since many substances of interest are not able to bind to metals, we later took up the use of more generalized types of binding forces.

A. ACYL TRANSFERS WITHIN MIXED METAL COMPLEXES

Metal cations are in general Lewis acids, so one problem in the design of synthetic catalysts is to avoid a "short circuit." That is,

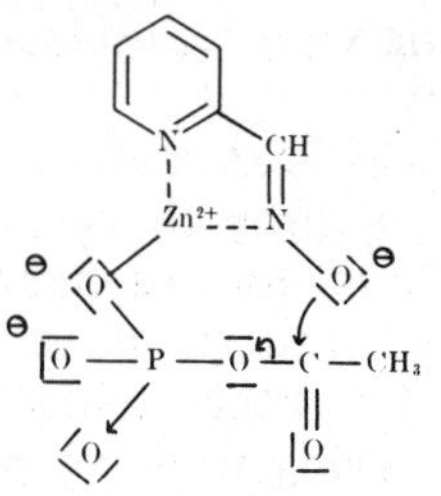

Figure 9.   The intracomplex attack by zinc pyridinecar-
boxaldoxime complex on bound acetyl phosphate.

one generally wants to incorporate a base or nucleophile along with
the metal ion, and one wants to avoid a direct interaction between
the two, which would neutralize them. In a sense a catalytic reaction
can be thought of as an electrical discharge through the substrate,
which would be vitiated by a direct short circuit interaction (this is
a highly figurative description, not to be taken too literally).

To solve this problem we developed (27) the use of the oxime of
pyridine-2-carboxaldehyde, PCA, as a metal ligand, since the coordination of the two nitrogen atoms leads to a geometry that prevents the oxygen atom from coordinating also. In our earliest study
(27) we found that the zinc complex of PCA was indeed an effective
catalyst for the hydrolysis of an acetate ester of a coordinating substrate, the acetate ester of 8-hydroxyquinoline sulfonate. Appropriate studies demonstrated that a mixed complex was formed and
there was then acetyl transfer within the complex. The product was
then relatively reactive toward further hydrolysis, so the entire process led to turnover catalysis and an accelerated hydrolysis of the
substrate. In subsequent work we demonstrated that this same zinc–
PCA complex was effective in hydrolyzing acetyl phosphate (28)
(Fig. 9) and Cooperman (29) demonstrated that it could be used to
catalyze some phosphate transfer reactions.

One limitation of such systems is their requirement that the substrate be a good ligand for metals. In most metalloenzymes binding
of the substrate may in part involve coordination to the metal, but
much of the binding comes from the normal sorts of interactions
common to all enzymes, including in particular hydrophobic interactions. Thus we constructed a model system (30) incorporating the
metal–PCA nucleophile–electrophile combination, but with a cyclodextrin as a binding group so that this characteristic metal re-

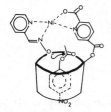

Figure 10. A metal-binding group attached to cyclodextrin permits catalyzed hydrolysis of *p*-nitrophenyl acetate, which is not a metal ligand but binds in the cyclodextrin cavity.

action could be performed on substrates that have no particular metal affinity.

We attached a metal binding group to cyclodextrin by acyl transfer, and then used this to bind a Ni^{2+}–PCA catalytic group. The resulting mixed complex, which was the first substance described as an artificial enzyme (30), was able to bind *p*-nitrophenyl acetate into the cyclodextrin cavity and then catalyze its hydrolysis by acetyl transfer and subsequent deacetylation (Fig. 10). The resulting catalyst could be inhibited by other substances that bind into the cavity, and showed the expected selectivity for substrates that could utilize the cyclodextrin binding site. In other work related to this (31) we used the original combination of cyclodextrin with a metal-binding group to bind Cu^{2+}, and showed that the resulting complex could use both the cyclodextrin and the metal to achieve two binding interactions with the *p*-nitrophenyl ester of glycine, and its subsequent hydrolysis. We and others have attached other metal binding groups to cyclodextrins, in order to achieve such mixed binding and catalytic reactions.

B. CYCLODEXTRINS AND DERIVATIVES AS ENZYME MIMICS

The last examples discussed can be thought of as modified metal systems or as cyclodextrin derivatives. In general, the cyclodextrins have excited wide interest in the construction of enzyme mimics because of their attractive properties. They are easily available, and form reasonably strong complexes with substrates that can fit into their cavities. The binding of a substrate can itself induce chemistry catalyzed or promoted by the hydroxyls of cyclodextrin, and in addition it is relatively easy to modify the cyclodextrins and introduce other catalytic groups, as in the cyclodextrin–metal systems just described.

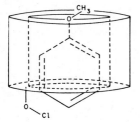

Figure 11. Cyclodextrin catalyzes the selective para chlorination of anisole by an intracomplex mechanism.

Unmodified cyclodextrin can perform some very interesting catalyzed reactions, using binding and the hydroxyl groups of the cyclodextrin. The earliest example of a selective synthetic reaction directed by cyclodextrin binding was our study (32,33) of the chlorination of anisole. In free solution anisole reacts with the mild chlorinating agent hypochlorous acid, HOCl, to afford both ortho- and para-chlorinated products. When a cyclodextrin is included in the reaction mixture, the substrate binds into the cyclodextrin cavity and the chlorination process becomes quite selective. The most striking effect is seen with α-cyclodextrin, consisting of six glucose molecules in a ring. Anisole fits rather tightly into this cavity, and the geometry is such as to prevent chlorination of the bound material by simple HOCl in solution. However, the HOCl reacts rapidly with hydroxyl groups of the cyclodextrin in a reversible equilibrium to form a glucose hypochlorite, and this species is able to chlorinate the bound substrate (Fig. 11). The ortho positions of anisole are quite inaccessible in the complex, and chlorination is directed selectively to the para position.

A detailed kinetic study (33) showed that this system also accelerates the chlorination reaction; the extent of acceleration is a function of the reaction conditions. One of the simplest effects is that α-cyclodextrin, with a small tight cavity, is more effective than is β-cyclodextrin at catalyzing the chlorination. This is as expected if it is important not simply to bind a substrate into the cavity, but to bind it with some degree of geometric definition in the complex. The more interesting finding was that the catalyzed chlorination showed a first-order dependence on HOCl concentration, as expected from the mechanism suggested, but the reaction in free solution depended on the second power of the HOCl concentration. That is, chlorination in free solution involves the species Cl_2O, which is in equilibrium with two molecules of HOCl. Because the two processes have different kinetic orders, the rate ratios depend on the precise concentration of HOCl being considered. At the relatively high concentrations that were used in most studies the cyclodextrin reaction was only five times faster than the reaction without cyclodextrin, but this rate advantage would of course increase if the concentration of HOCl were diminished.

This selective chlorination reaction has the potential to be practical, but for a truly practical process the necessity to separate the product from a solution of cyclodextrin is not attractive. For this reason we examined the properties of a cyclodextrin polymer (34) prepared from cyclodextrin and epichlorhydrin. We found that this material is able to perform a highly selective chlorination of anisole, producing greater than 99% of the para isomer, in a reaction in which the reagents are run down a chromatograph tube packed with the polymer. This kind of thing, related to the use of immobilized enzymes, has the potential to be of real practical interest.

Further studies on this reaction (33) have helped to clarify the mechanism. Various other substrates were examined in order to establish that the general predictions of the mechanism are correct. More importantly, it was shown that cyclodextrin does not catalyze the attack of a reagent (a diazonium cation) that could not be delivered by this special mechanism involving the hydroxyl group. Finally, the particular hydroxyl group that is involved, or at least that can be involved, was established by the finding (34) that α-cyclodextrin-modified with 12 added methyl groups is a very effective selective catalyst for this chlorination reaction. The methylation involved all of the primary C-6 hydroxyl groups, and all of the C-2 hydroxyl groups on the secondary side. The fact that the product was still a highly effective chlorinating reagent meant that the C-3 hydroxyl could play the catalytic role.

In aqueous solution one would expect the primary methyl groups to cluster and close the cavity, forming a pocket out of the previous tube shape. If such a closed cyclodextrin is used in a molecular model the anisole fits into this pocket in such a way as to bring its para position precisely into contact with a chlorine on the C-3 oxygen. Thus molecular models make it clear why this reaction was so effectively catalyzed in a selective fashion.

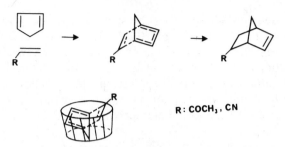

R : COCH₃ , CN

Figure 12.   The diene and dienophile of a simple Diels–Alder reaction are bound together into a cyclodextrin cavity, catalyzing the reaction.

It occurred to us that cyclodextrins might be able to bind two molecules into the same cavity, promoting their reaction. This is related to the general function of many enzymes in binding and gathering together two different reactants, either substrates or substrates and coenzymes. Molecular model building made it seem likely that some simple Diels–Alder reactions would be catalyzed by β-cyclodextrin, since a simple diene such as cyclopentadiene and a simple dienophile such as acrylonitrile could both fit into the cyclodextrin cavity (Fig. 12). Indeed this prediction proved to be correct (35). Cyclodextrin catalyzes the reaction of cyclopentadiene with simple dienophiles that should fit into the cavity, provided the β-cyclodextrin is used. With α-cyclodextrin, which has a smaller cavity, we observed inhibition instead. Apparently the diene could bind into the cavity, but the dienophile was then excluded. When larger diene and dienophile systems were used, even the β-cyclodextrin cavity was too small for both of them and inhibition was seen.

The reaction rates for the catalyzed reactions were several powers of 10 faster than typical bimolecular reaction rates in most solvents, in which Diels–Alder reactions are normally conducted, but interestingly a special effect was seen with the solvent water. We observed that water strongly catalyzed many of these Diels–Alder reactions by a hydrophobic effect. That is, in water solution the hydrocarbon sections of the two reactants were bound together in order to escape water contact, and this resulted in a Diels–Alder reaction inside a different cage, a cage constructed of solvent alone. Studies with salts that increase and decrease hydrophobic effects confirmed that this was indeed the phenomenon being observed (36). The result of this is that in the comparison of the cyclodextrin-catalyzed reaction with a reaction in water in the absence of cyclodextrin, the special effect of cyclodextrin binding was to some extent masked by the fact that it was competing with an alternative mode of binding in pure water, aggregation because of the hydrophobic effect.

In spite of this the cyclodextrin reactions were 10 or more times faster than the water reactions, and they also showed substrate selectivities. That is, the cyclodextrin reactions were only catalyzed in cases in which both the diene and the dienophile could fit in the cavity, in contrast to the situation in water in which many Diels–Alder reactions are apparently accelerated. Interestingly, the Diels–Alder reaction performed in water alone is, in the cases we examined, faster than the reaction without any solvent, and it shows (37) a very high product selectivity for products that have a minimum volume. Apparently the hydrophobic packing of the two components not only promotes their interaction, but promotes the formation of particular products in a synthetically attractive reaction. Since the Diels–Alder reaction inside a cyclodextrin cavity is in a chiral environment, one might hope for optical selectivity in the product. However, we have seen only very moderate such effects to date.

One of the real challenges in enzyme model chemistry is to achieve the rates of typical enzymatic reactions. Of course intramolecular reactions, in which neighboring groups attack, can often have very high velocities and be quite suitable models for enzymatic reaction rates. However, it was important to demonstrate that molecular complexing could also lead to high velocities. For this reason we were concerned that the cyclodextrin reactions that had been studied typically gave very modest rate accelerations, of the order of 10- to 100-fold. A classic study was performed by Bender and his co-workers some years ago (38), in which they examined acylation of cyclodextrin hydroxyl groups by bound phenyl esters (Fig. 13). The reaction is reminiscent of the acetylation of chymotrypsin by p-nitrophenyl acetate, the first step in the enzymatic hydrolysis of this unnatural substrate. In spite of a lot of geometric manipulation of the substrates, the maximum acceleration by cyclodextrin seen in the early work by the Bender group was only 250-fold. We set

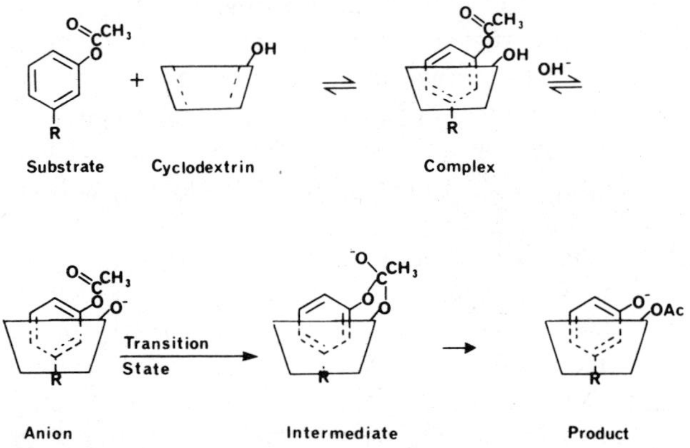

Figure 13.   Acetyl transfer in a cyclodextrin–substrate complex.

out to see if we could understand the reason for this and get the accelerations up into the enzymatic rate region.

It is important to understand clearly the actual velocity comparison being made. With cyclodextrin present at a given pH one of the good substrates examined by the Bender group, *m*-nitrophenyl acetate, undergoes a reasonably rapid acetyl transfer to a cyclodextrin hydroxyl group. In the absence of cyclodextrin at the same pH the substrate undergoes a hydrolysis (acetyl transfer to solvent) at approximately 1% of the cyclodextrin rate. We would say that the deacetylation of the substrate has been accelerated by 100-fold in this comparison. Of course, in simple solution an overall hydrolysis has occurred, while with cyclodextrin present, only the first step of a possible two-step hydrolysis sequence has occurred. Thus such acetyl transfer steps can by no means be considered to be catalytic. However, for a two-step enzymatic acceleration of hydrolysis to be fast, each individual step must be fast as well. Thus the first challenge was to learn how to accelerate such intracomplex acylation reactions; the second challenge would be to cleave the intermediates rapidly in the next step.

Molecular model building (39) gave us the clue to the problem with substrates that had been examined so far. In the best cases seen by the Bender group such as *m-t*-butylphenyl acetate, models showed that the substrate could bind nicely into the cavity in such a way that the acetyl group to be transferred was in contact with a cyclodextrin hydroxyl. However, when we made a model of the tetrahedral intermediate involved in such a transfer, we found that the m-t-butylphenyl group pulled up significantly out of the cavity. Although such models are not terribly quantitative, we estimated that as much as 75% of the original deep binding into the hydrophobic cavity was lost in the tetrahedral intermediate. If this were true it was not surprising that the reaction rate was modest, since the critical thing in catalysis is to bind the transition state, not the starting material. In fact, if binding is lost on going from starting material to transition state this will slow the intracomplex reaction, which is the process being studied. The situation is more complicated when substrate and catalyst are at low concentrations, and the velocity is less than V_{max}.

Our first approach to this problem was to modify cyclodextrin itself (39,40). As we have already described, the anisole chlorination reaction was clarified by methylating the cyclodextrin so as to produce a cavity with a floor, converting it into a pocket of relatively well-defined dimensions. A somewhat different floor was constructed on the cyclodextrin system by preparing β-cyclodextrin heptatosylate, substituted entirely on the primary carbon atoms, and reacting this with methylamine or ethylamine. The resulting compounds were then formylated to retain their water solubility but get rid of the charge. Molecular models suggested that the methyl groups or ethyl groups of these derivatives should cluster together and invade the cavity to some extent, since the same hydrophobic binding that operates to put substrates into the cyclodextrin cavity should certainly operate to put hydrocarbon segments of the molecule into the cavity if possible (Fig. 14). Studies on binding of these modified cyclodextrins to adamantanecarboxylate confirmed this interpretation. The adamantane derivative showed changes in binding properties indicating that there was indeed a floor on the cyclodextrin cavity.

As expected, the cyclodextrin derivatives with these intrusive floors were more reactive in the acetyl transfer process from bound m-t-butylphenyl acetate. Such a floor of course blocks the primary face of the cavity, and substrates are forced to bind only into the reactive secondary side of the cyclodextrin. In addition, the intru-

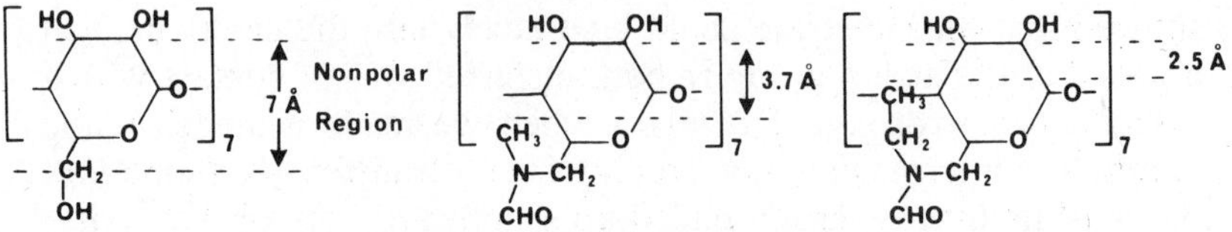

Figure 14. Flexible capping of the cyclodextrin ring. The methyl and ethyl groups intrude into the cavity, making it shallower but adding a new hydrophobic floor.

sion of methyl or ethyl groups into the cavity makes it shallower; the binding geometry of the substrate is closer to the binding geometry of the tetrahedral intermediate. The transition state must closely resemble the tetrahedral intermediate, so such an adjustment in binding should be relevant to reaction rates.

The findings with these capped cyclodextrin derivatives confirmed this picture. The $k_{complex}/k_{un}$ was increased by 10-fold with the methyl capped compound, and by 20-fold with the ethyl capped compound, in the acetyl transfer reaction with m-nitrophenyl acetate. Interestingly, the binding constant was essentially the same when the methyl capped material was used. The shallower cavity, which should bind more poorly, was compensated by the presence of a new hydrophobic floor. As expected from this, the compound with ethyl groups bound the substrate more poorly by a factor of about five. It still had a new hydrophobic floor, but now had an even shallower cavity as the ethyl groups were able to intrude further up into the cyclodextrin. Similar changes in rate and binding constants were observed with the substrate m-t-butylphenyl acetate.

Of course there are two approaches to trying to optimize the fit of a substrate and a cavity, so that the transition state for a chemical reaction within the complex will be well bound. The easier approach is to try to find other substrates in which a geometric problem does not occur during the acyl transfer. We looked at this situation with a series of p-nitrophenyl esters of acyl groups, in which the hydrophobic character of the acyl groups was sufficient to guarantee that they were the molecular fragments bound into the cavity. We hoped to optimize the structure of such substrates to the point at which we could get truly large rate accelerations of acyl transfer processes. The studies (39,41–43) will be described briefly in order of increasing effectiveness of the geometries, which is not the historical order in which the materials were prepared and studied.

In general with these more complex substrates, solubilities required us to work in mixed $DMSO/H_2O$ solvents. Our comparisons are to uncatalyzed hydrolysis reactions at the same pH in the same medium. In earlier studies we had found (44) that cyclodextrin will bind many hydrocarbon substrates in highly polar organic solvents such as DMSO, although up to that point it was believed by workers in the field that water was the unique solvent that promoted such complexing. We also found that the uncatalyzed hydrolysis reaction of typical substrates in $DMSO/H_2O$ was 25-fold or so faster than was the equivalent hydrolysis reaction with the same buffers in H_2O alone. Thus when we discuss the acceleration of some of the reactions we may refer both to the acceleration relative to the $DMSO/H_2O$ uncatalyzed process and also to the even larger acceleration that is seen relative to a pure water solution with the same buffer present. This latter comparison with reaction rates in pure water is the one normally done for enzymatic processes themselves, even though the interior of an enzyme may well resemble a polar organic solvent or mixed solvent. In any case, the reaction rates relative to processes in pure water are relevant to the reader in evaluating what has actually been achieved in velocity. A small part of that rate increase has come from a change of medium.

Some increase in $k_{complex}/k_{un}$ was observed with a cinnamic acid ester of p-nitrophenol in which the cinnamate carried a m-t-butyl group and a methoxyl group (Fig. 15). However, molecular models suggested that in this case the m-t-butylphenyl group still bound deeply into the cavity and lost some of this binding on going to the transition state. In order to check this point we attached a projection on the bound phenyl group (Fig. 15) that prevented it from binding so deeply. This led to an improved rate ratio, since it moved the binding geometry of substrate closer to that of the transition state, just as the intrusive floors had done.

Molecular models suggested that adamantanepropiolic acid p-nitrophenyl ester (Fig. 15) should bind the adamantane group on the face of the cavity, rather than down into it, and some of our studies with adamantane derivatives had confirmed this idea. Thus we expected that the geometry for acyl transfer in this case might be a lot closer to the original geometry of the bound substrate; the results

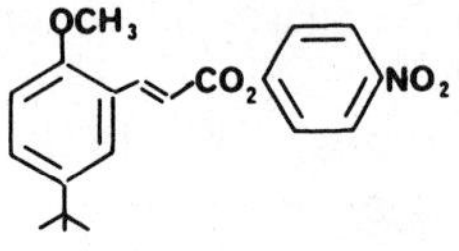

Figure 15. Some substrates for β-cyclodextrin acylation. The extra projections on 9 and 11 help adjust the geometry, leading to improved rates.

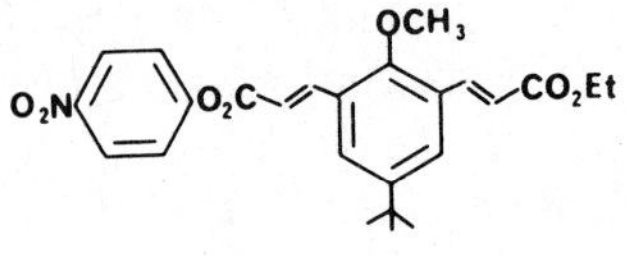

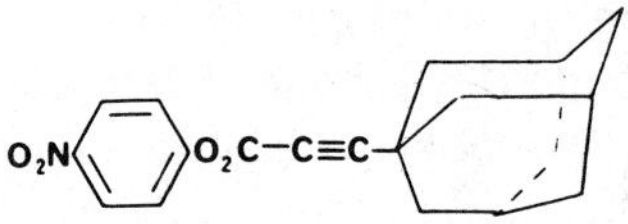

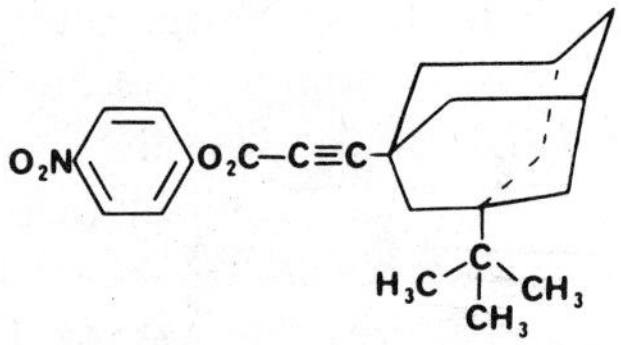

actually observed were quite surprising. The rate ratio increased to 2150-fold, and the compound showed extremely strong binding. As it turned out, something very interesting was happening.

The adamantane nucleus can bind only partially into the cyclodextrin cavity, mainly being occluded on the secondary face. Under these circumstances the substrate can derive better overall binding by having the side chain carrying the ester group pass down through the middle of the cyclodextrin cavity, rather than simply stick off into solution so as to perform the acyl transfer reaction with the secondary hydroxyls. In the better bound geometry in which the chain passes down through the cavity, the ester group is of course no longer within reach of the secondary hydroxyls and it is not within reach of the primary hydroxyls either. Thus one would expect this better geometry to represent an unproductive complex. Two kinds of evidence showed that this was indeed the case. Most strikingly, use of the capped cyclodextrin with an intrusive methyl floor prevented this alternative mode of binding, since the floor prevented penetration of the side chain. The ratio $k_{complex}/k_{un}$ improved to 14,000, but the compound was bound nine times more weakly. In the productive geometry for reaction, no extra binding contribution appears from the binding of the side chain and the adamantane group

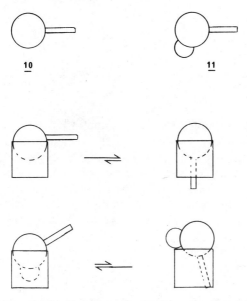

Figure 16.   Alternative binding geometries for substrates 10 and 11 (of Fig. 15).

itself does not penetrate far enough into the cavity to contact the methyl floor and pick up more binding interactions.

The second way to prevent this nonproductive complexing was to add a *t*-butyl group to the adamantane nucleus. Models showed (Fig. 16) that the *t*-butyl group could occupy the part of the cyclodextrin cavity that the adamantane did not fill, so it was expected that this compound would prefer the productive geometry in which the side chain projects over the secondary face and brings its carbonyl group near the reactive hydroxyl of cyclodextrin. Furthermore, if one tries to put this molecule into the nonproductive geometry that we deduced for the adamantyl compound without a *t*-butyl group, the *t*-butyl group interferes with this alternative nonproductive complex geometry. We indeed found that the reactivity ratio for this substance was 15,000, essentially the same as that of the adamantyl ester without a *t*-butyl group that had been pushed into the correct geometry by the use of a methyl floor. In this *t*-butyladamantyl case a methyl floor should be a problem, since the

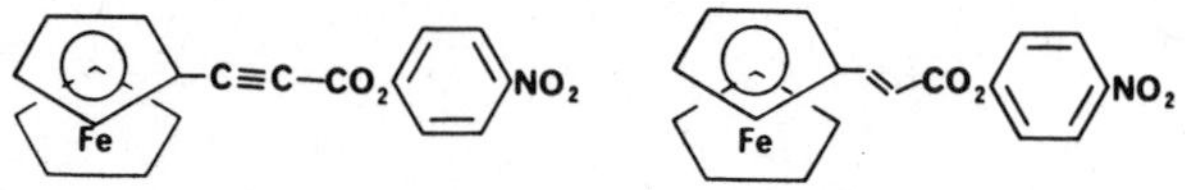

Figure 17.   Two excellent substrates derived from ferrocene.

cavity is fully occupied by the adamantyl nucleus and the *t*-butyl group. Indeed when we looked at the reaction of the *t*-butyladamantylpropiolic ester with the cyclodextrin capped with a methyl intrusive floor, we found that the rate ratio dropped, and a 2:1 cyclodextrin to substrate complex was formed.

The adamantyl system is essentially spherical, and the chain can adopt various orientations. When a *t*-butyl group is added it locks the conformation, but the chain makes an obtuse angle with the axis of the cyclodextrin cavity that is not quite ideal for binding of the tetrahedral intermediate. Ideally the side chain carrying a carbonyl group to be transferred to the cyclodextrin hydroxyls should come off at a 90° angle to the binding axis, or even an acute angle if possible. The 90° geometry was accessible with a series of derivatives of ferrocene; these led to the best substrates that have so far been examined for such cyclodextrin reactions.

In molecular models the ferrocene ring fits β-cyclodextrin very well, and our earlier studies (44) had demonstrated that it was extremely well bound. The *p*-nitrophenyl ester of ferrocenepropiolic acid showed a rate ratio that had risen to 140,000. If the side chain had the double bond in ferroceneacrylic ester, the rate ratio with β-cyclodextrin was 750,000 (Fig. 17). When the cyclodextrin derivative with an intrusive methyl floor was examined, binding was increased somewhat because of the new hydrophobic surface, but the acylation rate was decreased. The geometry is apparently not quite as good once the ferrocene substrate has been raised in the cavity. On the other hand, if we used a capped cyclodextrin derivative (Fig. 18) in which the floor does not intrude into the cavity it did not change the binding, and actually very slightly improved the rate.

The products of all these reactions are acylated cyclodextrins. When they are examined in the NMR spectrum they are seen to be a mixture of 2-acyl and 3-acyl compounds. Since acyl transfers between these two positions are expected to be very rapid, this evi-

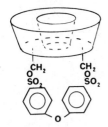

Figure 18. A capped cyclodextrin in which the floor does not intrude into the cavity.

dence is consistent with the idea that a single hydroxyl is the original attacking group, and the product then equilibrates; it does not require that both the C-2 and C-3 hydroxyls are equally reactive in the original process. Evidence bearing on this was obtained by our study (45) of a related sulfonate transfer reaction within a cyclodextrin complex; the sulfonate group cannot migrate from one hydroxyl to another. The product from *m*-nitrophenyl tosylate was exclusively the C-2 sulfonate. Thus we conclude that this reaction at C-2 is probably true for the acyl transfer reactions as well, and that the product mixture results from the expected equilibration.

Even better substrates have been obtained by restricting some of the freedom in the ferroceneacrylate system. In one study (42) a cyclohexane ring was fused on, and the rate ratio $k_{complex}/k_{un}$ increased to 3,200,000. The compound is a mixture of two enantiomers (Fig. 19), and one of them was more reactive than the other by a factor of 20. The absolute configuration of the more reactive enantiomer was established. This is quite a good chiral selectivity; this substrate not only shows a very high reaction rate but also a very high substrate selectivity with respect to enantiomers. An even better substrate (Fig. 19) is the corresponding cyclopentane derivative (43), which shows a 5,900,000-fold acceleration relative to hydrolysis in aqueous DMSO under the same conditions. Because of the solvent effect on the uncatalyzed reaction, this is 150,000,000 times as fast as the hydrolysis in pure water with the same buffer. This cyclopentane compound also showed an even higher chiral selectivity, with a 62-fold preference for one enantiomer over the other.

We also showed (43) that ruthenocene derivatives had no advantage over the ferrocene compounds. We found some systems in which nonproductive binding occurred, and blocked it by putting a

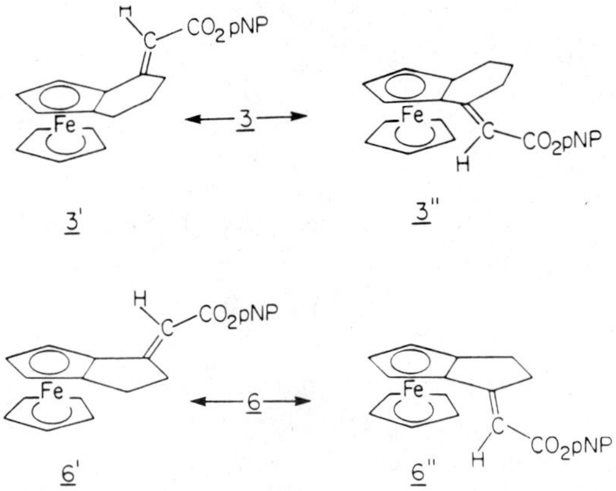

Figure 19.   Two substrates that show outstanding rates and stereoselectivities.

floor onto the cavity (43). The geometry achieved in the best cases was so good that even locking it with a covalent bond, so as to produce an intramolecular reaction rather than an intracomplex reaction, did not improve the rate.

Although the very high rates of these reactions might be interpreted to indicate that they are close to optimal, it does not seem likely that this is the case. The geometry is very good for a transition state leading to a tetrahedral intermediate, but it is by no means optimal for a transition state that comes after a tetrahedral intermediate, on the way to the final product. Reaction of an ester with the hydroxyl group of a cyclodextrin requires a 90° twist. That is, the hydroxyl must originally attack perpendicular to the plane of the ester group, but it must end up *in the plane* of the product ester (Fig. 20). This geometric requirement may explain the observation that the high rate accelerations we see with a good leaving group, such as the p-nitrophenoxide ion, become much smaller when poorer leaving groups are used, and the transition state comes later. The solution to this problem will require enough flexibility in the system that it can move to follow the reaction coordinate, not simply hold an original rigid geometry. Substrates with this semirigid character have not yet been examined.

Our ideas about the geometric changes required when various

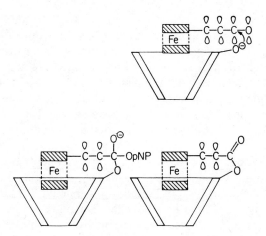

Figure 20.   Acylation of a cyclodextrin hydroxyl requires an overall 90° rotation of the carbonyl group.

substrates go from their original bound geometry to the geometry of the tetrahedral intermediate have of course all been based on model building, and supported by the correctness of various predictions from this geometric model. It is gratifying that studies (46) of the effect of pressure on these reactions also confirm the predictions of the models. Equilibria and rates change under pressure as a reflection of the volume changes involved in the equilibration or rate processes. Thus we were able to deduce the volume change both when a ferrocene-based substrate bound and also when it then underwent acyl transfer. Similar data were obtained for the poorer cinnamate ester substrate. The data confirmed our deduction that the great rate advantage of the ferrocene case corresponded to only a small geometric change on going to the transition state, with little loss of binding, while in the cinnamate case much more of the binding geometry was lost on proceeding to the transition state of the reaction.

We have referred to the problem that substrates can bind in either a productive or a nonproductive geometry. This is of course a general problem with an uncapped cyclodextrin, since even very simple substrates have the opportunity to bind with their top on either the primary or the secondary side of the cyclodextrin molecule. Thus one might wonder whether it is important to mount functional groups, designed to catalyze reactions within cyclodextrin complexes, on the secondary face or the primary face of the molecule. We had observed (47) that a metal binding group could be mounted on either the secondary or the primary face, and that the resulting metal complexes would catalyze reactions of bound substrates more or less equivalently. A similar conclusion arose from our study (48) of the catalytic abilities of the three individual cyclodextrin phosphates.

We developed specific syntheses for cycloheptaamylose 2-, 3-, and 6-phosphoric acids (Fig. 21). These were examined as acid catalysts for the hydrolysis of bound p-nitrophenyl tetrahydropyranyl ether, and it was found that only the 3- isomer was a catalyst for this hydrolysis. This reaction is suppressed by binding of the substrate to simple cyclodextrin, so the catalytic group must overcome the decrease in rate that results from moving the substrate into the less polar cyclodextrin environment.

At high pH these three phosphate esters were ionized and could act as basic catalysts for reactions of bound substrates. The process looked at was base-catalyzed tritium exchange in p-t-butylphenacyl alcohol (Fig. 21). The buffer-catalyzed enolization of this compound, which leads to tritium exchange, was actually accelerated by binding to simple β-cyclodextrin. All three isomers of the cyclodextrin phosphate were even better at catalyzing this reaction, and with similar rates. This indicates again that substrates can generally bind in both possible orientations into a cyclodextrin, and be accessible to catalytic groups located in either the primary or the secondary face. Thus in the synthesis of simple cyclodextrin catalysts the mounting of catalytic functional groups on either of the two faces of the molecule can be useful. Since substrates can generally bind in either orientation, a defined orientation of the substrate can only be achieved by modifying the cavity, as by capping it to produce a cup-shaped molecule.

Ribonuclease is one of the best understood enzymes. Two imidazole rings and an ammonium ion of lysine cooperate in the two-step sequence by which ribonucleic acid, RNA, is hydrolyzed. The intermediate in the hydrolysis is a cyclic phosphate derived from the substrate itself, so the enzyme can act as a catalyst for either of these steps without becoming inactivated by covalent bonding of

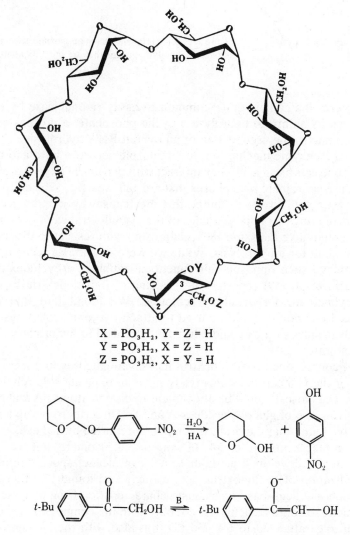

$$X = PO_3H_2, \quad Y = Z = H$$
$$Y = PO_3H_2, \quad X = Z = H$$
$$Z = PO_3H_2, \quad X = Y = H$$

Figure 21. The three β-cyclodextrin phosphates and the reactions they catalyze.

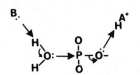

Figure 22.   Linear displacement at phosphorus can be assisted by catalytic groups that are not 180° apart.

an intermediate. It is quite common to assay ribonuclease by examining its ability to hydrolyze a cyclic phosphate, a process corresponding to the second step of an overall RNA hydrolysis.

In the enzyme mechanism one imidazole acts as a base and the other one acts as an acid; they interact with two apical oxygen atoms of a five-coordinate phosphorus that are 180° apart. It seemed to us very attractive to try to mimic this mechanism by mounting two imidazole rings on opposite sides of the cyclodextrin cavity, so that they could have the same 180° relationship with respect to the transition state for a hydrolysis. While we were developing techniques to achieve such opposite side functionalization of a cyclodextrin, Tabushi et al. (49) reported the bridging of β-cyclodextrin by diphenylmethane-4,4'-disulfonyl chloride. We found (50) that his bridged compound could be used to attach two imidazole rings on opposite sides of the cyclodextrin cavity, linked to the primary carbon atoms.

Of course since β-cyclodextrin has seven glucoses in a ring, no two of the C-6 carbon atoms really have an opposite side relationship. The closest would be attachment to carbon atoms 6A and 6D, involving C-6 of glucose residues A and D. On further investigation (51) we found that the bridged compound reported by Tabushi et al. was in fact a mixture of 6A, 6D and 6A, 6C isomers, but we (51) and he (52) developed methods to achieve cleaner specific difunctionalization of cyclodextrin. Interestingly, we found (51) that the bifunctional acid–base mechanism characteristic of the enzyme ribonuclease could be achieved not only with the compound carrying imidazoles attached to 6A and 6D, but also with the isomer with imidazoles attached at 6A and 6C, which are by no means on opposite sides of the cavity. This can be understood when it is realized (Fig. 22) that the catalytic mechanism requires that the oxygen atoms have a 180° relationship with each other, but not necessarily the imidazole groups. Their interaction with protons binding to electron pairs on the oxygen atoms can be achieved if the catalytic groups themselves are as close as those in the 6A, 6C isomer.

Of course it would be of interest to use such a catalyst to cleave RNA itself or a simple nucleotide derived from RNA, but molecular model building made it fairly clear that the geometry would not be ideal. For this reason we went to a synthetic substrate, a cyclic phosphate, with the correct geometry to interact with this catalyst. Part of the price of choosing the cyclodextrin nucleus for binding is the necessity to select substrates that fit it, rather than arbitrary substrates of interest.

The most interesting results (50) were obtained with the cyclic phosphate derived from 4-*t*-butylcatechol. The *t*-butylphenyl ring was known to bind into a β-cyclodextrin cavity, and molecular models suggested that one of the imidazoles in the molecular complex could act as a base to deliver the water molecule for hydrolysis while the other one, as an imidazolium cation, could act as an acid to protonate the leaving group. In fact, this bifunctional mechanism was used by the catalyst, as demonstrated most clearly by the pH–rate profile (Fig. 23).

The observed rate of hydrolysis of the substrate–catalyst complex as a function of pH followed almost exactly the calculated curve, assuming that both imidazole rings had their normal titration pKs (50). At high pH the catalyst was a little more effective than expected by this mechanism, indicating that the acidic group was helpful but not absolutely required.

The contrast of this behavior with that of the enzyme is sensible. In the enzyme reaction the oxygen that must be protonated is on a saturated carbon atom and would not form a very stable anion, while in our substrate it is a phenoxide anion that does not absolutely require a proton to be supplied. As expected from this finding, cyclodextrin monoimidazole is also a base catalyst for the cleavage of the cyclic phosphate group, but it is not as effective as the bifunctional catalyst with two imidazole rings. The process is in fact catalysis, since the product can dissociate from the cyclodextrin and turnover is observed. However, there is product inhibition of the catalytic reaction, with the product bound about as effectively as the starting material is bound.

Molecular models suggested that this mechanism should be specific with respect to the cleavage reaction (Fig. 24). The enzymatic

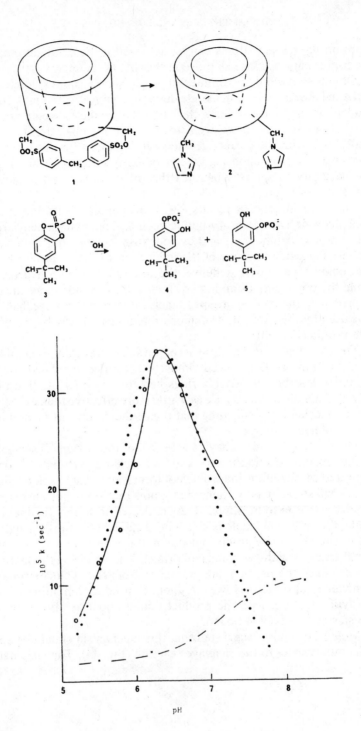

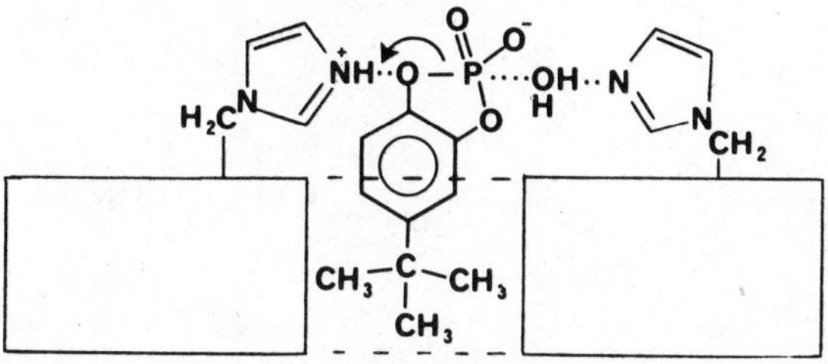

Figure 24. The catalytic mechanism, deduced from molecular models and the pH–rate data, that explains the selective formation of product **5** (Fig. 23).

process cleaves a 2,3-cyclic phosphate to leave the phosphate group on C-3, and we expected a related selectivity with our model system. This was actually observed (50). The cleavage of the substituted catechol cyclic phosphate occurred largely so as to leave the phosphate group attached to position 2; less than 10% of the product had the phosphate attached to position 1. By contrast, simple chemical hydrolysis of this substrate with various buffers, including imidazole buffers, gives esentially random cleavage of the phosphate ester. The major product in the catalyzed reaction is derived by delivery of water from close to the cyclodextrin cavity. Cleavage in the other direction would require that the water approach from a more remote direction.

We have also reported (51) the synthesis of another series of catalysts, in which the imidazole ring is attached by a flexible link in such a way that it can reach further out into solution. As expected from this, the species can deliver water from the more remote direction, cleaving the cyclic phosphate in the other way (Fig. 25). We had originally thought that it was selective for such an alternative mode of cleavage, but reinvestigation of the process indicates that the original product identification was in error. Instead this more

Figure 23. The cyclodextrin bisimidazole 2 and its precursor 1. Substrate 3 is hydrolyzed, in a complex with 2, with a pH–rate profile (solid line) closely fitting the theoretical curve (dotted line). The dashed line is the curve for catalysis by cyclodextrin monoimidazole. The catalyzed hydrolyses of 3 selectively form product 5, in contrast to simple hydrolysis that forms a 4,5 mixture.

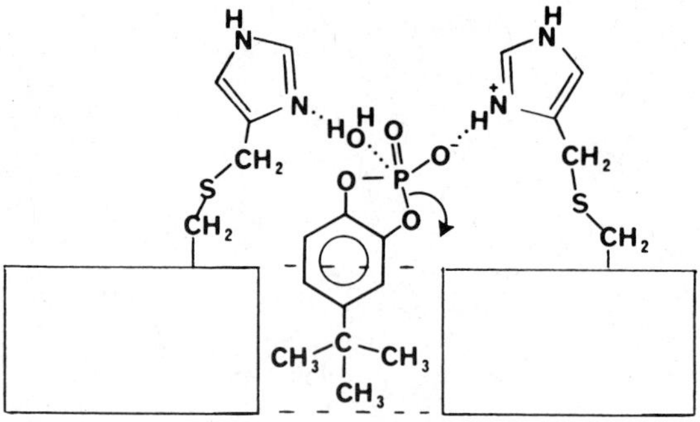

Figure 25.   A catalyst with less defined geometry, which can cleave substrate **3** (Fig. 23) to both **4** (shown) and **5**.

flexible bifunctional catalyst, which can deliver water from either a remote or a nearby position, gives pretty much random cleavage of the cyclic phosphate in either of the possible directions. In order to achieve a selective *reversal* of the original cleavage direction, one would presumably need a rigid base that could not come close to the cavity.

It is interesting that the pH–rate profile (51) for this more flexible catalytic system was also bell shaped, indicating acid–base cooperative catalysis, but with a significant difference from the profile of the previous catalyst. The kinetic pH–rate plot did not correspond to titration pKs for the catalyst itself, but instead showed that the proton is more strongly bound in the transition state than it was in the simple catalyst. For this reason we believe that the acid group in this case is coordinated to the phosphate anion, not to the leaving oxygen. It thus serves the function that is sometimes assigned to the lysine ammonium group. A full model for this enzyme would carry a third catalytic group, and would have the imidazoles more immobilized than they are in either of these simple model systems.

The original model system (50) showed good selectivity with respect to substrate and also good selectivity in product formation. We set out to try to incorporate such selectivity into a model reaction that performs useful synthetic chemistry, and in which selectivity was desperately needed.

The enzyme reactions for which pyridoxal phosphate or pyridoxamine phosphate are coenzymes are many, and sometimes with the same substrate. For instance, pyridoxal phosphate can be the cofactor in the conversion of serine to hydroxypyruvate by transamination, to pyruvate and ammonia by dehydration, to glycine and formaldehyde by reverse aldol reaction, to tryptophan by reaction with indole, and so on. The enzymes for each of these processes are able to impose selectivity on a chemically versatile process, selecting the substrate, the chemistry to be performed, and the geometry of the products. We set out to impose such selectivity in artificial enzymes incorporating pyridoxal and pyridoxamine systems.

Our first approach to this area was simply to attach pyridoxamine to β-cyclodextrin, so that we could get substrate selectivity by using cyclodextrin binding (53). The CH_2OH group of pyridoxamine was converted to a CH_2SH group, and this was used to form a thioether by displacement on β-cyclodextrin-6-tosylate. The attachment of the cyclodextrin ring added two kinds of selectivity to the transamination reaction (Fig. 26).

With simple pyridoxamine, we found (53) that pyruvic acid, phenylpyruvic acid, and indolepyruvic acid all had comparable reactivities, when studied in competition. Pyridoxamine was able to convert these ketoacids to the corresponding amino acids at essentially the same rate. If a molecular of cyclodextrin was added to this reaction mixture the results were similar, although in this case the aromatic keto acids were a little less reactive because they were to some extent bound to the cyclodextrin. However, when the cyclodextrin group was *attached* to the pyridoxamine the aromatic keto acids were considerably more reactive.

In our earliest study (53) we reported that the indole compound, for instance, forms as much tryptophan in 10 min with the pyridoxamine–cyclodextrin molecule as it forms in 30 h with simple pyridoxamine alone (Fig. 27). This would correspond to a 200-fold rate increase, but of course this is a rather rough kinetic method. In later work (54) we examined the actual initial rates carefully, and found that the true acceleration is of the order of 70-fold. In the 30 h required for the pyridoxamine reaction there are side reactions that diminish the overall yield. Even the improved kinetic runs were done under conditions of partial saturation, so they do not represent the maximum rate advantage for the cyclodextrin–pyridoxamine re-

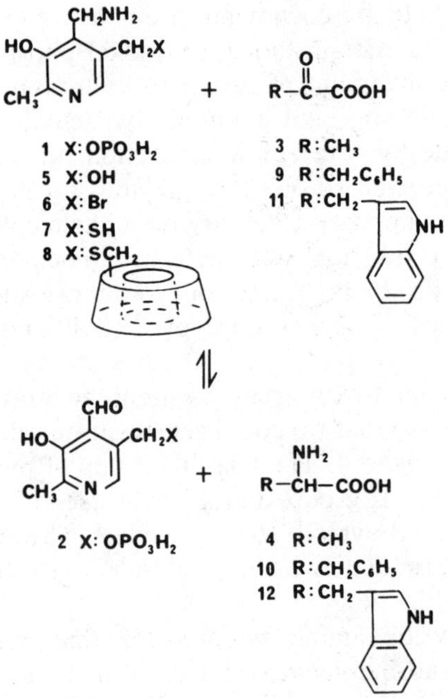

Figure 26.   Conversion of pyridoxamine (5) to the cyclodextrin derivative 8 through intermediates 6 and 7. The cyclodextrin derivative selectively transaminates aromatic keto acids 9 and 11, which bind to the cyclodextrin unit.

agent. It seems fair to say that an acceleration of the order of 10^2 results for substrates that can use the binding site, but not for those that cannot.

We found that simple substrates such as pyruvic acid had essentially the same reactivity with pyridoxamine whether the cyclod-

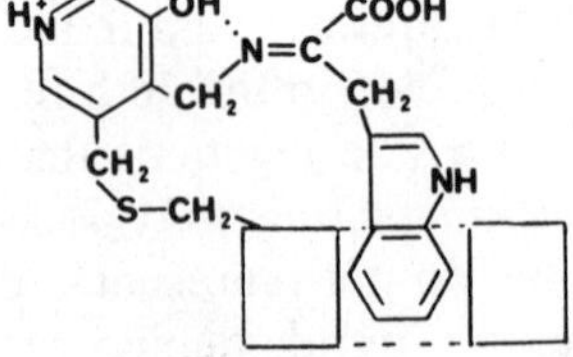

Figure 27.   A transamination intermediate with cyclodextrin binding.

extrin binding group was attached or not. As a result, when the artificial transaminase enzyme was presented with a mixture of pyruvic acid and indolepyruvic acid, it showed typical enzymatic selectivity in forming tryptophan with no detectable formation of alanine. On standing, the cyclodextrin–pyridoxal that was formed was able to reverse the reaction and set up an equilibrium, in which both amino acids were present. Other studies showed that we can achieve as many as 10 turnovers: A keto acid is converted to the corresponding amino acid, and a different amino acid is then sacrificed in the reverse direction to achieve the overall transamination process.

This molecule also showed stereoselectivity. In contrast to the situation with simple pyridoxamine, the pyridoxamine–cyclodextrin hybrid produces amino acids inside a chiral cavity, and this influences the stereochemistry of the product. We found (55) a 5:1 preference for the formation of L-phenylalanine by this catalyst. Thus the asymmetric cavity of this artificial transaminase enzyme achieves a rather good selectivity in the formation of the optically active amino acid. The stereoselectivity is expected in general, but is not obvious in detail from an examination of the molecular models. We shall discuss later another rationally directed approach to the stereochemistry of such transamination processes.

As part of the general concern about the geometry of binding of substrates into an uncapped cyclodextrin cavity, we wanted to examine a case in which pyridoxamine is attached to the secondary side of the cyclodextrin molecule. Although the work we have described previously indicated that both faces should be suitable for the attachment of such functional groups, it seemed worthwhile to see whether the stereochemistry and rate accelerations would differ when the secondary side attachment was used. In general a more rigid attachment is possible on the secondary face, since the free rotation of the C–CH$_2$ groups is not present. The stereochemistry is also more obviously expressed on the secondary face, since attachment would be directly to an asymmetric carbon atom. We had found (45) a good procedure for preparing the C-2 tosylate of β-cyclodextrin, by a tosyl transfer reaction, so we used this material (Fig. 28). We found that this C-2 tosylate derivative of cyclodextrin rapidly closed to the 2,3-epoxide, and that this epoxide could be opened with various thiols (55).

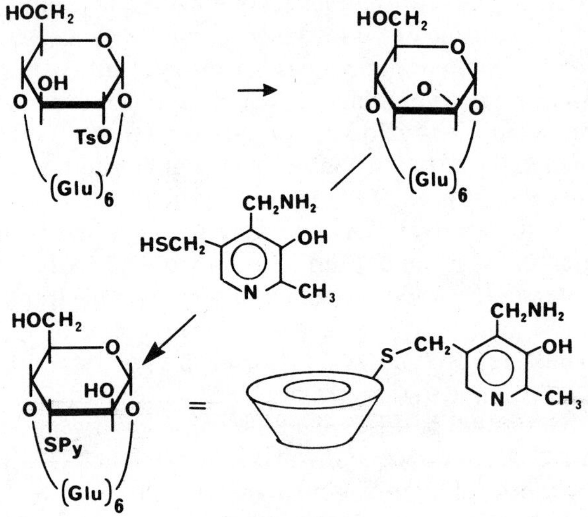

Figure 28. Conversion of β-cyclodextrin-2-tosylate to a pyridoxamine derivative attached at C-3.

The general rule in sugar chemistry is that epoxides open by axial attack of the nucleophile. This would require that the 2,3-epoxide of β-cyclodextrin open with attack at C-3, and this attack would have to come from *inside* the cavity of the molecule. One might wonder whether the normal rule would still apply in this case, but apparently it did. NMR data showed that the displacement reaction on the epoxide indeed occurred at C-3. This should have resulted in a cyclodextrin molecule with an axial 2-hydroxyl pointing away from the cavity and an axial pyridoxamine pointing into the cavity. However, the NMR data also showed that the pyridoxamine unit is attached as an equatorial substituent.

There are really only two ways in which this could happen. One would be a ring opening of the epoxide at C-3 with retention of configuration, totally out of the question unless there are special structural factors not present in this system. The other choice, which is most likely, is that the original attack was axial; the chair glucose ring on which the two axial groups were attached then underwent a chair flip in its conformation, which would make these groups both equatorial in the final stable conformation. Models show that such a chair flip in one of the glucose rings produces an indentation into the cavity of the cyclodextrin, but it does not decrease the size of the cavity to the point at which an indole ring could not fit. Thus in this molecule we expected again to see the preferential binding of aromatic keto acids. The artificial transaminase with pyridoxamine attached to the secondary face of the cyclodextrin should also show both substrate selectivity and product stereoselectivity.

This new pyridoxamine derivative did again show a preference for indolepyruvic acid or phenylpyruvic acid (55), but the preference was only about half as large as that with our original molecule in which the pyridoxamine was attached to the primary side. The stereochemical results were curious. In contrast to the previous artificial transaminase, this new one showed essentially no detectable stereoselectivity in the synthesis of phenylalanine. However, it did show about a 2:1 preference for one enantiomer in the synthesis of tryptophan, comparable to the magnitude of the preference that had been exhibited by the previous reagent, but now the preference was in the reverse direction. That is, the primary pyridoxamine–cyclodextrin compound showed a 2:1 preference for the formation of L-tryptophan, while the secondary cyclodextrin–pyridoxamine compound showed a 2:1 preference for the formation of the D-tryptophan. This reversal of stereochemistry is interesting, but it again emphasizes the desirability of a rational control of stereochemistry (to be discussed) if such methods are to be used for stereoselective synthesis of amino acids.

The original argument for using cyclodextrin as a binding group was that it was conveniently available, and would permit us to learn quickly what could be achieved with the combination of catalytic groups and binding groups. However, at this point it seemed desirable to generalize the chemistry a bit, and look at a different binding cavity that also used hydrophobic binding forces and that also would therefore be able to operate in polar solvents such as water.

A number of synthetic macrocyclic systems have been made that have cavities, but in only a few cases is there real evidence that these cavities persist in solution and that substrates bind into them. In other cases the possibility has not yet been excluded that substrates simply bind to the face of a synthetic molecule; in some cases it has been demonstrated that indeed this is what happens. For this

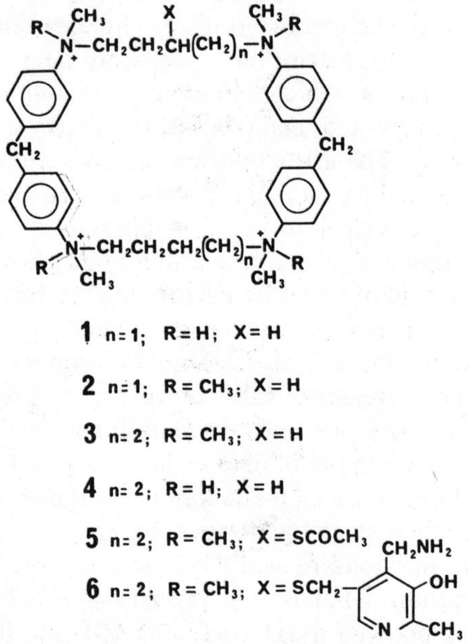

1 n = 1; R = H; X = H

2 n = 1; R = CH₃; X = H

3 n = 2; R = CH₃; X = H

4 n = 2; R = H; X = H

5 n = 2; R = CH₃; X = SCOCH₃

6 n = 2; R = CH₃; X = SCH₂—

Figure 29. Known macrocyclic binding rings (**1** and **3**) and our related quaternary salts (**2** and **4**). The pyridoxamine derivative **6**, prepared from **5**, shows substrate selectivity similar to that of the cyclodextrin–pyridoxamine catalysts.

reason we were very attracted to cavities that had been introduced by Koga and his co-workers (56), building on some earlier work by Stetter.

Koga's group had studied some tetraamines (Fig. 29) that were soluble in strong acid solution and that bound hydrocarbon pieces such as naphthalene rings and durene into the cavity of the molecule. NMR methods had been used to establish the position of the substrate in the cavity, and an X-ray structure determination of a solid complex showed that at least one of these substrates was inside the cavity, as deduced from the NMR work. We modified the Koga compounds in a simple way (54), alkylating the nitrogen atoms so that the molecule would be soluble in water at any pH, and did a preliminary study to show that the resulting species was fundamentally the same as the one studied by Koga. Binding constants, and

NMR spectra of bound molecules, were so similar that it was completely clear that our modified cavity would indeed bind molecules into this interior.

We then built a derivative of this system (Fig. 29) in which a functional group was attached to one of the chains comprising a side of the ring, so that groups such as pyridoxal could be attached (54). Molecular models showed that the transition state for a transamination reaction should also be able to fit this species pretty much as it had the cyclodextrin derivatives, with a phenyl or indole ring bound into the cavity. In fact we found that this molecule was just about as good as either of the cyclodextrin derivatives with respect to substrate selectivity. That is, it showed a significant preference for aromatic keto acids compared with small aliphatic ones that could not utilize the binding interaction. Of course this synthetic cavity is not chiral, so it has no possibility of producing optical activity in the product amino acids. However, in general it seems likely that totally synthetic binding cavities will become quite important in artificial enzymes because they can be tailored to fit a particular substrate of interest. The cyclodextrins come with a built in shape, which can only partly be altered by chemical modification.

Of course a transaminase enzyme, or more generally any enzyme that catalyzes transformations using pyridoxal or pyridoxamine, consists of more than a binding site. It is of interest to add the extra catalytic groups used by such enzymes into our artificial enzyme systems. In transamination, a critical catalytic group is the basic group of the enzyme that removes a proton from one carbon atom and transfers it to another carbon atom. There is reasonable evidence that such transfers do occur with a single basic group, which in transamination must remove the proton from the CH_2 group of the pyridoxamine unit and move the proton to the α carbon of the product amino acid (Fig. 30). If we are to imitate such enzymes in detail, we certainly must incorporate a basic group of this kind.

Our first work on this problem addressed the question of whether we could get evidence that a basic group in a simple enzyme model could indeed perform the proton transfer required for transamination. We studied this (57) by attaching a series of bases to the C-5 CH_2 group of pyridoxamine, the same group that we had used previously to attach pyridoxamine to various binding rings (Fig. 31). We developed reaction conditions in which the pyridoxamine re-

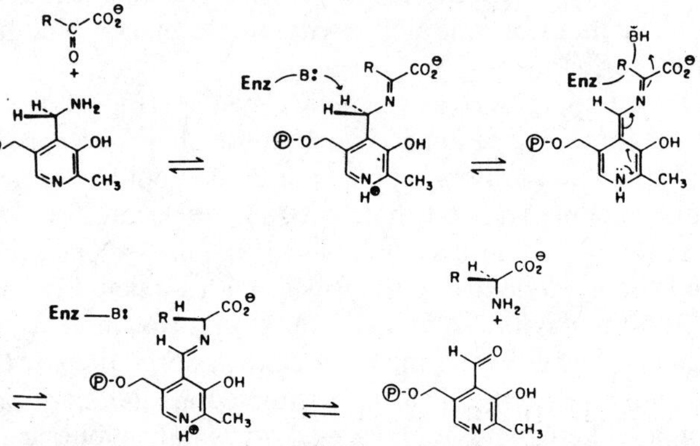

Figure 30. An enzyme basic group performs a proton transfer during transamination.

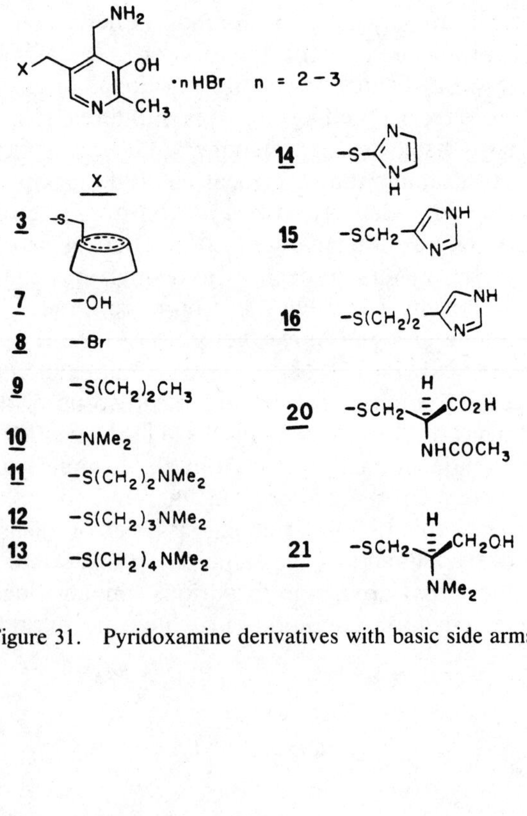

Figure 31. Pyridoxamine derivatives with basic side arms.

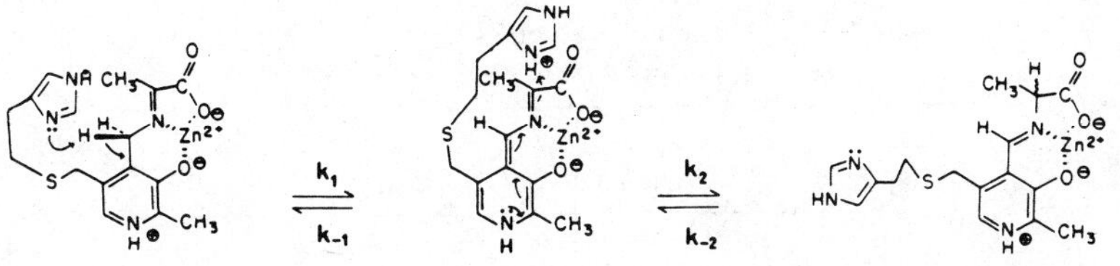

Figure 32.   The proton transfer steps in transamination.

acted rapidly with a keto acid to form the corresponding ketimine intermediate, and the rate-limiting step was the isomerization of this ketimine to the aldimine formally derived from pyridoxal and the amino acid (Fig. 32). This isomerization is a key step in transamination, in which the proton is removed from one carbon atom and placed on another (the overall process of course requires catalysis of the other steps as well). With a kinetic system that would let us observe this key proton transfer, we were in a position to evaluate our ability to catalyze it with bases held in a well-defined position within the molecule.

As Table I shows (57), the most effective catalysts for this isomerization were not those with a basic group held near the original CH_2 group of the pyridoxamine unit. Instead, the most effective catalytic systems had a basic nitrogen at the end of a rather long chain, which would permit it to reach out to the α carbon of the developing amino acid. This seemed to indicate strongly that the base was doing more than simply remove the proton from the pyridoxamine CH_2 unit; it was also moving that proton out to the carbon atom of the developing amino acid.

Unfortunately we could not get direct evidence on this. Fast exchange processes in the course of the reaction were such that we could not directly detect transfer of a hydrogen isotope from one of these carbons atoms to the other, although we looked carefully for such internal transfer. However, in general longer more flexible side arms should lead to slower reaction rates unless that extra length and flexibility is required for the catalyzed process. Molecular models suggested that the optimum system observed, with one sulfur and three methylene groups linked to a dimethylamino base, or with

TABLE I
Rates of Conversion of Ketimines (e.g., 17) to Aldimines
(e.g., 19) at pH 4.00 in Methanol (30.0°C)[a]

Compound	Side chain	$k_{obsd}(s^{-1})$[b]	Relative rate
9	SPr	8.7×10^{-6}	1.0
7	OH	1.2×10^{-5}	1.4
10	NMe$_2$	1.3×10^{-4}	15.0
11	S(CH$_2$)$_2$NMe$_2$	2.3×10^{-4}	26.0
12	S(CH$_2$)$_3$NMe$_2$	3.3×10^{-4}	38.0
13	S(CH$_2$)$_4$NMe$_2$	1.1×10^{-4}	13.0
14	S(Im)	5.4×10^{-5}	6.0
15	SCH$_2$Im	1.1×10^{-4}	13.0
16	SCH$_2$CH$_2$Im	6.8×10^{-4}	78.0
20	N-acetylcysteine	9.6×10^{-5}	11.0
21	N,N-dimethylcysteinol	2.3×10^{-4}	26.0[c]

[a] Structures shown in Figs. 31 and 32. pH as read with a glass electrode. The pHs were unchanged at the end of the reaction.

[b] Standard deviation within each run less than 1%; duplicate runs usually within 1%, with a few within 10%.

[c] With α-oxovaleric acid as substrate, which is ~20% slower than pyruvic acid.

one sulfur and two methylene groups linked to an imidazole, were of the correct length to be able to put the proton onto the developing α carbon of the product amino acid.

The kinetics as a function of pH showed that indeed we were dealing with a species carrying one extra proton, presumably on the pyridine ring. Thus the mechanism that seemed most sensible for the catalyzed process we were observing was one in which the flexible chain curled up to permit the base to remove a proton from the CH$_2$ group of the pyridoxamine in the ketimine intermediate, and then the chain straightened out to permit the resulting protonated base to put the proton on the remote carbon atom so as to form the aldimine intermediate. This interpretation was strengthened by the finding that an optically active side chain basic group was able to induce some optical activity in the product, as expected if it was indeed delivering the proton to the product carbon atom.

Emboldened by this, we constructed another model system (58).

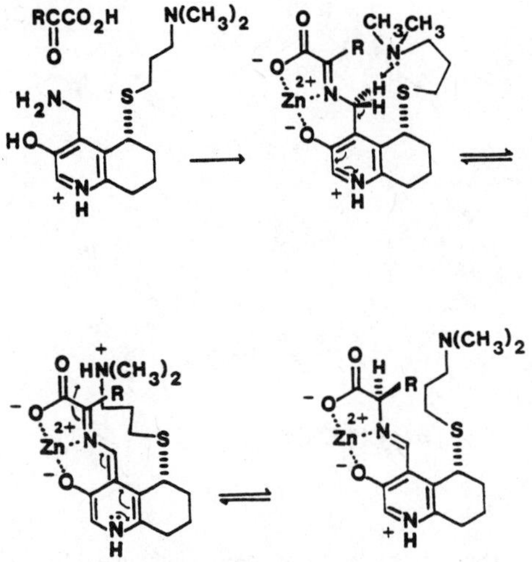

Figure 33. A basic side arm that directs transaminations with very high optical inductions.

A chain carrying a basic group was to be held in such a way that it could reach only one face of the reaction intermediate in transamination, so that this chain should be able to give good optical induction in the formation of the product. To achieve the required specification of geometry, we had the chain come off a ring fused to the pyridoxamine unit, and we used a metal ion to bind the carboxyl group of the ketimine intermediate so that that intermediate had a well-defined orientation also (Fig. 33). A somewhat laborious synthesis produced the molecule, optically resolved, in which one of the optical isomers examined had the chain held in such a way that it could reach the *re* face of the intermediate. This system produced optically active amino acids with as much as 96% of the expected D isomer. When a simple alkyl chain, not carrying a basic group, was mounted in the same stereochemical position, it shielded the *re* face to some extent, leading to a small but real preference for the formation of the L-amino acid product.

These results have stimulated our current work. We are combining a pyridoxal or pyridoxamine coenzyme, a binding group, and a catalytic base unit in an optically selected sense so as to achieve

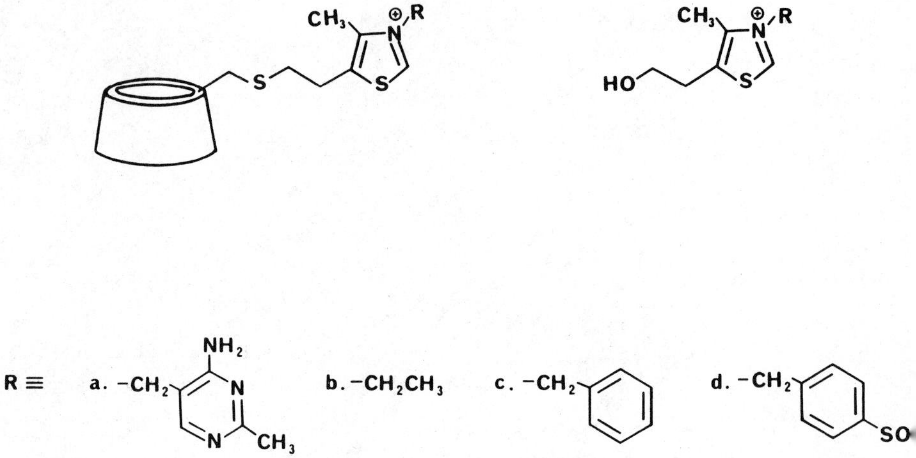

Figure 34. Thiamine and related compounds linked to cyclodextrin.

some of the most attractive aspects of selectivity seen in the actual enzyme catalyzed reactions with these cofactors.

In other recent work (59), we have attached thiamine to β-cyclodextrin (Fig. 34), and have examined the ability of this binding group to promote some selective reactions for which thiamine pyrophosphate acts as a coenzyme. We found that the cyclodextrin will indeed promote the preferential reaction of a thiazolium salt, such as thiamine or related compounds, with substrates that can bind to the cavity. The magnitudes of the acceleration are modest, similar to those we have described for the pyridoxamine system. However, in this work we also found that the resulting intermediate, bound into a cyclodextrin cavity, may not be particularly reactive toward a second bulky reagent. Thus the cyclodextrin unit can promote reactions such as the thiamine-catalyzed oxidation of aldehydes, but it is not a good promoter for the thiamine catalyzed coupling of the two aldehyde molecules to form a benzoin derivative. Similarly, in some of these systems groups that are part of the coenzyme can bind into the cavity, and diminish the total reactivity of the molecule. In the conversion of small catalyst systems into enzyme mimics by the attachment of binding groups, it is critical to pay attention to problems that may arise from binding of pieces of the catalyst into the cavity, or from binding of reaction intermediates that might block further transformations.

In principle the attachment of a binding group to a cofactor could be used to hold a second reagent near an intermediate, rather than promote the formation of the intermediate. We have observed such an effect in the reaction of serine with indole to form tryptophan (60). Pyridoxal catalyzes this reaction in the absence of an enzyme, but very poorly. In the enzyme tryptophan synthetase, pyridoxal phosphate promotes this reaction by catalyzing the dehydration of serine to a reactive intermediate that then couples with a bound indole. We have duplicated this feature of the enzyme reaction by the use of the molecule (Fig. 26) in which a pyridoxal ring is attached as a thioether to the primary side of β-cyclodextrin. We find (60) that this species is a better catalyst for the synthesis of tryptophan, although of course it will be fully optimized only when the additional catalytic groups are present. However, the improved catalysis of tryptophan synthesis by this synthetic tryptophan synthetase must reflect the simultaneous binding of serine to the pyridoxal unit and of indole into the cyclodextrin cavity. This promotion of the coupling of two separate reagents by binding both of them is also reminiscent of the catalysis of the Diels–Alder reaction described earlier.

IV. Remote Oxidation and Related Reactions

A major purpose in the construction of artificial enzymes and enzyme models is to learn how to perform new chemistry. The remarkable selectivity of enzyme reactions is very much worth imitating in new synthetic reactions, developed to improve the tools of chemistry. Nowhere is this more true than in the ability of enzymes to carry out reactions at completely unactivated saturated carbon atoms, while ignoring other sections of a molecule that would be much more reactive in any normal chemical process (Fig. 35). Enzymes can oxidize methyl groups on saturated carbons atoms of steroids, even though such methyl groups are probably the least reactive part of any organic molecule. Such attack on these inactive carbon atoms can occur in steroid molecules that have much more reactive other regions.

$$CH_3(CH_2)_{16}CO_2H \xrightarrow[\text{oxidation}]{\text{enzymatic}}$$
stearic acid

$$CH_3(CH_2)_7CH=CH(CH_2)_7CO_2H$$
oleic acid

Figure 35. An example of regioselective enzymatic attack on unactivated C–H groups.

Certainly the secret of such selectivity is geometric. That is, the enzyme binds the substrate in such a way that an intrinsically non-selective functional group, capable of attacking pretty much any kind of C–H bond, is held so as to attack the desired bond and no other. It seemed to us that this general style, which must certainly be true regardless of the details of the chemistry involved, was a sufficiently powerful general approach that we should try to imitate it. We wanted to learn how to perform geometrically controlled selective reactions.

Much of our work was directed at selectively functionalizing un-activated C–H bonds, but some of it was aimed at performing se-lective reactions on particular double bonds while ignoring others. Since we have written an extensive review of this work a few years ago (2), we shall simply outline the principal findings and recent progress, and show how this area can be developed in the direction of true artificial enzymes.

A. BENZOPHENONE PHOTOCHEMISTRY

The first question was whether it would indeed be possible to direct the attack on an unactivated C–H bond by using simple geo-metric control. To examine this question, we attached a rigid ben-zophenone molecule to a flexible chain and then examined the result of photolysis of this species (Fig. 36) (61). We found that indeed the photoexcited benzophenone molecule would attack only those C–H bonds that were geometrically accessible, but that the flexibility of the chain meant that a fair range of such bonds could actually be reached. To improve the selectivity of such processes, two ap-proaches were used. In one of them we used more rigid substrates, while in the other approach we tried to impose additional constraints to immobilize the flexible substrates.

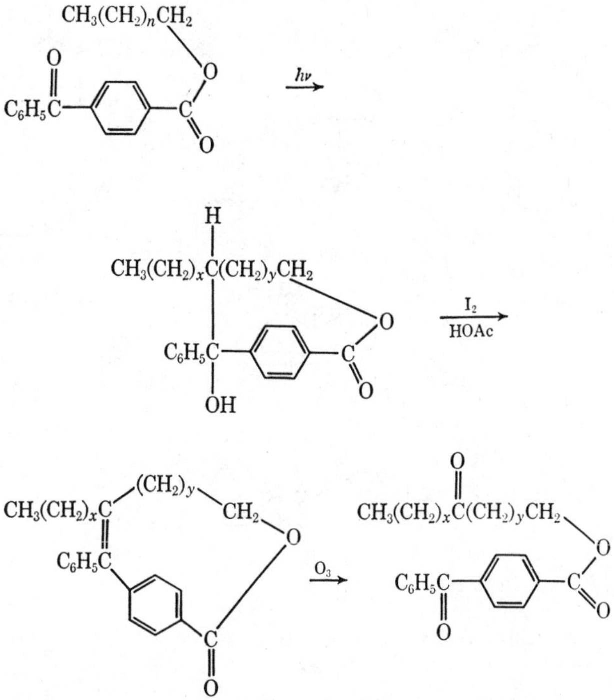

Figure 36. Geometrically directed functionalization of a flexible chain. The flexibility results in only moderate selectivity.

The rigid substrates selected were steroids. These are molecules of considerable interest, and so much of their chemistry has already been worked out that we were confident about our ability to discover easily the nature of any transformations performed on them. We found that when a benzophenone reagent was attached to a steroid, it would perform C–H attack, on irradiation, only at positions that were geometrically accessible as judged from molecular models (62–66). Sometimes several such positions could be reached, but in other interesting cases highly selective reactions occurred.

One of the most striking was the finding (Fig. 37) that we could introduce a double bond between carbon-14 and carbon-15 in ring D of the steroid, starting with a rigid benzophenone attached at C-

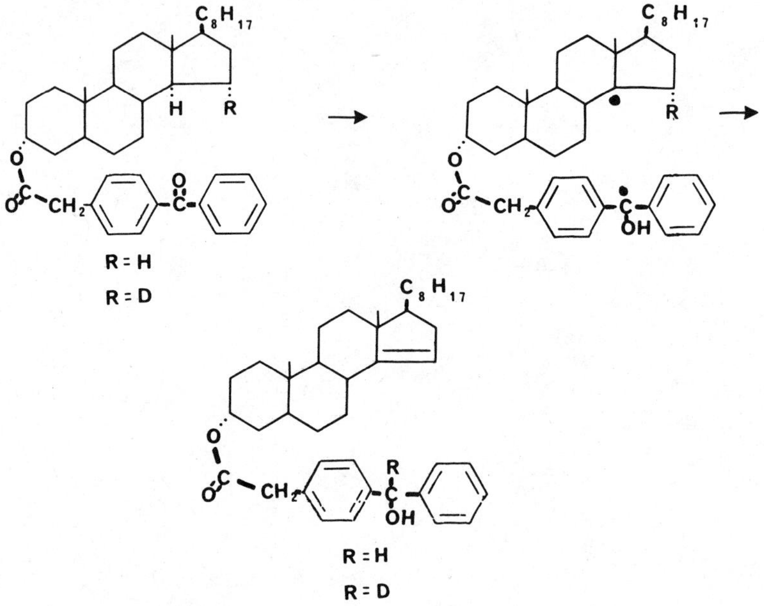

Figure 37. A highly selective geometrically controlled functionalization of a rigid steroid substrate.

3 in ring A of the steroid. This highly selective reaction was predictable from the geometry of the system. Labeling studies demonstrated that the process did indeed involve the geometrically controlled removal of two hydrogen atoms on the α face of the steroid, from carbon atoms 14 and 15. The benzophenone group was attached to the α face of the steroid, so it performed its selective chemistry on that face. In very recent work (67) we have shown that a similar benzophenone ester can be attached on the β face of the steroid, and that it will give predictable attack at geometrically accessible positions on that face. By adjusting the geometry of the benzophenone derivative, it has been possible to move the positions of functionalization in a controlled and predictable way.

Much of this selectivity comes from the fact that both the substrate and the reagent are relatively rigid species. Not many conformations are available to them, and only a few positions can be reached in an intramolecular attack. We examined the possibility that a flexible long chain substrate might also be attacked selectively if the chain was incorporated into an organized structure such as a micelle or bilayer membrane. We found interesting changes in the positions attacked (68,69), reflecting the effect of such organized structure, but in no case was the organization sufficient to give us highly selective reactions that might be synthetically useful in producing single products.

Of course in an enzyme it is likely that the flexible chains of molecules such as stearic acid are completely immobilized by binding into the enzyme cavity; the selective conversion of stearic to oleic acid, for instance, involves attack on a chain that is bound in such a way as to have lost all of its flexibility. We achieved a related immobilization of an otherwise flexible chain by incorporating two binding points (70). That is, we were able to achieve rather good selectivity in the attack on a long chain dicarboxylic acid, with a carboxyl at each end, by stretching this chain across a rigid benzophenone reagent and then attacking the middle of this stretched out chain by irradiating the benzophenone unit (Fig. 38).

The interactions used to stretch out such a difunctional substrate involved complexing. Fairly good selectivity was achieved using hydrogen bonding, in which each carboxyl shared a proton with a carboxylate ion of the reagent. Even better definition of geometry was obtained by using a reagent with two positively charged groups located in such a position that they could form ion pairs with the

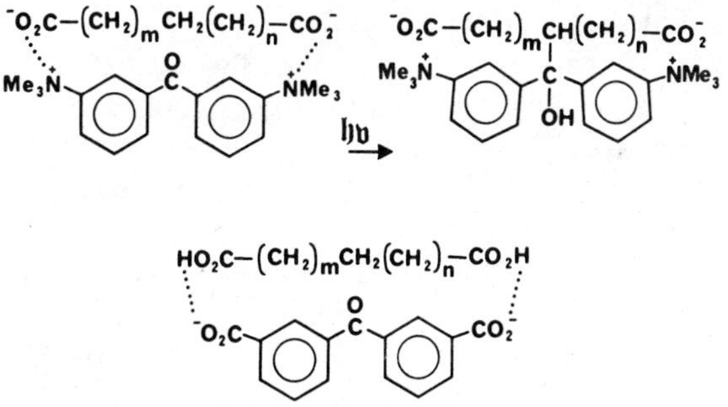

Figure 38. Selective functionalization of flexible substrates immobilized by two binding interactions.

carboxylate anions of the substrate. When the substrate was precisely long enough to stretch out between these two positively
charged points of the reagent, a very selective reaction in the center
of the molecule resulted. When the chain was made longer, it had
several conformations available to it and the reaction became less
selective.

With the less selective reactions observed here, or in the flexible
chain (71), micelle, or membrane systems, a considerable amount
of information about conformations of the chains was obtained from
analyzing the products of reaction. Useful as such information is,
the major goal of this work is to learn how to perform highly selective
reactions by freezing in a single conformation of the substrate.

B.   TEMPLATE-DIRECTED CHLORINATIONS

The finding that we could achieve selective attack on substrates
by a reagent that was simply complexed, using hydrogen bonding
or ion pairing, certainly helped move the processes toward enzyme
imitation. However, the benzophenone species that performed the
selective functionalization was a reagent, not a catalyst. Catalysis
was achieved by developing directed free radical halogenation processes, specifically chlorination reactions.

A free chlorine atom is sufficiently reactive that it can attack
essentially any C–H bond, although it shows the normal preference
for tertiary hydrogen atoms over secondary hydrogen atoms. The
primary hydrogen atoms of a CH_3 group are least reactive. We reasoned that if we could direct such free radical chlorinations in some
way, using geometric control, we might be able to introduce the
selectivity that they normally lack.

As one approach (Fig. 39), we looked at the selectivity of chlorination reactions produced by the attachment of a rigid phenyliodine
dichloride unit to a substrate (71–73). In a free radical chain we
expected that hydrogen removal from the substrate would involve
a radical derived from this species, which corresponds more or less
to a chlorine atom attached to the iodine of an iodobenzene unit.
Data in the literature suggested that a species of this type could
remove hydrogen atoms, and that it showed a considerable preference for attack on tertiary C–H groups. Some chemical selectivity
of this sort was an advantage. The rudimentary geometric control
we were applying involved simply the attachment of a rigid reagent

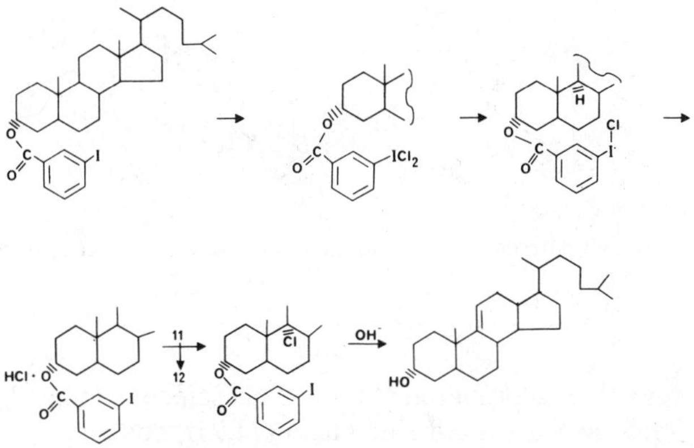

Figure 39. Geometrically controlled chlorination of a steroid. The double bond produced can be used to synthesize corticosteroids.

to a rigid substrate at one point, and a number of degrees of conformational freedom were still available.

We found (74) that the linked reagent–substrate derived by making an ester of 3-α-cholestanol with *m*-iodobenzoic acid, and then attaching the chlorines to the iodine, was able to perform a highly selective free radical chlorination at C-9 of the steroid (Fig. 39). This is a tertiary position, but not otherwise activated. It was of considerable interest, since it let us readily produce the 9,11 double bond that is of importance in the synthesis of corticosteroids. Molecular models showed that this was indeed the tertiary C–H bond that should have been attacked with this reagent, and this was also consistent with distance calculations based on known X-ray structures of relevant compounds.

When the reagent was changed in geometry, to *p*-iodophenylacetic acid, the reaction was again quite selective (Fig. 40), but this time selective for chlorination at C-14. The movement of the iodine from the meta to the para position, and the introduction of an extra CH₂ group, lengthened the reagent by two bonds. The result was to bite the steroid two bonds further away from the point of attachment. Steroids chlorinated at C-14 are of interest in the synthesis of cardiac-active substances. Finally, the use of an even longer reagent, derived from biphenyl, put the iodine out so far that we were able

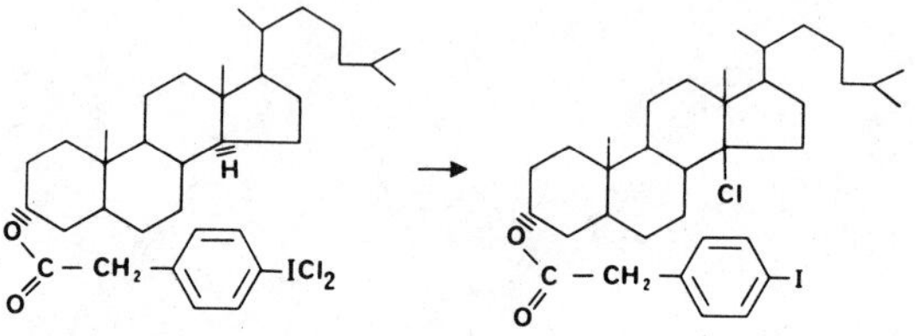

Figure 40. With a different reagent, the same steroid of Fig. 39 can be chlorinated at C-14.

to get a selective chlorination at C-17. This selective attack has been used to remove the steroid side chain (74,75), converting available sterols to useful intermediates.

A number of other geometric modifications were studied, but at the same time the chemistry was changed. It was awkard to attach an iodoaryl group and then convert it to the corresponding dichloride, and sometimes this conversion was a problem because of the reactivity of other groups in the molecule. We decided that the same intermediate radical, with a chlorine atom attached to the iodine, should be formed if one carried out a simple chlorination of a substrate carrying an iodoaryl group (Fig. 41). That is, we expected that a chlorine atom, either free or complexed to some other carrier species, should preferentially react with the iodine atom. The chlorine would then be transferred under geometric control to the correct C–H bond, rather than simply attack the substrate at random. This proved to be the case. A process was invented (74,76), radical relay chlorination, in which the geometrical control was furnished by an iodoaryl template molecule that was temporarily attached to the substrate. This template species captured an incoming chlorine atom, delivered it to a C–H group, and was thus restored to its original condition. Obviously such a template relay process has the potential to be catalytic; a catalytic version of this was indeed developed later.

One might wonder whether the iodine atom is unique in its potential to capture and deliver a chlorine atom. We have found that the sulfur of a diarylsulfide (77) and the sulfur of a thiophene ring (78) are both able to perform this function. Undoubtedly there are other species that also could be used.

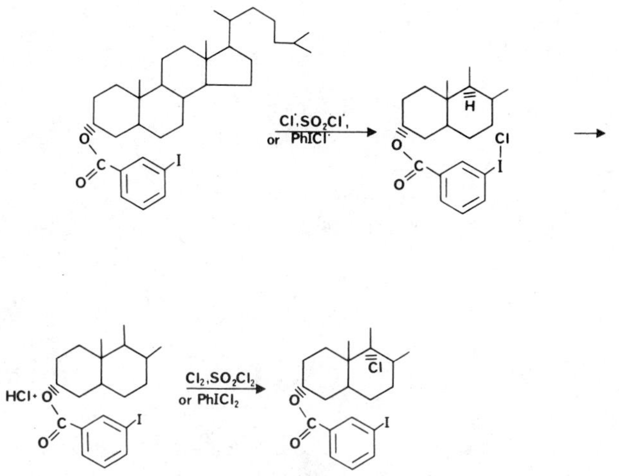

Figure 41. The radical relay mechanism, in which a template catalyzes and directs a chlorination.

One advantage of the radical relay method is that it was possible to use a small excess of chlorinating agent in order to drive the reaction to completion, since the chlorine need no longer be present in stoichiometric amount relative to the substrate. Radical relay halogenation has precisely the same selectivity as that of the previous system involving attached ICl_2 groups, and it has generally replaced the previous method. Two developments have moved this toward a catalytic process that might be closer to an enzyme mimic.

In one scheme (78) we attached one template to three steroids, and showed that the template could functionalize all of them (Fig. 42). That is, an incoming chlorine atom was captured by the template and delivered to one of the steroids, but the template was then regenerated and able to direct attack on the other steroids as well. Interestingly, the geometric selectivity of these methods is good enough that there was not any further attack on an already chlorinated steroid in some other position, so the reaction was quite clean.

The second process involved the use of complexing between templates and substrates to direct this reaction (79). The interaction examined so far is ion pairing, in which an iodophenyl group carries

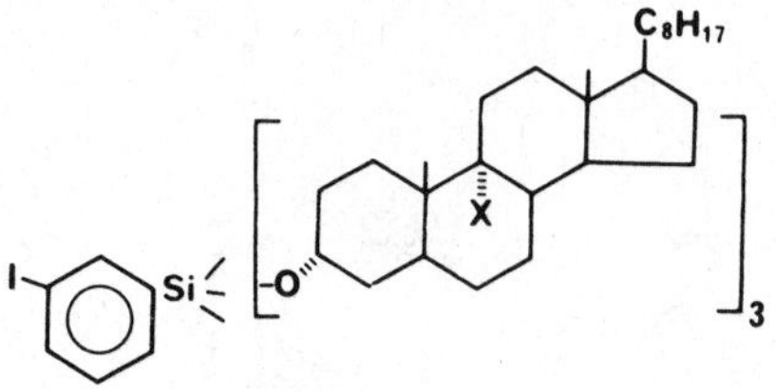

1 X : H

3 X : Cl

Figure 42. The same template can direct the conversion of all three steroids in substrate **1** to product **3**.

a positive charge and the substrate carries a negative charge or vice versa. Reasonably good selectivity of halogenation was seen in these cases also, although the geometric control was not quite as good as with direct ester bonding. A single ion pair is not enough of a restriction on the movement of substrate and catalyst relative to each other. Current work is aimed at developing these techniques using more complex binding interactions. We hope to achieve better geometric control, but retain the feature of a temporary association in a catalytic process.

C. DIRECTED EPOXIDATION

Although the most demanding challenge is the functionalization of saturated C–H groups under geometric control, it is also of some interest to learn how to select among several more reactive groups by geometric methods. One approach was based on the work by Sharpless (80) and others demonstrating that a double bond near a hydroxyl group can be selectively epoxidized. A metal coordinates to the hydroxyl group so as to deliver an oxidizing reagent to the nearby double bond. The Sharpless work all involved systems in which the hydroxyl group was in fact attached to a carbon atom that was a neighbor to the double bond to be attacked, but we reasoned that simple proximity in space should be enough.

We examined a case of a steroid with two double bonds, in which one of the double bonds was near a hydroxyl group and the other was not (81). With the normal Sharpless conditions the double bond

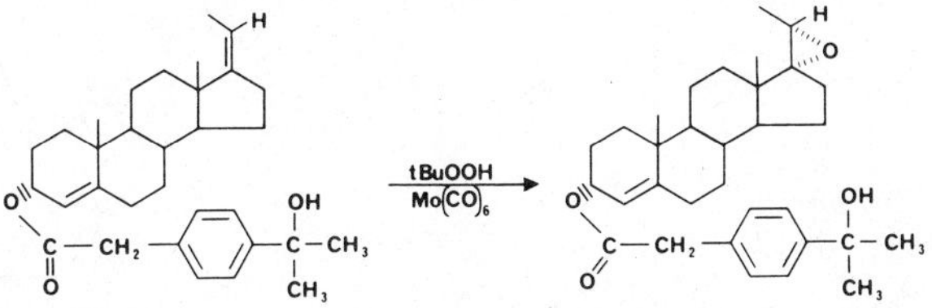

Figure 43. A template directs the regioselective and stereoselective oxidation of a double bond. Without the attached ester the OH group directs oxidation of the other double bond.

in ring A, near the hydroxyl group, was epoxidized. However, when we attached a template to that hydroxyl group (Fig. 43) we carried a new OH out in space into a region near the remote double bond. Now we found that the metal and peroxide reagent used to form the epoxide selectively attacked this remote double bond and with full stereoselectivity. The reaction was very sensitive to the precise structure of this template, as expected if geometric control is determining the reaction. Furthermore, the process was catalyzed by this template; with templates of a different structure that could not deliver a metal near the remote double bond, no attack on either of the double bonds was observed.

Such selective oxidation of particular double bonds in polyenes could also be of interest in a catalytic process. This would require that the template simply be temporarily complexed to the substrate, rather than attached as a covalent ester.

Although we have emphasized the selectivity aspects of this chemistry, the template directed chlorination and epoxidation are in fact accelerated processes. Thus, in the presence of an iodoaryl group steroids can be chlorinated under reaction conditions in which they are recovered unchanged if the iodine atom is not present. Similarly, the epoxidation of a double bond by a metal coordinated to a template hydroxyl group is accelerated, and if such a hydroxyl group is missing or is the wrong geometric location the substrate is recovered unchanged. As we discussed at the beginning of this review, selectivity and increased velocity are different aspects of the same fundamental behavior. With improved geometric control we can expect both the velocities and selectivities to increase.

REFERENCES

1. For a review, cf. Breslow, R., *Science* **218**, 532–537 (1982).

2. For a review, cf. Breslow, R., *Acc. Chem. Res.*, **13**, 170 (1980).

3. Breslow, R., *Chem. Ind. (London)* 28 (1956).

4. Breslow, R., *J. Am. Chem. Soc.*, **79**, 1762 (1957).

5. Breslow, R., *Chem. Ind. (London)*, 893 (1957).

6. Breslow, R., *J. Am. Chem. Soc.*, **80**, 3719 (1958).

7. Breslow, R. and McNelis, E., *J. Am. Chem. Soc.*, **81**, 3080 (1959).

8. Breslow, R., *CIBA Foundation Study Group II*, J.A. Churchill Ltd., p. 65, 1961.

9. Breslow, R., *Ann. N.Y. Acad. Sci.*, **98**, 445, (1962).

10. Breslow, R., and McNeilis, E., *J. Am. Chem. Soc.*, **82**, 2394 (1960).

11. Cf., ref. (12), footnote 3.

12. Breslow, R. and Khanna, P.L., *J. Am. Chem. Soc.*, **98**, 1297 (1976).

13. Breslow, R. and Katz, I., *J. Am. Chem. Soc.*, **90**, 7376 (1968).

14. Breslow, R. and Wernick, D.L., *J. Am. Chem. Soc.* **98**, 259 (1976).

15. Breslow, R. and Wernick, D.L., *Proc. Natl. Acad. Sci. USA*, **74**, 1303–1307 (1977).

16. Breslow, R., Chin, J., Trainor, G., and Hilvert, D., *Proc. Natl. Acad. Sci. USA*, **80**, 4585–4589 (1983).

17. Chin, J. and Breslow, R., *Tetrahedron Lett.*, **23**, 4221–4224 (1982).

18. Breslow, R., McClure, D.E., Brown, R.S., and Eisenach, J., *J. Am. Chem. Soc.*, **97**, 194 (1975).

19. Breslow, R., Fairweather, R., and Keana, J., *J. Am. Chem. Soc.*, **89**, 2135 (1967).

20. Breslow, R. and Schmir, M., *J. Am. Chem. Soc.*, **93**, 4960 (1971).

21. Breslow, R. and McAllister, C., *J. Am. Chem. Soc.*, **93**, 7096 (1971).

22. A. Schepartz, unpublished work.

23. Groves, J.T. and Chambers, R.R., *J. Am. Chem. Soc.*, **106**, 630 (1984).

24. Tang, C.C., Davalian, D., Huang, P., and Breslow, R., *J. Am. Chem. Soc.*, **100**, 3918 (1978).

25. Breslow, R., Hunt, J., Smiley, R., and Tarnowski, T., *J. Am. Chem. Soc.*, **105**, 5337–5342 (1983).

26. Brown, R.S., Salmon, D., Curtis N.J., and Kusuma, S., *J. Am. Chem. Soc.*, **104**, 3188 (1982) and references therein.

27. Breslow, R. and Chipman, D., *J. Am. Chem. Soc.*, **87**, 4195 (1965).

28. Malmin, J., Ph.D. thesis, Columbia University, 1969.

29. Reviewed by B. Cooperman, in *Metal Ions in Biological Systems*, Vol. 5, Chap. 2, H. Higel, Ed., 1976.

30. Breslow, R. and Overman, L.E., *J. Am. Chem. Soc.*, **92**, 1075 (1970).

31. Breslow, R., in *Bioinorganic Chemistry* (*Advances in Chemistry Series*), American Chemical Society, Washington, D.C. 1971.

32. Breslow, R. and Campbell, P., *J. Am. Chem. Soc.*, **91**, 3085 (1969).

33. Breslow, R. and Campbell, P., *Bioorg. Chem.*, **1**, 140 (1971).

34. Breslow, R., Kohn, H., and Siegel, B., *Tetrahedron Lett.*, 1645 (1976).

35. Breslow, R. and Rideout, D., *J. Am. Chem. Soc.*, **102**, 7816 (1980).

36. Breslow, R., Maitra U., and Rideout, D., *Tetrahedron Lett.*, **24**, 1901–1904 (1983).

37. Breslow, R. and Maitra, U., *Tetrahedron Lett.*, **25**, 1239–1240 (1984).

38. van Etten, R.L., Sebastian, J.F., Clowes, G.A., and Bender, M.L., *J. Am. Chem. Soc.*, **89**, 3242 (1967).

39. Breslow, R., Czarniecki, M.F., Emert, J., and Hamaguchi, H., *J. Am. Chem. Soc.*, **102**, 762 (1980).

40. Emert, J. and Breslow, R., *J. Am. Chem. Soc.*, **97**, 670 (1975).

41. Czarniecki, M.F. and Breslow, R., *J. Am. Chem. Soc.*, **100**, 771 (1978).

42. Breslow, R. and Trainor, G., *J. Am. Chem. Soc.*, **103**, 154 (1981).

43. Breslow, R., Trainor, G., and Ueno, A., *J. Am. Chem. Soc.*, **105**, 2739 (1983).

44. Siegel, B. and Breslow, R., *J. Am. Chem. Soc.*, **97**, 6869 (1975).

45. Ueno, A. and Breslow, R., *Tetrahedron Lett.*, **23**, 3451–3454 (1982).

46. le Noble, W.J., Srivastava, S., Breslow, R., and Trainor, G., *J. Am. Chem. Soc.*, **105**, 2745 (1983).

47. Chao, Y., Ph.D. thesis, Columbia University, 1972.

48. Siegel, B., Pinter, A., and Breslow, R., *J. Am. Chem. Soc.*, **99**, 2309 (1977).

49. Tabushi, I., Shimokawa, H., Shimizu, N., Shirakata, H., and Fujita, K., *J. Am. Chem. Soc.*, **98**, 7855 (1976).

50. Breslow, R., Doherty, J., Guillot, G., and Lipsey, C., *J. Am. Chem. Soc.*, **100**, 3227 (1978).

51. Breslow, R., Bovy, P., and Lipsey Hersh C., *J. Am. Chem. Soc.*, **102**, 2115 (1980).

52. For a review, cf. Tabushi, I., *Acc. Chem. Res.*, **15**, 66 (1982).

53. Breslow, R., Hammond, M., and Lauer, M., *J. Am. Chem. Soc.*, **102**, 421 (1980).

54. Winkler, J., Coutouli-Argyropoulous, E., Leppkes, R., and Breslow, R., *J. Am. Chem. Soc.*, **105**, 7198–7199 (1983).

55. Breslow, R. and Czarnik, A.W., *J. Am. Chem. Soc.*, **105**, 1390 (1983).

56. Odashima, K., Itai, A., Iitaka, Y., and Koga, K., *J. Am. Chem. Soc.*, **102**, 2504 (1980).

57. Zimmerman, S.C., Czarnik, A.W., and Breslow, R., *J. Am. Chem. Soc.*, **105**, 1694–95 (1983).

58. Zimmerman, S.C. and Breslow, R., *J. Am. Chem. Soc.*, **106**, 1490–1491 (1984).

59. Hilvert, D. and Breslow, R., *Bioorg. Chem.*, **12**, 206–220 (1984).

60. Weiner, W., Winkler, J., Zimmerman, S., Czarnik, A., and Breslow, R., *J. Am. Chem. Soc.*, **107**, 4093 (1985).

61. Breslow, R. and Winnik, M.A., *J. Am. Chem. Soc.*, **91**, 3083 (1969).
62. Breslow, R. and Baldwin, S.W., *J. Am. Chem. Soc.*, **92**, 732 (1970).
63. Breslow, R. and Scholl, P.C., *J. Am. Chem. Soc.*, **93**, 2331 (1971).
64. Breslow, R. and Kalicky, P., *J. Am. Chem. Soc.*, **93**, 3540 (1971).
65. Breslow, R., Baldwin, S., Flechtner, T., Kalicky, P., Liu, S., and Washburn, W., *J. Am. Chem. Soc.*, **95**, 3251 (1973).
66. Wife, R.L., Prezant, D., and Breslow, R., *Tetrahedron Lett.*, 517 (1976).
67. Breslow, R., Maitra, U., and Heyer, D., *Tetrahedron Lett.*, **25**, 1123–1126 (1984).
68. Breslow, R., Kitabatake, S., and Rothbard, J., *J. Am. Chem. Soc.*, **100**, 8156 (1978).
69. Czarniecki, M.F. and Breslow, R., *J. Am. Chem. Soc.*, **101**, 3675 (1979).
70. Breslow, R., Rajagopalan, R., and Schwarz, J., *J. Am. Chem. Soc.*, **103**, 2905 (1981).
71. Breslow, R., Rothbard, J., Herman, F., and Rodriguez, M.L., *J. Am. Chem. Soc.*, **100**, 1213 (1978).
72. Breslow, R., Dale, J.A., Kalicky, P., Liu, S.Y., and Washburn, W.N., *J. Am. Chem. Soc.*, **94**, 3276 (1972).
73. Breslow, R., Corcoran, R.J., Dale, J.A., Liu, S., and Kalicky, P., *J. Am. Chem. Soc.*, **96**, 1973 (1974).
74. Breslow, R., Corcoran, R.J., Snider, B.B., Doll, R.J., Khanna, P.L., and Kaleya, R., *J. Am. Chem. Soc.*, **99**, 905 (1977).
75. Snider, B.B., Corcoran, R.J., and Breslow, R., *J. Am. Chem. Soc.*, **97**, 6580 (1975).
76. Breslow, R., Corcoran, R.J., and Snider, B.B., *J. Am. Chem. Soc.*, **96**, 6791 (1974).
77. Breslow, R., Wife, R.L., and Prezant, D., *Tetrahedron Lett.*, 1925 (1976).
78. Breslow, R. and Heyer, D., *J. Am. Chem. Soc.*, **104**, 2045 (1982).
79. Breslow, R. and Heyer, D., *Tetrahedron Lett.*, **24**, 5039–5042 (1983).
80. Sharpless, K.B. and Michaelson, R.C., *J. Am. Chem. Soc.*, **95**, 6136 (1973) et seq.
81. Breslow, R. and Maresca, L.M., *Tetrahedron Lett.*, 623 (1977).

SUPEROXIDE DISMUTASES

By IRWIN FRIDOVICH, *Department of Biochemistry,
Duke University Medical Center, Durham, North
Carolina 27710*

CONTENTS

I. Introduction

The superoxide radical (O_2^-) is a frequently encountered intermediate of the reduction of dioxygen and it poses a threat to living cells, much as does H_2O_2. Metalloenzymes, called superoxide dismutases, SODs, provide a defense against O_2^- and are found in virtually all organisms. These enzymes, properly called superoxide/superoxide oxidoreductases, catalyze the conversion of O_2^- to H_2O_2 + O_2 and operate close to the diffusion limit. A decade has passed since the last review on SODs appeared in *Advances in Enzymology* (1). Interest in these enzymes has grown steadily and rapidly. We shall now survey some of the work fueled by this interest.

II. Assays

Any continuous source of O_2^- can be coupled with any indicating scavenger of this radical to provide the basis for assaying SOD. The essence of such assays is competition between SOD and the indicating scavenger for the available O_2^-. The SOD thus inhibits the rate of the indicating reaction and one unit of the enzyme is arbitrarily defined as the amount that causes 50% inhibition under specified conditions. A linear relationship between the concentration of SOD and the ratio of the uninhibited to the inhibited rates (2) is useful in calculating the amount of enzyme that would be needed to achieve precisely 50% inhibition. In the first SOD assays to be described, O_2^- was generated either by the xanthine oxidase reaction or by the cathodic reduction of dioxygen, and it was detected by its ability to reduce cytochrome c or tetranitromethane, or to oxidize epinephrine (3).

Nitroblue tetrazolium, NBT coupled with a photochemical source of O_2^- (4), has been used to provide an assay for SOD applicable either in free solution or to polyacrylamide gels (5). The insolubility of the purple formazan produced by the reduction of NBT is crucial for the latter application. Since SOD inhibits the reduction of NBT by O_2^-, it signals its presence on gels by causing an achromatic band against the purple background of formazan. Users of NBT must maintain an awareness that this compound can both detect the presence of O_2^- and mediate its production (6,7) because the univalent reduction of NBT yields a tetraazoinyl radical that can reduce dioxygen to O_2^-.

Autoxidations in which O_2^- serves as a chain propagating species can combine the functions of generation and of detection of O_2^-. This is particularly the case when the autoxidation produces a chromophoric product. The inhibition of such autoxidations by SOD provides the basis for several convenient assays for this enzyme. Thus, epinephrine autoxidizes to adrenochrome in the pH range above 9.0 and SOD inhibits (8). The accumulation of adrenochrome can be followed either in the visible or in the mid UV (9). Pyrogallol is stable in acid solution, but autoxidizes in neutral to alkaline solution to yield strongly absorbing products and SOD inhibits. This provides an assay that is superior to that dependent upon epinephrine since it can be performed at pH 7.9 (10), whereas the epinephrine autoxidation assay is useful only at higher pH. 6-Hydroxydopamine is yet another compound that autoxidizes rapidly in the neutral pH range to yield colored products. SOD at 0.1 μg/mL caused 50% inhibition of this autoxidation (11). Autoxidations might be expected to be sensitive to multiple interferences, yet the pyrogallol autoxidation assay was found to be unaffected by glutathione (11a). Hydroxylamine, NH_2OH, is oxidized to NO_2^- by O_2^-, and NO_2^- can be converted, by way of nitrous acid, to colored diazo coupled products. This has been made the basis of an SOD assay, by using the xanthine oxidase reaction as a source of O_2^- (12,13). Linear logit plots of such inhibition data provide a useful means of analysis (14). At pH 10.2 NH_2OH autoxidizes to NO_2^- and SOD inhibits (15), providing yet another autoxidation assay for this enzyme. The most recently described and in some ways the most interesting of the autoxidation assays for SOD involves the use of hematoxylin (16). The rate of autoxidation of hematoxylin increases 118-fold as the pH is increased from 6.60 to 8.90. Furthermore, SOD inhibited this autoxidation below pH 8.0, but activated above pH 8.0; allowing assays based either upon dimution or upon acceleration of the rate of autoxidation.

Inhibition, by SOD, of an autoxidation in which O_2^- serves as a chain propagator, is easily understood. The basis for augmentation of autoxidation by SOD is less obvious. For the case of hematoxylin autoxidation the relevant overall reaction is

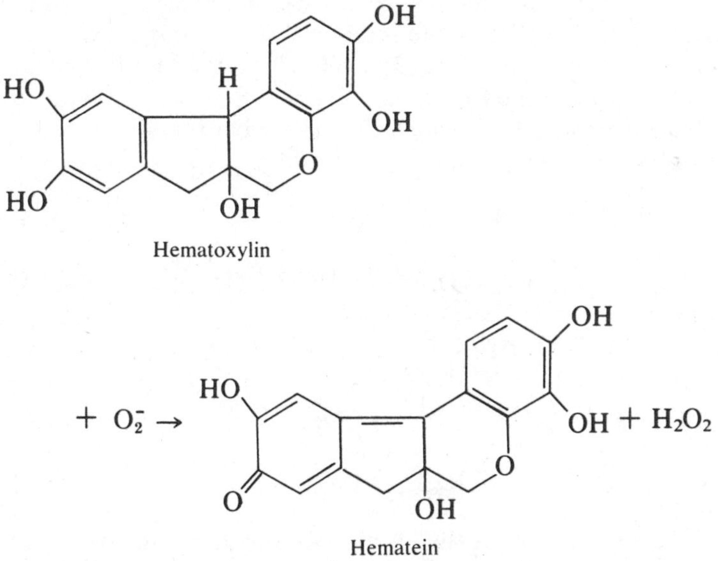

Hematoxylin

(II.a)

Hematein

If we represent hematoxylin as He–OH we may suppose that the autoxidation below pH 8.0 is propagated by O_2^- as follows:

$$He\text{---}O^- + O_2 \rightleftharpoons He\text{---}O\cdot + O_2^- \qquad \text{(II.b)}$$

$$O_2^- + He\text{---}O^- + 2H^+ \rightleftharpoons He\text{---}O\cdot + H_2O_2 \qquad \text{(II.c)}$$

$$2He\text{---}O\cdot \rightleftharpoons He\text{=}O + He\text{---}O^- \qquad \text{(II.d)}$$

SOD will have the effect of lowering the concentration of O_2^- and will therefore decrease the rate of the chain propagating Reaction (II.c). At pH above 8. we suppose that Reaction (II.b) is very rapid, while Reaction (II.c) is correspondingly slower. In that case the autoxidation is no longer effectively a chain reaction because the chain propagating step [Reaction (II.c)] is slower than the initiation step [Reaction (II.b)]. The overall rate of the autoxidation will then depend upon the rate of Reaction (II.b) which, being an equilibrium with O_2^- as a product, will be displaced to the right by SOD. It remains to be seen whether this plausible explanation will prove correct.

The oxidation of dianisidine, photosensitized by riboflavin, is also augmented by SOD and this has been used as the basis of assays for SOD, applicable either in free solution (17) or on polyacrylamide gel electropherograms (18). The explanation advanced for this effect of SOD (17) included the following reactions. Fl denotes riboflavin and DH_2 dianisidine. Fl* indicates electronically excited riboflavin, while hν represents a photon of visible light.

$$Fl + h\nu \rightarrow Fl^* \qquad \text{(II.e)}$$

$$Fl^* + DH_2 \rightarrow FlH\cdot + DH\cdot \qquad \text{(II.f)}$$

$$FlH\cdot + O_2 \rightarrow Fl + H^+ + O_2^- \qquad \text{(II.g)}$$

$$DH\cdot + O_2^- + H^+ \rightarrow DH_2 + O_2 \qquad \text{(II.h)}$$

$$2DH\cdot \rightarrow D + DH_2 \qquad \text{(II.i)}$$

In this scheme of reactions the O_2^- made during the autoxidation of the flavin semiquinone [Reaction (II.g)] reduces the dianisidine radical [Reaction (II.h)] and thus decreases the net yield of the colored oxidation product D by Reaction (II.i). SOD, by lowering $[O_2^-]$ decreases the rate of Reaction (II.h) leaving more DH· for dismutation to product [Reaction (II.i)].

The xanthine oxidase/cytochrome c method of assaying for SOD continues to be widely used and has been modified to both increase sensitivity and/or to minimize interferences. Sensitivity is limited by the competition between SOD and ferricytochrome c for O_2^- made by the xanthine oxidase reaction, and by the molar extinction change that accompanies the reduction of cytochrome c. The balance of this competition can be tipped further in favor of SOD by lowering the concentration of ferricytochrome c and by raising the pH. Elevation of pH, in the range 7–10, markedly decreases the rate of reaction of O_2^- with ferricytochrome c while having a lesser effect on the rate of reaction of O_2^- with SOD. The molar extinction change can be increased by following the reduction of cytochrome c at 418 nm, where $\Delta\epsilon_M = 70,000\ M^{-1}\ cm^{-1}$, rather than at 550 nm where it is $21,000\ M^{-1}\ cm^{-1}$. Salin and McCord (19) have utilized all of these devices and, in applying their assay method to tissue extracts, used CN^- to selectively inhibit CuZnSOD and thus to allow assay of both MnSOD and CuZnSOD.

Cytochrome c reductases and oxidases constitute potential sources of interference with all assays dependent upon the reduction of this cytochrome. This can be minimized by using acetylated (20,21) or succinylated (22) cytochrome c, which is reducible by O_2^- but is not a good substrate for the reductases or oxidases. Since the reaction of cytochrome c with anionic reductants is electrostatically facilitated by the net positive charge on the protein, acylated cytochrome c reacts more slowly with O_2^- than does the native cytochrome (23). This contributes increased sensitivity to SOD assays based upon acylated cytochrome c. The greatest achievable sensitivity can be arrived at by entirely eliminating the competition between SOD and the indicating scavenger for the O_2^-. This was done by delaying the addition of cytochrome c for several minutes after the xanthine oxidase reaction had been initiated. Addition of cytochrome c was then accompanied by a burst of reduction, whose magnitude was a measure of the steady state level of O_2^- achieved by the balance between the rate of production of O_2^- and its removal by the spontaneous dismutation (24). SOD present in the reaction mixture lowered the steady state level of O_2^- and hence the burst of reduction seen when cytochrome c was added. This method allowed the measurement of picomolar levels of SOD (25).

The number of assays that have been devised for measuring the activity of SODs seems almost to equal the number of investigators who study these enzymes. A few examples of the ingenuity expended in this enterprise will suffice. O_2^- is produced during the cathodic reduction of dioxygen and when the cathode is coated with a hydrophobic compound, such as triphenylphosphine oxide, O_2^- diffuses from the cathode. In the presence of SOD this O_2^- is dimuted to $O_2 + H_2O_2$ close to the cathode, thus increasing the level of O_2 in the vicinity of the electrode. This increases the current flow and this increase in the polarographic current may be used to assay SOD (26). Phenylhydrazine autoxidizes at pH 10.2 and reduces NBT. The inhibition of this NBT reduction by SOD has also been made the basis of an assay (27). O_2^- absorbs in the ultraviolet and the rate of decay of this absorbance can be used to assay SOD. Because the decay of O_2^- by spontaneous dismutation is quite rapid, this method has most often been used in the short time frame of pulse radiolysis (28–35). However, by working at an elevated pH in scrupulously clean glassware the spontaneous dismutation may be slowed sufficiently to allow the use of KO_2 as the source of O_2^- and then to assay SOD in terms of the rate of decay of the UV absorbance of this O_2^- (36).

The spontaneous dismutation of O_2^- depends upon a source of protons. Stable solutions of superoxide salts may therefore be prepared in dry nonprotic solvents such as DMSO (dimethyl sulfoxide). A very simple assay for SOD was based upon measuring the reduction of NBT by O_2^- added as a solution of KO_2 in DMSO (37). KO_2 is not very soluble in DMSO, but this problem can be circumvented by solubilizing the cation, and therefore the salt, with a crown ether such as 18-crown-6 (38). Alternately, tetraalkyl ammonium salts of O_2^- , which are soluble in DMSO, may be used (39). The water soluble carotenoid *crocin* is belached by O_2^- and the inhibition of this bleaching by SOD provides yet another assay for its activity (4). CuZnSOD broadens the NMR spectrum of ^{19}F and this has been proposed as an assay for this particular SOD (41). Pyrogallol, which in autoxidizing produces O_2^- , has been combined with NBT to assay

SOD in tissue extracts (41a). Reactions leading to the chemiluminescence of luminol are initiated by O_2^-. Luminol therefore emits light when in the presence of the xanthine oxidase reaction and the inhibition of this luminescence by SOD provides another assay (42). Certain hydroxylamine derivatives are oxidized to stable nitroxides by O_2^-, providing an ESR assay for O_2^- production or for SOD activity (43). Horseradish peroxidase is converted to an inactive ferrooxy form by O_2^-. Since O_2^- is produced during the autoxidation of rifamycin SV, the oxidation of this antibiotic by horseradish peroxidase is accelerated by SOD and this can be used as an assay for SOD (44). The SOD content of suspensions of bacteria have been assayed after permeabilizing the cells with toluene (45). DMSO, made alkaline by addition of aqueous NaOH, autoxidizes and builds up a steady state level of O_2^-. This has been used, in conjunction with cytochrome c, as the basis for a very simple assay (46).

Extracts of tissues or cells will ordinarily contain more than one type of SOD, and means for individually assaying SODs in such mixtures have been devised. These usually depend upon reagents that inhibit or inactivate one of the SODs, while having no effect on the others. Thus cyanide inhibits CuZnSOD, but not FeSOD or MnSOD (29,47,48). One may use 5 mM CN$^-$ to specifically suppress the activity of CuZnSOD in extracts (49). Hydrogen peroxide on the other hand, inactivates CuZnSOD (31,48,50) and FeSOD (51), but has no effect on MnSOD (51,52). CuZnSOD exhibits virtually the same level of activity at pH 10.0 as at pH 7.8, whereas the activity of MnSOD is substantially suppressed at the higher pH. Assaying tissue extracts at both of these pH values allows calculation of the amount of both of these SODs (53). Alternately, MnSOD activity can be entirely eliminated by 2% sodium dodecyl sulfate, which has no effect on CuZnSOD (54).

Immunoassays, responsive to specific constellations of antigenic determinants rather than to catalytic activity, have been developed for each of the types of SOD. A radioimmunoassay developed for the CuZnSODs from man, cow, or rat, was sensitive to 5 ng of CuZnSOD in lysates of human erythrocytes (55). Normal human blood was found to contain 854 ± 100 ng SOD/mg hemoglobin. This is equivalent to approximately 256 mg SOD/L packed erythrocytes. Blood from individuals with trisomy 21 exhibited a 50% overdosage of the CuZnSOD (56). A radioimmunoassay has also been developed for the MnSODs from man, cow, and rat. Interspecies cross reactivities were examined. The CuZnSODs of these species did not cross react, but the MnSODs did so (57). The use of radioimmunoassays for CuZnSOD in the diagnosis of disease has been considered (58).

Enzyme-linked immunoassays have also been developed. Such a method for the human CuZnSOD was useful in the range 0.05 to 10.0 ng of this enzyme and was used to show elevation of CuZnSOD in both serum and in urine in patients suffering from renal disease (59). An enzyme-linked immunoassay was developed for human MnSOD (60). It was used to show that serum MnSOD is markedly elevated during liver disease, while urinary MnSOD was high in patients with nephrotic syndrome and low with those who were hypertensive. This immunoassay was compared with ordinary activity assays. When applied to normal human lung the two methods were in agreement but in adenocarcinoma an immunoreactive, but enzymically inactive, MnSOD was detected (61). This finding may correlate with the observations of very low MnSOD in cancerous tissues or in transformed cells (61a–61d).

There are clearly many assay methods that can be applied to the SODs. Each has its advantages, yet each has potential pitfalls and cannot be applied blindly. It is a useful exercise to consider some of these potential sources of artifact. We begin with the xanthine oxidase/cytochrome c method. Any substance that inhibits xanthine oxidase would also inhibit the reduction of cytochrome c and might thus be mistaken for an SOD. A control in which the sample was tested for any effect on the rate of conversion of xanthine to urate, in the absence of cytochrome c, would expose this problem. Catalysis of the dismutation of O_2^-, while inhibiting cytochrome c reduction, would have no effect on the conversion of xanthine to urate.

Commercially available preparations of cytochrome c are occasionally contaminated with SOD. When this happens the contaminant is most often the very stable CuZnSOD, which is inhibitable by cyanide. This problem can be diagnosed by adding 1.0-mM CN$^-$ to the xanthine oxidase/cytochrome c reaction mixture. If CN$^-$ increases the rate of reduction of cytochrome c then the CuZnSOD may be assumed to be present. Given that the molecular weight of the CuZnSOD is ~32,500 dalton, while that for cytochrome c is

~12,500 dalton the contaminant can be removed by gel exclusion chromatography.

Commercially available xanthine oxidase is frequently contaminated with lactoperoxidase, which can catalyze the oxidation of ferrocytochrome c by H_2O_2. Since both H_2O_2 and ferrocytochrome c accumulate during the reaction, lactoperoxidase would catalyze a progressively more rapid peroxidation of ferrocytochrome c, which would give the appearance of a progressive decline in the rate of accumulation of ferrocytochrome c. This problem can be solved by adding catalase to prevent the accumulation of H_2O_2. It would, of course, be necessary to ascertain that the catalase used was free of SOD (62).

Native xanthine oxidase cannot directly reduce cytochrome c, but this reduction can be mediated by a variety of electron carrying compounds. Dioxygen can serve in this way (63) and so can a variety of dyes and quinones (64,65). It follows that xanthine oxidase contaminated with quinones could cause the reduction of cytochrome c by a route independent of O_2^-. This situation is easily diagnosed. Thus, the presence of quinones will make some part of the rate of cytochrome c reduction impervious to inhibition by SOD. The presence of reductants, such as ascorbate, in the sample, will also provide an O_2^--independent pathway for the reduction of cytochrome c. A graphical method for detecting the occurrence and extent of such problems has been described (66).

A certain means of detecting O_2^--independent pathways of cytochrome c reduction, by the xanthine oxidase reaction, is to exclude dioxygen and a certain means of removal of low molecular weight quinones and reductants is by dialysis of samples prior to assay. Yet, dialysis is not a solution to all problems. Thus, whereas native xanthine oxidase will not catalyze the direct reduction of cytochrome c, the deflavoxanthine oxidase does so (67). Flavin adenine dinucleotide, FAD, is labilized when xanthine oxidase is exposed to very high concentrations of salts. Freezing aqueous solutions concentrates solutes such as enzymes and salts in the domains between the water crystals. Some loss of FAD occurs during prolonged frozen storage of xanthine oxidase. Deflavination is accompanied by the appearance of oxygen-independent and SOD-insensitive reduction of cytochrome c.

Finally, one must consider a subtle interference by compounds that cannot directly reduce cytochrome c, but which can react with O_2^- to yield a product that does reduce this cytochrome. When present in the assay mixture such a compound will neither prevent nor accelerate the reduction of cytochrome c by the xanthine oxidase reaction. What it does is to displace O_2^- as the reductant of cytochrome c and thus to replace SOD inhibitable by SOD-insensitive reduction of the cytochrome. It thus gives the appearance of making SOD less effective as an inhibitor of the reduction of cytochrome c. In effect, such compounds appear to be inhibitors of SOD. Pamoic acid acts in this way (68).

The xanthine oxidase/cytochrome c assay for SOD is thus clearly susceptible to artifact, yet it served to guide the first isolations of CuZnSOD (3), MnSOD (69), and FeSOD (70) and it continues to be widely used. In the use of this, or any other assay for SOD, there is no substitute for understanding the reactions involved and for doing the controls needed to expose potential problems. It might be well, for the neophyte, to begin with very simple and direct assays such as the one in which a DMSO solution of a superoxide salt is added to buffered NBT (37). Methods for assaying SODs have been reviewed previously (71).

III. Isoenzymes and Electromorphs

Application of a common selection pressure to a varied biota over a long period of time is apt to elicit parallel evolutionary adaptations. It is therefore not surprising that the oxygenation of the biosphere by early photosynthesis led to the appearance of more than one type of SOD. We now find three distinct types of SODs, representing two independent lines of evolutionary descent. The first of these is the copper-and zinc-containing SOD, CuZnSOD, characteristically found in the cytosol of eukaryotic cells; the second is the manganese-containing SOD, MnSOD, found in many bacteria and in the matrix of bacteria; the third is an iron-containing SOD, FeSOD, found primarily in bacteria and in a few plants. MnSODs and FeSODs from a variety of sources show extensive amino acid sequence homology and clearly represent one line of descent; while the CuZnSODs show no sequence homology with the MnSODs and FeSODs and result from an independent evolutionary history. Nevertheless, all of the SODs catalyze the same reaction, that is, the dismutation of O_2^- , and do so at nearly identical rates; which approach the diffusion limit.

The discovery, isolation, and characteristics of these enzymes has been described (1,72) and Steinman (73) has written a thorough account of their protein chemistry and of the relationship of structure to their function. CuZnSODs have been isolated from a wide range of organisms including yeast (74), *Neurospora crassa* (75), spinach (76), wheat germ (48), swordfish liver (77), chicken liver (78), and bovine blood (3). In all cases a homodimeric enzyme with a molecular weight of ~32,000 dalton and containing one Cu(II) and one Zn(II) per subunit is obtained. The subunits are stabilized by an intrachain disulfide bond but associate by noncovalent forces (79–81). CuZnSODs are generally very stable enzymes, tolerating exposure to organic solvents during purification and retaining activity in 8.0 M urea (82,83) or in 2% sodium dodecyl sulfate, SDS (54). The CuZnSOD of yeast is not as stable towards either urea (84) or SDS (74) as is the bovine enzyme.

The CuZnSOD from bovine erythrocytes is resolved into three electromorphs on polyacrylamide gels (5), while the corresponding enzyme from chicken liver is composed of four electromorphs that copurify (78). The chicken liver CuZnSOD can be separated by isoelectric focusing into six electromorphs whose isoelectric points range from 6.75 to 5.35 (85). Three CuZnSOD electromorphs with pI = 4.0, 4.5, and 5.0 were noted in several mouse tissues. Their relative abundances were 1:6:23. These results were explained in terms of two types of subunit of pI = 4.0 and 5.0 with relative abundances of 0.14:0.86 (86). These subunits, designated A and B could associate to yield AA, AB, and BB, which would correspond to the observed electromorphs. Hybridization of CuZnSODs, from different species, by exchange of subunits, has been observed (87). Two of the electromorphs of bovine CuZnSOD, with pI = 5.2 and 4.9, have been isolated by preparative isoelectric focusing and failed to show any differences in molecular weight, metal content, antigenic reactivity, or in optical and in ESR spectra. There was however a difference in thermal lability (88). These electromorphs were considered to have arisen by posttranslational modification.

Exposure of isolated bovine CuZnSOD to Fe(III)–EDTA plus

ascorbate has been reported to generate electromorphs similar to those seen in erythrocyte lysates. These authors (89) suggest that oxidation of nonessential amino acid residues was responsible for generating the observed heterogeneity. Yet, the patterns of electromorphs of CuZnSOD in rat and mouse appear to be tissue specific (90) and in some plants some of the CuZnSOD electromorphs have been shown to be distinct gene products since they differed in amino acid composition (48,91). A genetic variant of human CuZnSOD with a pI of 4.85, compared to pI = 4.7 for the normal form, has been described (92). This variant is fully active. There is a substantial literature dealing with genetic variants of an activity called tetrazolium oxidase (93–96). Tetrazolium oxidase is, in fact, superoxide dismutase (97,98).

A high molecular weight (135,000 dalton) tetrameric CuZnSOD has been found in human tissues. Because this enzyme constitutes 90% of the total SOD of blood plasma, but only 7% of the SOD of lung, it is being called extracellular SOD or ECSOD (99,100). This enzyme appears to be a glycoprotein, in keeping with the suggestion that it is primarily an extracellular enzyme. Although there is very little total SOD in extracellular fluids, when compared with the amount found within cells, that amount may be critically important in moderating the effects of O_2^- liberated from activated phagocytic cells.

B. MANGANESE-CONTAINING SUPEROXIDE DISMUTASES

MnSOD was first isolated from *Escherichia coli* (69) and has since been obtained from a wide range of organisms including *Streptococcus mutans* (101), chick liver mitochondria (78), *Pleurotus olearius* (102), *Bacillus stearothermophilus* (103), yeast (104), human liver (105), *Thermus aquaticus* (106), *Porphyridium cruentum* (107), *Pisum sativum* (108), maize (109), and several species of Mycobacteria (110–113). MnSODs have a subunit weight of ~23,000 dalton with one atom Mn(III) per subunit. These enzymes are frequently dimeric, although the MnSODs from mitochondria, the mycobacteria, and the thermophilic bacteria are tetrameric. In the case of the *E. coli* MnSOD, three electromorphs, which copurify, have been detected (114).

C. IRON-CONTAINING SUPEROXIDE DISMTUASES

FeSODs were first isolated from *E. coli* (70) and subsequently from numerous other microbial sources including photobacteria

(115), a blue green alga (51), *Chlorobium thiosulfatophilum* (116), *Chromatium vinosum* (47), *Pseudomonas ovalis* (117), *Anacystis nidulans* (118), and *Methanobacterium bryantii* (118a). Although long thought to be restricted to bacteria, FeSODs have now been found in three families of higher plants, i.e., the ginko tress, water lilies, and mustards (119). Forty three other plant families were examined and found not to contain detectable FeSOD and there is, as yet, no explanation for this spotty distribution of FeSOD among higher plants. With few exceptions all FeSODs described to date are dimeric and like the MnSODs have a subunit weight of ~23,000 dalton with one atom of metal per subunit. One exception is the tetrameric FeSOD from *Mycobacterium tuberculosis* (120). Another tetrameric FeSOD was obtained from *Rhodococcus bronchialis*. It was described as containing 2.2 Mn and 0.9 Fe/tetramer (121). Five electromorphs of FeSOD have been noted in *Euglena gracilis* and the two most abundant of these with pI = 4.2 and 4.3 have been isolated (122).

IV. Structure and Evolutionary Relationships

Amino acid sequences are available for members of all three classes of SODs. These are of interest both for the light they shed on the evolution of these enzymes and because they facilitate interpretation of X-ray diffraction data. The bovine CuZnSOD was the first to be completely sequenced (79–81). When the corresponding enzyme from *Saccharomyces cerevisiae* was examined (123,124) it was seen to exhibit 55% homology with the bovine enzyme. Given the great divergence between yeast and cow this degree of homology indicates a high degree of evolutionary conservatism. Comparison of the yeast and bovine CuZnSOD sequences with those of the MnSOD and FeSOD indicates no homology beyond that expected by chance, suggesting independent evolutionary origins for the CuZnSOD as opposed to the FeSOD/MnSOD family of enzymes. Human (125–127) and horse (128) CuZnSOD sequences have also been examined and they exhibit the expected homology to the bovine enzyme.

CuZnSOD is ordinarily found in eukaryotes, not in prokaryotes. Yet, there are two bacteria that do contain this type of SOD. These are *Photobacterium leiognathi* (129) and *Caulobacterium crescentis*

(130). Since the former is a symbiont found in a specific light producing gland of the pony fish, a gene transfer from the host fish to the bacterium was considered (131). Statistical comparisons of amino acid compositions of SODs from many organisms appeared to support this concept of gene transfer across the eukaryote/prokaryote abyss (132). Confidence in this proposal was subsequently weakened when an amino acid sequence for the *P. leiognathi* CuZnSOD became available and showed only 25% sequence homology with the eukaryotic CuZnSODs (133). Yet, those residues that are conserved in the *P. leiognathi* CuZnSOD are those essential for maintaining the structure of the active site. These include the six histidine residues and the one aspartate that constitute the ligands of the Cu(II) and Zn(II) and that arginine which is positioned close to the Cu(II) and which has been shown to be important for catalysis. Moreover, prediction of secondary structure from the amino acid sequence indicates that the *P. leiognathi* and the eukaryotic CuZnSODs have virtually identical structures (134). Finally, when allowances are made for frequently encountered amino acid replacements, the homology between the *P. leiognathi* and the eukaryotic SODs rises to 50% (134). Gene transfer remains the most likely explanation for the CuZnSOD in *P. leiognathi*. But what can be said about the CuZnSOD in the free-living *C. crescentis*? It is best to defer further speculation until its amino acid sequence has been determined.

MnSODs and FeSODs from a variety of organisms have been sequenced. These data support the view that these enzymes evolved from a common ancestral protein. This conclusion is based on the extensive homologies seen among them, whatever their source. Complete amino acid sequences have been reported for the MnSODs from *Bacillus stearothermophilus* (135), *E. coli* (136), and *S. cerevisiae* (137). Steinman, in analyzing the sequence of the *E. coli* MnSOD (136) predicted a compact structure, no extensive β sheets and histidine ligands to the Mn(III). These prognostications proved to be very close to the mark (138). Comparisons of the sequences of the MnSODs from *B. stearothermophilus* and *E. coli* with each other showed 59% identity and a great deal of homology with the known FeSOD sequences. The variability among these enzymes was of special interest since it did not seem to fit the accepted taxonomies (139).

Partial amino acid sequences are easily determined and are available for the FeSODs from *Desulfovibrio desulfuricans* (140), *Spirulina platensis*, *Plectonema boryanum*, and *P. leiognathi* (141) and for the MnSODs from *Thermus aquaticus, Rhodopseudomonas spheroides*, chicken liver mitochondria, and human liver mitochondria (141). Comparisons of the partial amino terminal sequences of the SODs from a number of bacteria (142,143) showed clear relatedness among MnSODs and FeSODs, but there were many amino acid replacements. Two closely positioned histidine residues were conserved in all of these amino terminal portions of SODs and predictions of secondary structure indicated that they would be located on the same helix.

The methods of X-ray diffraction have been applied to CuZnSODs (144–147), MnSODs (138,148–149), and FeSODs (150–154). The bovine CuZnSOD was the first to be studied in this way and has now been analyzed to 2 Å resolution. Its predominant feature is a cylinder composed of eight strands of antiparallel β-structure, which has been called the β-barrel. Two large nonhelical loops extend from the β-barrel, one from its top edge and the other from the bottom edge. These enclose the active site region. In the native dimeric molecule the two active sites are 34 Å apart which, given the size of the dimer, is nearly as far apart as they could possibly be. The Cu(II) at each active site is ligated to four imidazole rings in a distorted square planar arrangement. These imidizoles are derived from histidines 44, 46, 61, and 118. The Zn(II) is joined to the Cu(II) by a bridging ligand, which is the imidazolate of His 61. The remaining ligands to the Zn(II) are provided by His 69, His 78, and Asp 81. It is noteworthy that six of the eight histidine residues present in each subunit are found ligated to the metals.

The active site region has been examined in great detail (147). The Cu(II) lies at the bottom of a deep channel and is partially exposed to solvent. The Zn(II) is positioned below the Cu(II) and is completely buried in the protein structure. Of the atoms that form the active site channel 86% are conserved in the CuZnSODs that have been sequenced. In contrast only 41% of the atoms composing the remaining surface of the enzyme are conserved. It thus seems clear that the arrangement of atoms in the channel region is critically important for catalysis. Arginine 141 is positioned in the solvent access channel close to the Cu(II) such that an O_2^-, while hydrogen bonded to this arginine could ligate to the copper. When O_2^- is not present two water molecules bind in this position.

As expected from the amino acid sequence homologies, the structure for MnSOD and FeSOD, arrived at by X-ray crystallography (138), are congruent with each other and differ sharply from the structure of CuZnSOD. Thus, whereas β-structure is a predominant feature of CuZnSOD, it is virtually absent from MnSOD and FeSOD; which are rich in α helix. Moreover, the metal binding sites are far from the subunit contact face in CuZnSOD, but are close to that contact face in MnSOD and FeSOD. The tetrameric MnSOD of *T. thermophilus* has been analyzed to 2.4-Å resolution (138). This allowed identification of the metal ligands as histidines 26, 81, and 167 and aspartate 163. It also allowed dismissal of an earlier suggestion (153,154) that a particular region of electron density seen at lower resolution might be due to an unidentified prosthetic group. At 2.4-Å resolution this feature was seen to be due to tryptophan 130 (138). Analytical data for MnSODs and FeSODs have typically indicated between 0.5 and 1.0 atoms of the metal per subunit. X-ray crystallography indicates identical subunits with one metal binding site on each. The discrepancy between the results of atomic absorption spectrophotometry and the expectations from the structures arrived at by crystallography may be due to partial loss of metal during isolation or to metal replacements occurring *in vivo*.

V.   Mechanisms

With all of the SODs, removal of the active site metal results in loss of catalytic activity and its replacement results in restoration of that activity. Metal removal has been achieved by applying chelating agents at low pH or by combining low pH with chelating and chaotropic agents (3,106,135,155–157). In the case of FeSOD, metal removal was also achieved anaerobically at pH 11.0, by applying dithiothreitol plus EDTA (158,159).

In the CuZnSODs it is the copper that functions in electron transport during the catalytic cycle, whereas the zinc plays a secondary and perhaps primarily a structural role. Replacement of the Cu(II) by any other metal was thus associated with complete loss of activity. In contrast, the Zn(II) could be replaced by Co(II) (160), Hg(II) (82), Cd(II) (161), Cu(II) (162), or even by an empty site (163), without much loss of activity. The hydrophobic chelating agent diethyl dithiocarbamate removes Cu(II) from CuZnSOD and can be used to prepare the apoenzyme at neutral pH and without resort to chaotropic agents (164–166). Since the affinity of the apoenzyme for Cu(II) is very great, reconstitution with regain of activity could be used as the basis for an ultrasensitive assay for Cu(II) (167).

Most of the MnSODs and FeSODs that have been examined are entirely specific for the metal found in the native enzyme, in that its replacement results in loss of activity. Thus Mn(II) restored activity to the apoenzymes prepared from the MnSODs of *B. stearothermophilus* (168,169) and *E. coli* (157,170), but Co(II), Fe(II), Ni(II), Cu(II), and Zn(II) did not. All of these metals prevented reconstitution by Mn(II) and thus may be presumed to compete with Mn(II) for occupancy of the active site. Moreover, with a few of these metals, stoichiometric tight binding at the level of one atom metal per subunit could be shown and activity could subsequently be restored by removal of the foreign metal and its replacement by manganese.

The FeSOD of *P. ovalis* was as fastidious as the MnSODs just described. Thus, Fe(II) restored activity to the apoenzyme, but Mn(II), Cd(II), and Cr(II) did not (171). It is clear that these MnSODs and FeSODs have diverged sufficiently that only manganese will restore activity to the apoMnSOD and only iron to the apoFeSOD. Yet there are two reports of SODs in which either of these metals will impart catalytic activity. These are the SODs from *Bacterioides fragilis* (172,173) and *Propionibacterium shermanii* (174). It will be fascinating to compare the ligand fields of these SODs with those of the more discriminating MnSODs and FeSODs.

The blue color of CuZnSOD is bleached approximately 30% during the catalytic steady state (30,31). This was taken to indicate a ping–pong mechanism in which the active site Cu(II) is alternately reduced and reoxidized during successive encounters with O_2^-. It also indicates the approximate equality of the rate constants for the two half-reactions of this catalytic cycle. Given the very high turnover rate of this enzyme, which approaches the diffusion limit, saturation by substrate is not likely to be achieved. Indeed, variation of $[O_2^-]$ by pulse radiolysis gave no sign of saturation of CuZnSOD up to 0.5-mM O_2^- (29).

Each of the active sites in the homodimeric CuZnSOD appears to function independently. Early claims of an inhibitory interaction leading to a half-of-the-sites reactivity (175) were subsequently withdrawn (176,177). The possible importance of functional interactions between the subunits of the bovine CuZnSOD was explored by hybridizing native with diazo coupled and inactive enzyme. Native subunits exhibited identical activity whether coupled to other native subunits or to chemically modified and catalytically inactive subunits (178). This indicates that functional interactions between subunits are not essential for full expression of the catalytic activity. Nevertheless, the subunits of CuZnSOD have not yet been separated in a way that retains catalytic activity. This is largely due to the great stability of the subunit association, which is not disrupted even in 8.0 M urea (83).

Redox titration of CuZnSOD indicated the release of a basic group concomitant with reduction of the Cu(II) (179). This observation and the need to provide a mechanism for rapid conduction of protons to products, during the catalytic cycle, led to the proposal that the bridging imidazolate of histidine 61 was released from and religated to the Cu(II) during its successive reduction and reoxidation by O_2^- (52). The released imidazolate, being highly basic, would abstract a proton from the solvent and would then give up that proton during the second half of the catalytic cycle to form the product HO_2^-; which, upon release from the active site, would protonate to H_2O_2 in solution. This proposal is illustrated by the following scheme:

$$
\begin{array}{c}
\text{Im} \\
\diagdown \\
-Zn^{2+}-Im^- -Cu^{2+} \\
\diagup \quad \diagdown \\
\text{Im} \quad \text{Im}
\end{array}
\xrightarrow[+H^+]{+O_2^-}
\begin{array}{c}
\quad\quad\quad \text{Im} \\
| \quad\quad\quad \diagdown \\
-Zn^{2+}-ImH \quad Cu^+ \\
| \quad\quad\quad \diagup \quad \diagdown \\
\quad\quad\quad \text{Im} \quad \text{Im}
\end{array}
+ O_2
$$

$$
\Big\downarrow {\scriptstyle +O_2^-}
$$

$$
\begin{array}{c}
| \quad\quad\quad \diagdown \\
-Zn^{2+}-Im^- -Cu^{2+} + HO_2^- \\
| \quad\quad\quad \diagup \quad \diagdown
\end{array}
$$

Proton conduction is not rate limiting, in spite of the very rapid turnover rate of the catalytic process. Thus the enzymic dismutation of O_2^- was unaffected by changes in pH in the range 6.5 to 9.5 (28,29). Moreover, replacement of H_2O_2 by D_2O, at pH 10.0, did not slow the enzymic process (52).

The role of the bridging imidazolate in the catalytic mechanism has been probed experimentally. Thus in the enzyme bearing Co(II) in place of Zn(II), reduction of the Cu(II) perturbs the spectrum of the Co(II) in a way that resembles the effect of removal of the Cu(II) (180). Elimination of the effect of the copper upon the spectrum of the Co(II), by reduction of the Cu(II), was interpreted as an indication that the bridge between the two metals had been severed upon reduction of the copper. On the other hand SCN^- has been reported to bind to the Cu(II) of CuZnSOD, with displacement of the bridging imidazolate but without inactivation (181). Yet, the weight of available evidence does favor an active role for the bridging imidazolate in the catalytic mechanism.

Were the imidazolate bridging Cu(II) and Zn(II) to release one of these metals during the catalytic cycle, it would be the copper. This statement derives from the facts that it is the copper that is reduced during the catalytic cycle and from the sharp contrast between Cu(II) and Zn(II) when ligating multiple imidazole rings. Thus, with Cu(II) there are negative interactions between coordinated imidazole groups, whereas with Zn(II) there are positive interactions (182). It follows that release of one of the four imidazoles ligated to Cu(II) would be facilitated by the negative interactions among them; whereas release of any one of the three imidazoles ligated to Zn(II) would be hindered by the positive interactions they display. Extended X-ray absorption spectra of the reduced and of the oxidized forms of CuZnSOD supports the view that reduction is accompanied by release of the bridging imidazolate from the copper. The fine structure of these spectra changed in a way that indicated the presence of four imidazole ligands to the Cu(II) in the oxidized enzyme, but only three such ligands to the Cu(I) in the reduced enzyme (182a). X-ray crystallography of the reduced CuZnSOD may ultimately settle this fascinating question.

The SODs react with O_2^- at a rate that approaches the diffusion limit, yet the Cu(II) exposure to solvent at the active site represents no more than 0.1% of the surface area of the enzyme. Contemplation of this situation led Koppenol (183) to propose that electrostatic facilitation was important in guiding the anionic O_2^- to the active site of SOD. This proposal was tested by an examination of the effects of changes in the ionic strength on the activity of CuZnSOD (184,185). Increasing the ionic strength decreased the rate of the enzymic reaction, indicating that electrostatic facilitation was a factor in the catalyzed process. Moreover, acylation of the ε-amino groups of lysine residues on the enzyme inverted the response to ionic strength. These results suggested that lysine residues are important in providing for the electrostatic facilitation that had been proposed by Koppenol. MnSOD and FeSOD of *E. coli* were subsequently examined and also gave indications of electrostatic facilitation of the catalytic process by lysine residues (186).

The question of the importance of the positive charge of lysine residues on the activity of CuZnSOD has been approached in several ways. Carbamoylation or succinylation of lysine residues does diminish catalytic activity, but without perturbing the ESR spectrum of the Cu(II) (187). This indicates that the lysine residues play a role in accelerating the enzymic reaction without being close enough to the Cu(II) to interact with the unpaired electrons on that metal. Porcine CuZnSOD has a higher isoelectric point than does the bovine enzyme. Comparison of these two enzymes might shed some light on electrostatic influences on the catalytic process (188). The porcine enzyme showed a progressive decline in activity as the pH was raised above 7.5 without any corresponding change in ESR of the Cu(II). In contrast, the activity of the bovine enzyme was independent of pH until the pH exceeded 9.5 and there was then a parallel change in ESR signal. Succinylation of both enzymes drastically lowers their isoelectric points and causes a 90% loss of activity. After succinylation both the porcine and the bovine CuZnSODs were unaffected by pH until it exceeded 9.5. These results (188) suggest that a large net negative charge on the CuZnSOD is important for maintaining full activity and also that loss of the positive charge on specific lysine residues, by ionization above pH 9.5, causes loss of activity. This is in agreement with the concept that electrostatic guidance of the O_2^- is achieved by combining the repulsive effect of negatively charged residues with the attractive effect of cationic residues.

Getzoff et al. (189) utilized the highly detailed structural information developed from X-ray crystallographic data (147) to analyze the electrostatic field over the surface of the bovine CuZnSOD, with special attention to the active site region. They found an area of positive charge coincident with the long and deep solvent access channel to the active site Cu(II). They moreover noted that the electrostatic potential gradient would direct the anionic substrate O_2^- into the channel and down to the Cu(II) site. The amino acid residues that create this electrostatic field are sequence conserved in Cu-ZnSODs from very different species. Lysine 134 and glutamate 131 appeared to direct the long range approach of O_2^- to the channel, while arginine 141 seemed very important for local orienting effects upon the O_2^- in the depths of the channel. The papers describing this work (147,189) are graced by striking color-coded computer graphic illustrations. One hopes that this approach may soon be applied to MnSOD and FeSOD as the essential structural information becomes available.

The equilibrium for the dismutation of O_2^- to $H_2O_2 + O_2$ lies vastly in favor of the products (190). It has nevertheless been possible to demonstrate catalysis of the reverse reaction by CuZnSOD (191). This was done by using tetranitromethane to react with and thus trap O_2^-. Tetranitromethane was chosen because the rate constant for its reduction to the nitroformate anion by O_2^- is approximately $1 \times 10^9 \ M^{-1} \ \text{sec}^{-1}$ at 25°C. SOD catalyzed a dioxygen-dependent reduction of tetranitromethane by H_2O_2. Moreover, this process was a saturable function of the concentration of H_2O_2 and occurred at a rate that was proportional to the enzyme concentration. Elevation of the pH in the range 8.–10.5 increased the rate, indicating that HO_2^-, rather than H_2O_2, was the reactant for the reverse dismutation reaction. The reaction scheme advanced (191) to account for this process was

$$\text{E—Cu(II)} + HO_2^- \rightleftharpoons \text{E—Cu(I)—O}_2\text{H} \qquad \text{(V.a)}$$

$$\text{E—Cu(I)—O}_2\text{H} \rightleftharpoons \text{E—Cu(I)} + H^+ + O_2^- \qquad \text{(V.b)}$$

$$\text{E—Cu(I)} + O_2 \rightleftharpoons \text{E—Cu(II)} + O_2^- \qquad \text{(V.c)}$$

Reaction (V.c), the oxidation of the reduced enzyme by dioxygen, would be the slowest component of this overall process.

If HO_2^-, rather than H_2O_2, is the substrate for the reverse dismutation, then HO_2^- must also be the product of the forward dismutation. This has the added attraction that the production of HO_2^- requires that only one proton be supplied during the very rapid catalytic process, while the second is picked up from the solvent after the HO_2^- has left the active site. The reduction of the Cu(II) of CuZnSOD by H_2O_2 (or HO_2^-) is known to be a rapid process (3,174,192,193). Hydrogen peroxide causes a slow irreversible inactivation of CuZnSOD, which becomes more rapid as the pH is elevated (31,48,52,193), and which is associated with the modification of one histidine residue per subunit (193). Studies of the effect of pH on this inactivation and on the saturability of its rate with respect to $[H_2O_2]$, have demonstrated that HO_2^-, rather than H_2O_2, is the reactive species (194). It seems likely that inactivation of CuZnSOD by HO_2^- is a two-step process in that one HO_2^- reduces the active site copper to Cu(I), which in turn reduces a second HO_2^- into a bound and potent oxidant, at the oxidation level of $\cdot OH$ which then attacks one of the imidazole groups ligated to the copper. FeSOD is also subject to inactivation by H_2O_2 (51), but MnSOD is not, and this provides a basis for distinguishing these two enzymes in cell extracts.

Specific inhibitors for particular SODs would be exceedingly useful from both the experimental and practical points of view. An inhibitor specific for FeSOD would constitute a poison selective for pathogens containing FeSOD, while having a little effect on their hosts that do not contain FeSOD. This thought has occurred to several investigators and is being actively pursued. CuZnSODs are inhibited by CN^- (29) and all of the SODs are inhibited by azide (51,195) and by nitroprusside (195a), but these hardly qualify as specific inhibitors. Phosphate has been reported to weakly diminish the activity of CuZnSOD (192,196) presumably by binding to the conserved arginine residue in the active site crevice.

Quantum chemical studies of the interaction of O_2^- with a model of the active site of CuZnSOD have led to the proposal of a novel mechanism (196a). Rather than a ping–pong mechanism in which the active site Cu(II) is alternately reduced and reoxidized, while the products O_2^- and H_2O_2 are alternatively released; the new mechanism posits the binding of one O_2^- to the Cu(II) until a second O_2^- reduces that copper to Cu(I). The resting enzyme is thus thought to contain a water molecule ligated to the Cu(II) and hydrogen bonded to arginine 141 and O_2^- displaces this H_2O molecule. The Cu(II)–O_2^- is then reduced, by a second O_2^-, to a Cu(I)–O_2^- complex; within which charge redistribution leads to a Cu(II)-peroxide complex, which finally dissociates HO_2^-. Arginine 141 is assigned a critical role in stabilizing the Cu(II)–O_2^- complex by hydrogen bonding. Hopefully this entirely plausible mechanism will soon be subjected to experimental tests.

VI. Biosynthesis

Increases in cell content of SOD are often seen when conditions are imposed that increase intracellular production of O_2^-. Such inductions of SOD were seen in *Streptococcus faecalis* and in *E. coli* (197). Anaerobically grown *E. coli* contain FeSOD, while aerobically cultured cells contain FeSOD, MnSOD, and a hybrid SOD (198,199). Inductions of SOD by elevation of pO_2 have been reported to occur in rat lung (200–205), alevolar macrophages (206–209), cultured endothelial cells (210,211), *Photobacter leiognathi* (129), yeast (212), potato slices (213), *Bacterioides fragilis* (214), *Vibrio eltor* (215), and *Oscillatoria limnetica* (216) and in many other cells and tissues (217–224).

Inductions of MnSOD, at constant pO_2, demonstrate that O_2 is not itself responsible for derepression of SOD biosynthesis. Thus, dependence upon glucose, as an energy source, depressed MnSOD in *E. coli* (225). Glucose has since been seen to lower the MnSOD content of mouse heart (226). Enrichment of the culture medium with iron increased the FeSOD levels in *P. ovalis* (117) and in *E. coli* (227,228) while enrichment with manganese salts elevated MnSOD in these bacteria. Quinones, viologens, and dyes, which can enter the cells and there be reduced to autoxidizable products, increase the level of SOD in *E. coli* (199,229–232). Such compounds increase cyanide-resistant respiration and can be shown to mediate O_2^- production in NAD(P)H-enriched cell extracts. There is little doubt that such compounds divert electron flow from the normal cytochrome pathway, which produces little O_2^-, to truncated pathway but produces a great deal of O_2^-. As a first approximation one may take the increase in cyanide-resistant respiration as a measure of the increase in O_2^- production. In the case of *E. coli*, paraquat can increase CN^--resistant respiration 15-fold. Since glucose can meet the energy needs of cells by a process independent of oxygen utilization, the suppression of SOD by glucose suggests that a metabolite of dioxygen is involved in the induction of SOD biosynthesis. The very large inductions caused, at constant pO_2, by compounds that can increase O_2^- production, suggests that O_2^- may be the responsible metabolite of dioxygen.

There are observations that suggest that the controls on the biosynthesis of SODs are complex and interlocking. Thus, consider the following observations:

1. With the fungus *Dactylium dendroides* a deficiency of Cu(II) in the growth medium leads to a decline in CuZnSOD with a compensatory increase in MnSOD (233).

2. A manganese-deficient diet lowered MnSOD, while raising CuZnSOD, in rats, mice, and chickens (234).

3. Exposure of poplar leaves to SO_2 caused an elevation of SOD (235).

4. A prolonged deficiency of vitamin E elevated SOD in several tissues of the rat (236).

5. Treatment of rats with 2,4-dinitrophenol caused an elevation of liver MnSOD (237).

6. Maintenance of rats at half the normal pO_2 for one week elevated lung MnSOD (238).

Some of these results are easily explicable in terms of O_2^- production, but others are not. Thus, 2,4-dinitrophenol can mediate O_2^- production by being reduced to the nitroanion radical which, in turn, reduces dioxygen to O_2^- (239). Depletion of any one SOD by a nutritional deficiency of the corresponding metal, would lead to an elevation of the steady state level of O_2^- within the cell and this could then derepress the biosynthesis of the other SOD. This might account for the effect of Cu(II) deficiency on the SODs in *D. dendroides* (233). and of Mn(II) deficiency on SODs in rats, mice, and chickens (234). The induction of SOD in hypoxic rat lung is also understandable. Suppose that O_2^- production in lung is minimal at a pO_2 of 14-mm Hg and becomes elevated at both lower and at higher pO_2. This would lead to inductions of SOD in lung by hyperoxia and by hypoxia (238). We are left to puzzle over the induction of SOD in poplar leaves by SO_2 (235) and over the report that elevation of FeSOD in *E. coli*, by gene cloning, failed to prevent induction of MnSOD by oxygenation (240).

Studies of the effects of metal salts and of chelating agents on the SOD content of *E. coli* (227,241–243) are leading to a fuller appreciation of the complexity of the controls on the biosynthesis of these enzymes. Chelating agents with preference for iron elevated MnSOD and even caused the production of MnSOD by anaerobically grown cells. Fe(II) was clearly suppressing the production of active MnSOD, since growth in iron-poor medium elevated MnSOD, while iron enrichment lowered MnSOD (241,242). These results led to the suggestion that the MnSOD gene is controlled by an iron-containing repressor (241,242).

Examination of the effects of iron, manganese, chelating agents, and paraquat on the SODs in *E. coli* has led to a more complex hypothesis (243), which is illustrated in Fig. 1. We here suppose that the apoenzymes act as autogenous repressors (A.R.), as do the apoenzymes containing the wrong metal at the active site. In contrast, the active holoenzymes do not act as repressors. Chelating agents should then exert a biphasic effect, elevating MnSOD when used at low levels and then lowering MnSOD when added at higher concentrations. This expectation arises because chelating agents at low levels could relieve the competition by Fe(II), but at higher levels would also make the manganese unavailable to the apoenzyme. The chelators 8-hydroxyquinoline, *o*-phenanthroline, and ethylenediaminetetraacetic acid, EDTA, do exert such a biphasic effect. Autogenous repression by the apoenzymes and by the inactive, incorrectly substituted metalloenzymes would prevent the wasteful accumulation of inactive proteins, yet allow very dramatic increases in active enzyme as repression was lifted by insertion of the correct metal into the apoenzyme.

Competition between manganese and iron for the metal binding sites of these apoSODs has been demonstrated *in vitro* (156,157,168) and in the case of the *E. coli* enzymes only iron restores activity to

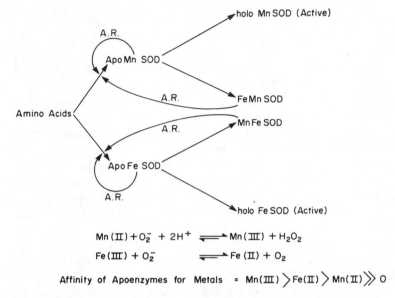

$$Mn\,(II) + O_2^- + 2H^+ \rightleftharpoons Mn\,(III) + H_2O_2$$

$$Fe\,(III) + O_2^- \rightleftharpoons Fe\,(II) + O_2$$

Affinity of Apoenzymes for Metals $= Mn(III) > Fe(II) > Mn(II) \gg 0$

Figure 1. Controls of the biosynthesis of SODs in *E. coli*: Hypothesis. We here suppose that the apo forms of both of the SODs made in *E. coli* act as autogenous repressors, A.R. It is also proposed, based upon metal replacement studies done *in vitro*, that the apoenzymes can bind either iron or manganese and that the active holoenzymes, formed when manganese is bound by the apoMnSOD or when iron is bound by the apoFeSOD, no longer act as autogenous regulators. Iron and manganese are thus thought to be in competition with each other and that competitive balance is tipped in favor of manganese when O_2^- oxidizes Mn(II), which binds less avidly than Fe(II) to Mn(III), which binds more tightly. This scheme fits available data, but has yet to be definitively tested.

apoFeSOD and only manganese to apoMnSOD. A key feature of the proposal shown in Fig. 1 is that Fe(II) binds more tightly to both apoSODs than does Mn(II), but that Mn(III) binds more tightly than Fe(II). O_2^- reduces Fe(III) to Fe(II), whether that Fe(III) be in cytochrome *c* (4), hemoglobin (244), or in an EDTA complex (245). In contrast, O_2^- oxidizes Mn(II) to Mn(III) (246–250). Anaerobic cells would contain Fe(II) and Mn(II) and the only active SOD produced would be FeSOD because Fe(II) successfully competes with Mn(II) for binding to the apoSODs.

Oxygenation and consequent O_2^- production would convert the

Mn(II) to Mn(III), while leaving the iron in the divalent state. Since Mn(III) shows greater affinity for the apoSODs, active MnSOD would now appear, while FeSOD would decline and this is what does happen. Chelating agents with preference for iron would tip the competitive balance between Fe(II) and Mn(II) and so allow active MnSOD to be made anaerobically. Once again, that is what is seen to occur. Increasing O_2^- production, as by addition of paraquat, would raise the fraction of cell manganese that is in the trivalent state, increasing its ability to compete with Fe(II) for the apoenzymes. Paraquat should thus induce MnSOD and should show a synergism with Mn(II), in this induction. This too is in accord with the actual behavior of the cells. Since Mn(III) can compete with Fe(II) for apoFeSOD, as well as for apoMnSOD, we might expect a decline in holoFeSOD in response to an increase in O_2^- production. This is seen. The observation that aeration decreases FeSOD, while elevating MnSOD, was also made with a strain of *E. coli* that overproduces FeSOD because of a multicopy plasmid bearing the FeSOD gene (240). The scheme in Fig. 1 thus accommodates the available data. Moreover the notion of autogenous repression by apoSODs is appealing, because it prevents wasteful production of an apoenzyme in amounts exceeding the supply of the prosthetic group, and it may be broadly applicable. It remains to be seen how well this scheme will survive the probing tests that are now being designed.

VII. Conclusions

The past decade has witnessed a satisfying growth and maturation in studies of the biology of the superoxide radical and in the enzymology of the superoxide dismutases. The very bulk of the pertinent literature is one measure of the work that has been done while the number of investigators currently engaged in this area is a measure of the work that remains to be done. Two journals devoted to this area are now being launched. One of these is *The Journal of Free Radicals in Biology and Medicine*, edited by K. A. Davies, and the second is *Advances in Free Radical Biology and Medicine*, edited by W. A. Pryor. It is hoped that these journals will make the writing of reviews such as this one an easier task in the years ahead.

References

1. Fridovich, I., *Adv. Enzymol.*, **41**, 35–97 (1974).

2. Sawada, Y. and Yamazaki, I., *Biochim. Biophys. Acta,* **327**, 257–265 (1973).

3. McCord, J.M. and Fridovich, I., *J. Biol. Chem.*, **244**, 6049–6055 (1969).

4. Ballou, D., Palmer, G., and Massey, V., *Biochem. Biophys. Res. Commun.*, **36**, 898–904 (1969).

5. Beauchamp, C. and Fridovich, I., *Anal. Biochem.*, **44**, 276–287 (1971).

6. Auclair, C., Torres, M., and Hakim, J., *FEBS Lett.*, **89**, 26–28 (1978).

7. Picker, S.D. and Fridovich, I. *Arch. Biochem. Biophys.* **228**, 155–158 (1984).

8. Misra, H.P. and Fridovich, I., *J. Biol. Chem.*, **247**, 3170–3175 (1972).

9. Sun, M. and Sigman, S., *Anal. Biochem.*, **90**, 81–89 (1978).

10. Marklund, S. and Marklund, G., *Eur. J. Biochem.*, **47**, 469–474 (1974).

11. Heikkila, R.E. and Cabbat, F., *Anal. Biochem.*, **25**, 356–362 (1976).

11a. Roth, E.E. and Gilbert, H.S., *Anal. Biochem.*, **137**, 50–53 (1984).

12. Elstner, E.F. and Heupel, A., *Anal. Biochem.*, **70**, 616–620 (1976).

13. Kobayashi, Y., Okahata, S., Tanabe, K., and Usui, T., *Hiroshima J. Med. Sci.*, **27**, 203–209 (1978).

14. Kobayashi, Y., Okahata, S., Tanabe, K., and Usui, T., *J. Immunol. Methods*, **24**, 75–78 (1978).

15. Kono, Y., *Arch. Biochem. Biophys,* **186**, 189–195 (1978).

16. Martin, J.P., *Fed. Proc.*, **43**, 2060 abst. (1984).

17. Misra, H.P. and Fridovich, I., *Arch. Biochem. Biophys.*, **181**, 308–312 (1977).

18. Misra, H.P., and Fridovich, I., *Arch. Biochem. Biophys.*, **183**, 511–515 (1977).

19. Salin, M.L. and McCord, J.M., *J. Clin. Invest.*, **54**, 1005–1009 (1974).

20. Azzi, A., Montecucco, C., and Richter, C., *Biochem. Biophys. Res. Commun.*, **65**, 597–603.

21. Buchanan, A.G. and Lees, H., *Can. J. Microbiol.*, **22**, 1643–1646 (1976).

22. Kuthan, H., Ullrich, V., and Estabrook, R.W., *Biochem. J.*, **203**, 551–558 (1982).

23. Finkelstein, E., Rosen, G., Patton, S.E., Cohen, M.S., and Rauckman, E.J., *Biochem. Biophys. Res. Commun.*, **102**, 1008–1015 (1981).

24. Hodgson, E.K. and Fridovich, I., *Biochim. Biophys. Acta,* **430**, 182–188 (1976).

25. Kirby, T.W. and Fridovich, I., *Anal. Biochem.*, **127**, 435–440 (1982).

26. Rigo, A., Viglino, P., and Rotilio, G., *Anal. Biochem.*, **68**, 1–8 (1975).

27. Misra, H.P. and Fridovich, I., *Biochemistry,* **15**, 681–687 (1976).

28. Klug, D., Rabani, J., and Fridovich, I., *J. Biol. Chem.*, **247**, 4839–4842 (1972).

29. Rotilio, G., Bray, R.C., and Fielden, E.M., *Biochim. Biophys. Acta,* **268**, 604–609 (1972).

30. Klug, D., Fridovich, I., and Rabani, J., *J. Am. Chem. Soc.*, **95**, 2786–2790 (1973).

31. Fielden, E.M., Roberts, P.B., Bray, R.C., and Rotilio, G., *Biochem. Soc. Trans.*, **1**, 52–53 (1973).

32. Pick, M., Rabani, J., Yost, F., and Fridovich, I., *J. Am. Chem. Soc.*, **96**, 7329–7333 (1974).

33. McAdam, M.E., Fox, R.A., Lavelle, F., and Fielden, E.M., *Biochem. J.*, **165**, 71–79 (1977).

34. Lavelle, F., McAdam, M.E., Fielden, E.M., Roberts, P.B., Puget, K., and Michelson, A.M., *Biochem. J.*, **161**, 3–12 (1977).

35. Fee, J.A., McClune, G.J., O'Neill, P., and Fielden, E.M., *Biochem. Biophys. Res. Commun.*, **100**, 377–384 (1981).

36. Marklund, S., *J. Biol. Chem.*, **251**, 7504–7507 (1976).

37. Henry, L.E.A., Halliwell, B., and Hall, D.O., *FEBS Lett.*, **66**, 303–306 (1976).

38. Valentine, J.S. and Curtis, A.B., *J. Am. Chem. Soc.*, **97**, 224–226 (1975).

39. Peters, J.W. and Foote, C.S., *J. Am. Chem. Soc.*, **98**, 873–875 (1976).

40. Montalbini, P., Koch, F., Burba, M., and Elstner, E.F., *Physiol. Plant. Pathol.*, **12**, 211–223 (1978).

41. Rigo, A., Viglino, P., Argese, E., Terenzi, M., and Rotilio, G., *J. Biol. Chem.*, **254**, 1759–1760 (1979).

41a. Minami, M. and Yoshikawa, H., *Clin. Chim. Acta*, **92**, 337–342 (1979).

42. Bensinger, R.E. and Johnson, C.M., *Anal. Biochem.*, **116**, 142–145 (1981).

43. Rosen, G.M., Finkelstein, E., and Rauckman, E.J., *Arch. Biochem. Biophys.*, **215**, 367–378 (1982).

44. Kono, Y., *J. Biochem.*, **91**, 1789–1794 (1982).

45. Whitelam, G.C. and Codd, G.A., *Anal. Biochem.*, **121**, 207–212 (1982).

46. Hyland, K., Voisin, E., Banoun, H., and Auclair, C., *Anal. Biochem.*, **135**, 280–287 (1983).

47. Kanematsu, S. and Asada, K., *Arch. Biochem. Biophys.*, **185**, 473–482 (1978).

48. Beauchamp, C.O. and Fridovich, I., *Biochim. Biophys. Acta*, **317**, 50–64 (1973).

49. Ysebaert-Vanneste, M. and Vanneste, W.H., *Anal. Biochem.*, **107**, 86–95 (1980).

50. Symonyan, M.A. and Nalbandyan, R.M., *FEBS Lett.*, **28**, 22–24 (1972).

51. Asada, K., Yoshikawa, K., Takahashi, M.-A., Maeda, Y., and Enmanji, K., *J. Biol. Chem.*, **250**, 2801–2807 (1975).

52. Hodgson, E.K. and Fridovich, I., *Biochemistry*, **14**, 5299–5303 (1975).

53. Salin, M., Day, E.D., and Crapo, J.D., *Arch. Biochem. Biophys.*, **187**, 223–228 (1978).

54. Geller, B.L. and Winge, D.R., *Anal. Biochem.*, **128**, 86–92 (1983).

55. Baret, A., Michel, P., Imbert, M.R., Morcellet, J.L., and Michelson, A.M., *Biochem. Biophys. Res. Commun.*, **88**, 337–345 (1979).

56. Del Villano, B.C. and Tischfield, J.A., *J. Immunol. Methods*, **29**, 253–262 (1979).

57. Baret, A., Schiavi, P., Michel, P., Michelson, A.M., and Puget, K., *FEBS Lett.*, **112**, 25–29 (1980).

58. Del Villano, B.C. and Tischfield, J.A., *Methods Enzymol.*, **74**, 359–370 (1981).

59. Nishimura, N., Ito, Y., Adachi, T., and Hirano, K., *J. Pharmacobio. Dyn.* **5**, 394–402 (1982).

60. Nishimura, N., Ito, Y., Adachi, T., Hirano, K., Sugiura, M., and Sawaki, S., *J. Pharmacobio. Dyn.* **5**, 869–876 (1982).

61. Iizuka, S., Taniguchi, N., and Makita, A., *J. Natl. Cancer Inst.*, **72**, 1043–1049 (1984).

61a. Yamanaka, N. and Deamer, D., *Physiol. Chem. Phys.*, **6**, 95–106 (1974).

61b. Sahu, S.K., Oberley, L.W., Stevens, R.H., and Riley, E.F., *J. Natl. Cancer Inst.*, **58**, 1125–1128 (1977).

61c. Oberley, L.W., Bize, I.B., and Sahu, S.K., *J. Natl. Cancer Inst.*, **61**, 375–379 (1978).

61d. Bize, I.B., Oberley, L.W., and Morris, H.P., *Cancer Res.*, **40**, 3686–3693 (1980).

62. Halliwell, B., *Biochem. J.*, **135**, 379–381 (1973).

63. McCord, J.M. and Fridovich, I., *J. Biol. Chem.*, **243**, 5753–5760 (1968).

64. Nakamura, S. and Yamazaki, I., *Biochim. Biophys. Acta*, **189**, 29–37 (1969).

65. McCord, J.M. and Fridovich, I., *J. Biol. Chem.*, **245**, 1374–1377 (1970).

66. Eldred, G.E. and Hoffert, J.R., *Anal. Biochem.*, **110**, 137–143 (1981).

67. Komai, H., Massey, V., and Palmer, G., *J. Biol. Chem.*, **244**, 1692–1700 (1969).

68. Hassan, H.M., Dougherty, H., and Fridovich, I., *Arch. Biochem. Biophys.*, **199**, 349–354 (1980).

69. Keele, B.B. Jr., McCord, J.M., and Fridovich, I., *J. Biol. Chem.*, **245**, 6176–6181 (1970).

70. Yost, F.J., Jr., and Fridovich, I., *J. Biol. Chem.*, **248**, 4905–4908 (1973).

71. Crapo, J.D., McCord, J.M., and Fridovich, I., *Methods Enzymol.*, **53**, 382–393 (1978).

72. Valentine, J.S. and Pantoliano, M.W., *Metal Ions Biol.*, **3**, 291–358 (1981).

73. Steinman, H.M., in *Superoxide Dismutase*, L.W. Oberley, Ed., Vol. I, CRC, Boca Raton, Fla., 1982 pp. 11–68.

74. Goscin, S.A. and Fridovich, I., *Biochim. Biophys. Acta*, **289**, 276–283 (1972).

75. Misra, H.P. and Fridovich, I., *J. Biol. Chem.*, **247**, 3410–3414 (1972).

76. Asada, K., Takahashi, M., and Nagate, M., *Agric. Biol. Chem.*, **38**, 471–473 (1974).

77. Bannister, J.V., Anastasi, A., and Bannister, W.H., *Comp. Biochem. Physiol.*, **56B**, 235–238 (1977).

78. Weisiger, R.A. and Fridovich, I., *J. Biol. Chem.*, **248**, 3582–3592 (1973).

79. Evans, H.J., Steinman, H.M., and Hill, R.L., *J. Biol. Chem.*, **249**, 7315–7325 (1974).

80. Steinman, H.M., Naik, V.R., Abernethy, J.L., and Hill, R.L., *J. Biol. Chem.*, **249**, 7326–7338 (1974).

81. Abernethy, J.L., Steinman, H.M., and Hill, R.L., *J. Biol. Chem.*, **249**, 7339–7347 (1974).

82. Forman, H.J. and Fridovich, I., *J. Biol. Chem.*, **248**, 2645–2649 (1973).

83. Malinowski, D.P. and Fridovich, I., *Biochemistry*, **18**, 5055–5060 (1979).

84. Barra, D., Bossa, F., Marmocchi, F., Martini, F., Rigo, A., and Rotilio, G., *Biochem. Biophys. Res. Commun.*, **86**, 1199–1205 (1979).

85. Connerdal, B., Keen, C.L., and Hurley, L.S. *FEBS Lett.*, **108**, 51–58 (1979).

86. Bloor, J.H., Holtz, D., Kaars, J., and Kosman, D.J., *Biochem. Genet.*, **21**, 349–364 (1983).

87. Tegelstrom, H., *Hereditas*, **81**, 185–198 (1975).

88. Civalleri, L., Pini, C., Rigo, A., Federico, R., Calabrese, L., and Rotilio, G., *Mol. Cell. Biochem.*, **47**, 3–9 (1982).

89. Mavelli, I., Ciriolo, M., and Rotilio, G., *Biochem. Biophys. Res. Commun.*, **117**, 677–681 (1983).

90. Crosti, N. and Sausa, P., *Biochem. Genet.*, **18**, 693–697 (1980).

91. Duke, M.V. and Salin, M.L., *Phytochemistry* **22**, 2369–2373 (1983).

92. Marklund, S., Beckman, G., and Stigbrand, T., *Eur. J. Biochem.*, **65**, 415–422 (1976).

93. Larsen, A.L. and Benson, W.C., *Crop. Sci.*, **10**, 493–495 (1970).

94. Oelshlegel, F.J., Jr., and Stahman, M.A., *Anal. Biochem.*, **42**, 338–341 (1971).

95. Baur, E.W. and Schorr, R.T., *Science*, **166**, 1524–1525 (1969).

96. Richmond, R.C. and Powell, J.R., *Proc. Natl. Acad. Sci., USA*, **67**, 1264–1267 (1970).

97. Lippitt, B. and Fridovich, I., *Arch. Biochem. Biophys.*, **159**, 738–741 (1973).

98. Sinet, P.M., in *Superoxide and Superoxide Dismutases*, A.M. Michelson, J.M. McCord, and I. Fridovich, Eds., Academic, London, 1977, pp. 458–465.

99. Marklund, S.L., Holme, E., and Hellner, L.,*Clin. Chim. Acta*, **126**, 41–51 (1982).

100. Marklund, S.L., *Proc. Natl. Acad. Sci., USA*, **79**, 634–638 (1982).

101. Vance, P.G., Keele, B.B., Jr., and Rajagopalan, K.V., *J. Biol. Chem.*, **247**, 4782–4786 (1972).

102. Lavelle, F., Durosay, P. and Michelson, A.M., *Biochimie*, **56**, 451–458 (1974).

103. Bridgen, J., Harris, J.I., and Northrop, F., *FEBS Lett.*, **49**, 392–395 (1975).

104. Ravindranath, S.D. and Fridovich, I., *J. Biol. Chem.*, **250**, 6107–6112 (1975).

105. McCord, J.M., Boyle, J.A., Day, E.D. Jr., Rizzolo, L.Z., and Salin, M.L., in *Superoxide and Superoxide Dismutases*, (Proc. EMBO Workshop, 1976), A.M. Michelson, J.M. McCord, and I. Fridovich, Eds., Academic, London, 1977, pp. 129–138.

106. Sato, S. and Harris, J.I., *Eur. J. Biochem.*, **73**, 373–381 (1977).

107. Misra, H.P. and Fridovich, I., *J. Biol. Chem.*, **252**, 6421–6423 (1977).

108. Fernandez, V.M., Sevilla, F., Lopez-Gorge, J., and Del Rio, L.A., *J. Inorg. Biochem.*, **16**, 79–84 (1982).

109. Baum, J.A. and Scandalios, J.G., *Arch. Biochem. Biophys.*, **206**, 249–264 (1981).

110. Chikata, Y., Kusunose, E., Ichihara, K., and Kusunose, M., *Osaka City Med. J.*, **21**, 127–136 (1975).

111. Kusunose, M., Noda, Y., Ichihara, K., and Kusunose, E., *Arch. Microbiol.*, **108**, 65–73 (1976).

112. Ichihara, K., Kusunose, E., Noda, Y., Chikata, Y., Kusunose, M., and Mori, T., *J. Biochem.*, *(Tokyo)*, **81**, 1427–1433 (1977).

113. Kusunose, E., Kusunose, M., Ichihara, K., and Izumi, S., *J. Gen. Appl. Microbiol.*, **26**, 369–372 (1980).

114. Clare, D.A., Blum, J., and Fridovich, I., *J. Biol. Chem.*, **259**, 5932–5936 (1984).

115. Henry, Y.A., Puget, K., and Michelson, A.M., *Fed. Proc.* **33**, 1321 (1974).

116. Kanematsu, S. and Asada, K., *FEBS Lett.*, **91**, 94–98 (1978).

117. Yamakura, F. *Biochim. Biophys. Acta*, **422**, 280–294 (1976).

118. Czeke, C., Horvath, L.I., Simon, P., Borbelly, G., Keszthelyi, L., and Farkas, G.L., *J. Biochem.* *(Tokyo)*, **85**, 1397–1404 (1979).

118a. Kirby, T.W., Lancaster, J.R., and Fridovich, I., *Arch. Biochem. Biophys.*, **210**, 140–148 (1981).

119. Bridges, S.M. and Salin, M.L., *Plant Physiol.*, **68**, 275–278 (1981).

120. Kusunose, E., Ichihara, K., Noda, Y., and Kusunose, M., *J. Biochem.* *(Tokyo)*, **80**, 1343–1352 (1976).

121. Ichihara, K., Kasaoka, I., Kusunose, E., and Kusunose, M., *J. Gen. Appl. Microbiol.*, **26**, 387–393 (1980).

122. Kanematsu, S. and Asada, K., *Arch. Biochem., Biophys.*, **195**, 535–545 (1979).

123. Johansen, J.T., Overballe-Peterson, C., Martin, B., Haseman, V., and Svendsen, I., *Carlsberg Res. Commun.*, **44**, 201–217 (1979).

124. Steinman, H.M., *J. Biol. Chem.*, **255**, 6758–6765 (1980).

125. Barra, D., Martini, F., Bossa, F., Rotilio, G., Bannister, J.V., and Bannister, W.H., *Biochem. Biophys. Res. Commun.*, **81**, 1195–1200 (1978).

126. Jabusch, J.R., Farb, D.L., Kerchensteiner, D.A., and Deutsch, H.F., *Biochemistry*, **19**, 2310–2316 (1980).

127. Barra, D., Martini, F., Bannister, J.V., Schinina, M.E., Rotilio, G., Bannister, W.H., and Bossa, F., *FEBS Lett.*, **120**, 53–56 (1980).

128. Ammer, D. and Lerch, K., *Dev. Biochem.*, **11A**, 230–236 (1980).

129. Puget, K. and Michelson, A.M., *Biochem. Biophys. Res. Commun.*, **58**, 830–838 (1974).

130. Steinman, H.M., *J. Biol. Chem.*, **257**, 10283–10293 (1982).

131. Fridovich, I., *Ann. Rev. Biochem.*, **44**, 147–159 (1975).

132. Martin, J.P. Jr., and Fridovich, I., *J. Biol. Chem.*, **256**, 6080–6089 (1981).

133. Steffens, K.J., Bannister, J.V., Bannister, W.H., Flohé, L., Gunzler, W.A., Kim, S.A., and Otting, F., *Hoppe-Seyler's Z. Physiol. Chem.,* **364,** 675–690 (1983).

134. Bannister, J.V. and Parker, M.W., *Proc. Natl. Acad. Sci., USA,* **82,** 149–152 (1985).

135. Brock, C.J. and Walker, J.E., *Biochemistry,* **19,** 2873–2882 (1980). Brock, C. J. and Walker, J. E., *Dev. Biochem.,* **11A,** 237–241 (1980).

136. Steinman, H., *J. Biol. Chem.,* **253,** 8708–8720 (1978).

137. Ditlow, C., Johansen, J.T., Martin, B.M., and Svendsen, I.B., *Carlsberg Res. Commun.,* **47,** 81–91 (1982).

138. Stallings, W.C., Pattridge, K.A., Strong, R.K., and Ludwig, M.L., *J. Biol. Chem.,* **259,** 10695–10698 (1984).

139. Walker, J.E., Auffret, A.D., Brock, C.J., and Steinman, H.M., *Dev. Biochem.,* **11A,** 212–222 (1980).

140. Bruschi, M., Hatchikian, E.C., Bonical, J., Bovier-Lapierre, G., and Couchoud, P., *FEBS Lett.,* **76,** 121–124 (1977).

141. Harris, J.E. and Steinman, H.M., in *Superoxide and Superoxide Dismutases,* A.M. Michelson, J.M. McCord, and I. Fridovich, Eds., Academic, London, 1977, pp. 225–230.

142. Harris, J.I., Auffret, A.D., Northrop, F.D., and Walker, J.E., *Eur. J. Biochem.,* **106,** 297–303 (1980).

143. Muno, D., Isobe, T., Okuyama, T., Ichihara, K., Noda, Y., Kusunose, E., and Kusunose, M., *Biochem. Int.,* **2,** 33–42 (1981).

144. Richardson, J.S., Thomas, K.A., Rubin, B.H., and Richardson, D.C., *Proc. Natl. Acad. Sci., USA,* **72,** 1349–1353 (1975).

145. Richardson, J.S., Thomas, K.A., and Richardson, D.C., *Biochem. Biophys. Res. Commun.,* **63,** 986–992 (1975).

146. Tainer, J.A., Getzoff, E.D., Beem, K.M., Richardson, J.S., and Richardson, D.C., *J. Mol. Biol.,* **160,** 181–217 (1982).

147. Tainer, J.A., Getzoff, E.D., Richardson, J.S., and Richardson, D.C., *Nature* **306,** 284–287 (1983).

148. Beem, K.M., Richardson, J.S., and Richardson, D.C., *J. Mol. Biol.,* **105,** 327–332 (1976).

149. Bridgen, J., Harris, J.I., and Kolb, E., *J. Mol. Biol.,* **106,** 333–335 (1976).

150. Powers, T.B., Slykehouse, T.O., Fee, J.A., and Ludwig, M.L., *J. Mol. Biol.,* **23,** 689–690 (1978).

151. Yamakura, F., Suzuki, K., Petsko, G.A., and Tsernoglou, D., *Dev. Biochem.,* **11A,** 242–253 (1980).

152. Yamakura, F., Suzuki, K., and Mitsui, Y., *J. Biol. Chem.,* **251,** 4792–4793 (1976).

153. Ringe, D., Petsko, G.A., Yamakura, F., Suzuki, K., and Ohmori, D., *Proc. Natl. Acad. Sci., USA,* **80,** 3879–3883 (1983).

154. Stallings, W.C., Powers, T.B., Pattridge, K.A., Fee, J.A., and Ludwig, M.L., *Proc. Natl. Acad. Sci., USA,* **80,** 3884–3888 (1983).

155. Brock, C.J., Harris, J.I., and Sato, S., *J. Mol. Biol.,* **107,** 175–178 (1976).

156. Puget, K., Lavelle, F., and Michelson, A.M. in *Superoxide and Superoxide Dismutases,* E.M. Michelson, J.M. McCord, and I. Fridovich, Eds., Academic, London, 1977, pp. 139–150.

157. Ose, D.E. and Fridovich, I., *Arch. Biochem. Biophys.* **194,** 360–364 (1979).

158. Yamakura, F. and Suzuki, S., *Biochem. Biophys. Res. Commun.,* **72,** 1108–1115 (1976).

159. Yamakura, F., *J. Biochem. (Tokyo),* **83,** 849–857 (1978).

160. Calabrese, L., Rotilio, G., and Mondovi, B., *Biochim. Biophys. Acta,* **263,** 827–829 (1972).

161. Beem, K.M., Rich, W.E., and Rajagopalan, K.V., *J. Biol. Chem.,* **249,** 7298–7305 (1974).

162. Fee, J.A. and Riggs, R.G., *Biochim. Biophys. Acta,* **400,** 439–450 (1975).

163. Valentine, J.S., Pantoliano, M.W., McDonnell, P.J., Burger, A.R., and Lippard, S.J., *Proc. Natl. Acad. Sci., USA,* **76,** 4245–4249 (1979).

164. Heikkila, R.E., Cabat, F.S., and Cohen, G., *J. Biol. Chem.,* **251,** 2182–2185 (1976).

165. Misra, H.P., *J. Biol. Chem.,* **254,** 11623–11628 (1979).

166. Cocco, D., Calabrese, L., Rigo, A., Marmocchi, F., and Rotilio, G., *Biochem. J.,* **199,** 675–680 (1981).

167. Stevenato, R., Viglino, P., Rigo, A., and Cocco, D., *Inorg. Chim. Acta,* **56,** L39–40 (1981).

168. Brock, C.J. and Harris, J.I., *Biochem. Soc. Trans.,* **5,** 1537–1539 (1977).

169. Brock, C.J. and Walker, J.E., *Dev. Biochem.,* **11A,** 237–241 (1980).

170. Ose, D.E. and Fridovich, I., *Arch. Biochem. Biophys,* **194,** 360–364 (1979).

171. Yamakura, F. and Suzuki, K., *J. Biochem. (Tokyo),* **88,** 191–196 (1980).

172. Gregory, E.M. and Dapper, C.H., *Arch. Biochem. Biophys,* **220,** 293–300 (1983).

173. Gregory, E.M. *Arch. Biochem. Biophys.* **238,** 83–89 (1985).

174. Meier, B., Barra, D., Bossa, F., Calabrese, L., and Rotilio, G., *J. Biol. Chem.,* **257,** 13977–13980 (1982).

175. Fielden, E.M., Roberts, P.B., Bray, R.C. Lowe, D.J., Mautner, G.N., Rotilio, G., and Calabrese, L., *Biochem. J.,* **139,** 49–60 (1974).

176. Cockle, S.A. and Bray, R.C., in *Superoxide and Superoxide Dismutases,* A.M. Michelson, J.M. McCord, and I. Fridovich, Eds., Academic, London, 1977, pp. 215–216.

177. Viglino, D., Rigo, A., Argese, E., Calabrese, L., Cocco, D., and Rotilio, G., *Biochem. Biophys. Res. Commun.,* **100,** 125–130 (1981).

178. Malinowski, D.P. and Fridovich, I., *Biochemistry,* **18,** 237–244 (1979).

179. Fee, J.A., and Di Corleto, P.E., *Biochemistry,* **12,** 4893–4899 (1973).

180. McAdam, M.E., Fielden, E.M., Lavelle, F., Calabrese, L., Cocco, D., and Rotilio, G., *Biochem. J.,* **167**, 271–274 (1977).

181. Strothkamp, K.G. and Lippard, S.J., *Biochemistry,* **20**, 7488–7493 (1981).

182. Nozaki, Y., Gurd, F.R.N., Chen, R.F., and Edsall, J.T., *J. Am. Chem. Soc.,* **79**, 2123–2129 (1957).

182a. Blackburn, N.J., Hasnain, S.S., Binstead, N., Diakun, G.P., Garner, G.D., and Knowles, P.F., *Biochem. J.,* **219**, 985–990 (1984).

183. Koppenol, W.H. in *Oxygen, Oxy-Radicals Chemistry and Biology* (Proc. Int. Conf., 1980), M.A.J. Rodgers, and E.L. Powers, Eds., Academic, New York, 1981, pp. 671–674.

184. Cudd, A. and Fridovich, I., *FEBS Lett.,* **144**, 181–182 (1982).

185. Cudd, A., and Fridovich, I., *J. Biol. Chem.,* **257**, 11443–11447 (1982).

186. Benovic, J., Tillman, T., Cudd, A., and Fridovich, I., *Arch. Biochem. Biophys.,* **221**, 329–332 (1983).

187. Cocco, D., Mavelli, I., Rossi, L., and Rotilio, G., *Biochem. Biophys. Res. Commun.,* **111**, 860–864 (1983).

188. Argese, E., Rigo, A., Viglino, P., Orsega, E., Marmocchi, F., Cocco, D., and Rotilio, G., *Biochim. Biophys. Acta,* **787**, 205–207 (1984).

189. Getzoff, E.D., Tainer, J.A., Weiner, P.K., Kollman, P.A., Richardson, J.S., and Richardson, D.C., *Nature,* **306**, 287–290 (1983).

190. Koppenol, W. H. and Butler, J., *FEBS Lett.,* **83**, 1–6 (1977).

191. Hodgson, E.K. and Fridovich, I., *Biochem. Biophys. Res. Commun.,* **54**, 270–274 (1973).

192. Rotilio, G., Calabrese, L., Bossa, F., Barra, D., Finazzi Agro, A., and Mondovi, B., *Biochemistry,* **11**, 2182–2187 (1972).

193. Bray, R.C., Cockle, S.A., Fielden, E.M., Roberts, P.B., Rotilio, G., and Calabrese, L., *Biochem. J.,* **139**, 43–48 (1974).

194. Blech, D.M. and Borders, C.L., Jr., *Arch. Biochem. Biophys.,* **224**, 579–586 (1983).

195. Misra, H.P. and Fridovich, I., *Arch. Biochem. Biophys.,* **189**, 317–322 (1978).

195a. Misra, H.P., *J. Biol. Chem.,* **259**, 12678–12684 (1984).

196. Mota de Freitas, D. and Valentine, J.S., *Biochemistry,* **23**, 2079–2082 (1984).

196a. Osman, R. and Basch, H., *J. Am. Chem. Soc.,* **106**, 5710–5714 (1984).

197. Gregory, E.M. and Fridovich, I., *J. Bacteriol.,* **114**, 543–548 (1973).

198. Gregory, E.M., Yost, F.J., and Fridovich, I., *J. Bacteriol.,* **115**, 987–991 (1973).

199. Hassan, H.M. and Fridovich, I., *J. Bacteriol.,* **129**, 1574–1583 (1977).

200. Crapo, J.D. and Tierney, D., *Clin. Res.,* **21**, 222 (1973).

201. Crapo, J.D. and Tierney, D.L., *Am. J. Physiol.,* **226**, 1401–1407 (1974).

202. Kimball, R.E., Reddy, K., Pierce, T.H., Schwartz, L.W., Mustafa, M.G., and Cross, C.E., *Am. J. Physiol.,* **230**, 1425–1431 (1976).

203. Autor, A.P., Frank, L., and Roberts, R.J., *Pediatr. Res.,* **10**, 154–158 (1976).

204. Stevens, J.B. and Autor, A.P., *J. Biol. Chem.*, **252**, 3509–3514 (1977).

205. Roberts, R.J., *J. Pediatr.* **95**, 904–909 (1979).

206. Rister, M. and Baehner, R.L., *Blood*, **46**, 1016 (1975).

207. Stevens, J.B. and Autor, A.P., *Lab. Invest.*, **37**, 470–478 (1977).

208. Simon, L.M., Liu, J., Theodore, J., and Robin, E.D., *Am. Rev. Respir. Dis.*, **115**, 279–284 (1977).

209. Nerurkar, L.S., Zeligs, B.J., and Bellanti, J.A., *Photochem. Photobiol.* **28**, 781–786 (1978).

210. Ody, C., Bach-Dieterle, Y., Wand, I., and Junod, A.F., *Exp. Lung Res.* **1**, 271–279 (1980).

211. Housset, B. and Junod, A.F., *Biochim. Biophys. Acta*, **716**, 283–289 (1982).

212. Gregory, E.M., Goscin, S.A., and Fridovich, I., *J. Bacteriol.*, **117**, 456–460 (1974).

213. Boveris, A., Sanchez, R.A., and Beconi, M.T., *FEBS Lett.*, **92**, 333–338 (1978).

214. Privalle, C.T. and Gregory, E.M., *J. Bacteriol.*, **138**, 139–145 (1979).

215. Ghosh, S. and Chatterjee, G.C., *J. Gen. Appl. Microbiol.*, **25**, 367–374 (1979).

216. Friedberg, D., Fine, M., and Oren, A., *Arch. Microbiol.*, **123**, 311–313 (1979).

217. Stowers, M.D. and Elkan, G.H., *Can. J. Microbiol.*, **27**, 1202–1208 (1981).

218. Moody, C.S. and Hassan, H.M., *Proc. Natl. Acad. Sci., USA*, **79**, 2855–2859 (1982).

219. Irwin, D., Subden, R.E., Meiering, A.G., Cunningham, J.D., and Fyfe, C., *Microbios*, **33**, 111–118 (1982).

220. Samah, D.A. and Wimpenny, J.W.T., *J. Gen. Microbiol.*, **128**, 355–360 (1982).

221. Natvig, D.O., *Arch. Microbiol.*, **132**, 107–114 (1982).

222. Smith, M.W. and Neidhardt, F.C., *J. Bacteriol.*, **154**, 344–350 (1983).

223. Nur, I., Okon, Y., and Henis, Y., *J. Gen. Microbiol.*, **128**, 2937–2944 (1982).

224. Martin, M.E., Strachan, R.C., Aranha, H., Evans, S.L., Salin, M.L., Welch, B., Arceneaux, J.E.L., and Byers, B.R., *J. Bacteriol.*, **159**, 745–749 (1984).

225. Hassan, H.M. and Fridovich, I., *J. Bacteriol.*, **132**,505–510 (1977).

226. Kasemset, D. and Oberley, L.W., *Biochem. Biophys. Res. Commun.*, **122**, 682–686 (1984).

227. Pugh, S.Y.R. and Fridovich, I., *Fed. Proc.*, **43**, 2057 (1984).

228. Pugh, S.Y.R., DiGuiseppi, J., and Fridovich, I., *J. Bacteriol.*, in press.

229. Hassan, H.M. and Fridovich, I., *J. Biol. Chem.*, **252**, 7667–7672 (1978).

230. Hassan, H.M. and Fridovich, I., *J. Biol. Chem.*, **253**, 8143–8148 (1978).

231. Hassan, H.M. and Fridovich, I., *Arch. Biochem. Biophys.*, **196**, 385–395 (1979).

232. Hassan, H.M. and Fridovich, I., *J. Bacteriol.*, **141**, 156–163 (1980).

233. Shatzman, A.R. and Kosman, D., *Biochim. Biophys. Acta*, **544**, 163–179 (1978).

234. de Rosa, G., Keen, C.L., Leach, R.M., and Hurley, L.S., *J. Nutr.*, **110**, 795–804 (1980).

235. Takahara, K. and Sugahara, K., *Plant Cell. Physiol.*, **21**, 601–612 (1980).

236. Chen, L.H., Thacker, R.R., and Chow, C.K., *Nutr. Rep. Int.*, **22**, 873–882 (1980).

237. Dryer, S.E., Dryer, R.L., and Autor, A.P., *J. Biol. Chem.*, **255**, 1054–1057 (1980).

238. Sjostrom, K. and Crapo, J.D., *Lab. Invest.*, **48**, 68–79 (1983).

239. Sealy, R.C., Swartz, H.M., and Olive, P.L., *Biochem. Biophys. Res. Commun.*, **82**, 680–684 (1978).

240. Nettleton, C.J., Bull, C., Baldwin, T.O., and Fee, J.A., *Proc. Natl. Acad. Sci., USA*, **81**, 4970–4973 (1984).

241. Moody, C.S. and Hassan, H.M., *Federation Proc.*, **43**, 1718 (1984).

242. Moody, C.S. and Hassan, H.M., *J. Biol. Chem.*, **259**, 12821–12825 (1984).

243. Pugh, S.Y.R. and Fridovich, I., *J. Bacteriol.*, in press.

244. Sutton, H.C., Roberts, P.B., and Winterbourn, C.C., *Biochem. J.*, **155**, 503–510 (1976).

245. Ilan, Y.A. and Czapski, G., *Biochim. Biophys. Acta*, **498**, 386–394 (1977).

246. Archibald, F.S. and Fridovich, I., *Arch. Biochem. Biophys.*, **214**, 452–463 (1982).

247. Homann, P.H., *Biochemistry*, **4**, 1902–1911 (1965).

248. Kono, Y., *J. Biochem.*, **91**, 381–395 (1982).

249. Kono, Y., Takahashi, M.A., and Asada, K., *Arch. Biochem. Biophys.*, **174**, 454–462 (1976).

250. Yamazaki, I. and Piette, L.H., *Biochim. Biophys. Acta*, **77**, 47–64 (1963).

BACTERIAL PROTEIN TOXINS WITH LATENT ADP-RIBOSYL TRANSFERASES ACTIVITIES

By CHUN-YEN LAI, *Central Research, Hoffmann-La Roche Inc., Nutley, New Jersey 07110*

CONTENTS

I. Introduction

A. BACTERIAL EXOTOXINS

In 1890 Von Behring and Kitasato demonstrated that rabbits and mice immunized with small amounts of culture filtrates of *Clostridium tetani* and *Corynebacterium diphtheriae* became resistant to infection by the respective bacteria. The sera from the immune animals were found to neutralize the toxicity of culture filtrates *in vitro* and to protect nonimmune animals from infection. These experiments provided conclusive evidence for the existence of toxic substances in the bacterial products that were the cause of disease states in animals. A great deal of research has since been directed to the isolation and identification of toxins from various pathogenic organisms. Toxoids from highly purified toxins have been found to be effective vaccines because of their specificities as immunogens. However, some highly infectious bacteria such as pneumococci were found not to produce toxins either *in vitro* or *in vivo*, their pathogenesis being different from those of toxin-producing bacteria. Other human and animal pathogens were shown to produce toxins that could not be correlated to the actual disease processes or caused illness only synergistically with other components of their metabolic products. These findings and perhaps more importantly, the development of antibiotics for treatment of infectious diseases slowed the studies of bacterial toxins considerably in the middle of this century. With the advances in protein biochemistry, immunology, and cell biology, interest in bacterial toxins has been renewed. In the past 20 years, several known toxins have been reinvestigated with newly available techniques, purified to homogeneity and the mechanism of actions studied at the molecular level.

The term bacterial toxins generally refers to toxic substances released by bacteria into culture medium or their external environment within the host. These are also called exotoxins to be distinguished from endotoxins, the toxic material found in and obtained by extraction of cell bodies of gram-negative bacteria. Bacterial endotoxins are usually lipopolysaccharides in nature, and have been shown to cause nonspecific body reactions that mimic those of bacterial sepsis but cause no ill effects on cultured animal cells. In fact lipopolysaccharide toxins have been found to stimulate B-lymphocyte differenciation, activate macrophages, and increase phagocytosis, and are sometimes considered more beneficial than harmful to the body (1–3). In contrast, bacterial exotoxins are proteins of defined molecular weights, and exert characteristic toxicities to animals that can be neutralized with specific antisera. They are traditionally named after the illness they cause, that is, diphtheria toxin, pertussis toxin, tetanus toxin, and so on (Table I). Of the many bacterial exotoxins identified, isolated, and studied in the past 90 years, only a few have been structurally characterized and their biochemical basis of action defined. In this chapter, discussion will be focused on the recent findings on these toxins and their mechanism of action. Excellent reviews on various studies on microbial toxins are available in books and monographs (4–6).

B. FUNCTIONAL DOMAINS IN BACTERIAL TOXINS—THE A–B MODEL

With advances in tissue culture techniques, studies on the mode of action of bacterial toxins at the cellular level became possible. By around 1960 it had become apparent that the toxicity of diphtheria toxin was due to the inhibition of protein synthesis leading to cell death. With the use of a cell-free protein synthesizing system, it was demonstrated that nicotanamide adenine dinuclotide, NAD^+, was required in the toxin's action, and that the enzyme involved in the translocation of the growing peptide chain on ribosome, elongation factor II, EF-2, became inactivated (7). In 1968 diphtheria toxin was discovered to be itself an enzyme that catalyzed the hitherto unknown reaction of transferring the ADP (adenosine diphosphate)-ribose moiety of NAD^+ onto an acceptor protein, in this case EF-2, and thereby caused its inactivation (8):

$$NAD^+ + EF\text{-}2 \rightarrow ADP\text{-ribosyl } EF\text{-}2 + \text{nicotinamide}$$

The discovery marked the first elucidation of the mechanism of action of bacterial toxin at the molecular level. In the years that followed, another unexpected fact became apparent. The intact diphtheria toxin was found to be virtually inactive enzymatically in the cell-free system; the ADP-ribosyl transferase activity came entirely from the NH_2-terminal proteolytic fragment of the toxin present in the preparations used in earlier experiments (9,10). This active fragment is about one third the size of the toxin and is called fragment A. Separated fragment A was fully active in catalyzing

TABLE I.
Samples of Bacterial Protein Toxins

Toxin	Approximate molecular weight	Biological activity	$LD_{50}(\mu g)$
Clostridium botulinum Botulinum types A and B	10^6 (dimer)	Neurotoxin	2×10^{-5} (i.p. mouse)
Clostridium tetani Tetanus toxin	6.7×10^4	Neurotoxin	3×10^{-5} (i.p. mouse)
Corynebacterium diphtheriae Diphtheria toxin	6.5×10^4	Cell death, dermonecrosis	0.04 (guinea pig)
Pseudomonas aeruginosa Exotoxin A	6.6×10^4	Burn infection, cell death	0.1 (mouse)
Bordetella pertussis Pertussigen	10^5 (subunits)	Promote leukocytosis and sensitivity to histamine	0.5 (mouse)
Virbrio cholerae Cholera toxin	8.4×10^4 (subunits)	Diarrheagenic	—
Escherichia coli Enterotoxins	Heat labile [10^5] Heat stable [4.4×10^3]	Diarrheagenic	—
Shigella dysenteriae Neurotoxin	8.2×10^4	Hemorrhage	2×10^{-3} (rabbit) 0.01 (mouse)
Streptococcus pyogenes Streptolysin O	2.7×10^4	Membrane damage	0.5 (i.v. mouse)

ADP-ribosylation of EF-2, but was found to be innocuous against whole cells. The finding suggested that the other region of the intact toxin was important for the translocation of the toxin through cell membrane into cytoplasm where fragment A was released to exert its action.

Subsequently, other bacterial toxins including *Pseudomonas aeruginosa* exotoxin A, cholera and *E. coli* enterotoxins, and recently, pertussis toxins have been found to catalyze ADP-ribosylation reaction (11–15). Interestingly, all of these toxins except pertussis toxin, whose molecular mechanism of action is still unclear, have been found to consist of two functional domains, A and B, in their molecular makeup. The A domain is the ADP-ribosyl transferase, which exerts the toxin's primary effect on the cell, but is inactive on association with the B domain. The B domain is responsible for the binding of toxin to the cell membrane after which the A domain is freed from the holotoxin via a proteolytic cleavage and/or disulfide reduction and enters the cell [for a review see ref. (16)]. Two plant toxins, abrin and ricin, and some glycoprotein hormones have also been shown to contain the two domain structure with similar modes of action. Bacterial protein toxins thus provide excellent models for the study of cell specificities, membrane structure and function, and a macromolecular entry mechanism, as well as the control of cell functions through protein modification.

The mechanism by which biologically active proteins and peptides enter the cells has been a subject of active research in recent years. Mammalian cells grown in culture have been shown to take up about 1/100 their volume of medium per hour at 30°C or 10% of their cell surface through endocytosis (16,17). Most of the endocytosed substances are, however, degraded and generally inactivated in the subsequent fusion of endocytotic vesicles to lysosomes. Although it is possible that active proteins enter into the cytoplasm via endocytosis, other modes of entry are generally considered more likely. This will be further discussed in Section II.F.

II. Diphtheria Toxin

A. DISEASE AND THE TOXIN—A HISTORICAL OVERVIEW

Diphtheria was a dreaded disease that killed up to 50% of the infected people until the late 19th century. The mortality rate was the highest with children of less than 4 years of age. Death usually occurred by suffocation from the characteristic greenish gray membrane formed in the pharynx of the patient. In 1884 Laeffler reported that in experimental animals killed by an injection of the pathogenic organism, C. *diphtheriae*, the bacteria were found localized at the site of innoculation, and suggested that a poison produced by the bacteria might be the direct cause of death in the disease. In 1888 Roux and Yersin succeeded in producing the symptom and the characteristic death of diphtheria in guinea pigs, rabbits, and pigeons by injection of a cell-free culture filtrate of diphtheria bacilli. The pathogenicity of the culture filtrate was confirmed 2 years later by the demonstration that specific antiserum made against the filtrate successfully protected the infected animals from the disease, and diphtheria became the model for the etiological studies of infectious disease. Development of formalin-treated diphtheria toxin (toxoid) as a vaccine in 1924 facilitated mass immunization against the disease and led to virtual eradication of diphtheria on earth (17).

In 1951 Freeman showed that an avirulent strain of C. *diphtheriae* could be made to produce toxin by infection with bacteriophage isolated from the toxigenic bacteria (18). Further studies by two other laboratories established the notion that the toxin production by the bacteria was due to its lysogeny with a corynephage carrying the toxin gene (19,20). The yield of toxin produced by the bacteria carrying (tox$^+$) prophage was found to vary from strain to strain, and was later correlated to the content of an iron-binding repressor protein produced by the host [for a review see ref. (17)]. The importance of iron deficiency in culture medium in the efficient toxin production had been known for a long time. It is interesting to note that the C. *diphtheriae* strain PW8, used in the commercial production of diphtheria toxin, was found to produce toxin amounting to as much as 35% of the total protein synthesized when cultured in an iron deficient medium containing succinate as a sole energy source (21). Genetic studies on the corynephage beta that carried *tox* gene also led to the discovery of CRMs (cross-reactive mutant proteins), the proteins produced by lysogenized bacteria that were serologically related to diphtheria toxin but had no or attenuated toxicity. These proteins have provided important tools for the studies of the structure-function relationship of diphtheria toxin (19).

In spite of the spectacular achievements in research for the control of diphtheria, studies on the mode of action of diphtheria toxin in the disease process progressed slowly. Degeneration of cardiac muscle fibers, and dimunition of response to biogenic amines by blood vessels were noted in animals dying of diphtheria before 1934. But it was the experiment of Straus and Hendee in 1959, in which diphtheria toxin was found to inhibit protein synthesis in cultured cells before changes in the cell morphology occurred, that provided a clue to the toxin's mechanism of action. The observation was confirmed by Collier and Pappenheimer in 1964 using a cell-free protein synthesizing system (22). Subsequently, the site of interaction with the toxin in the system was identified as EF-2, the enzyme that catalyzed translocation of the growing peptide chain on ribosome (7,23). These studies led to the discovery in 1968 that diphtheria toxin was itself an enzyme that catalyzed the ADP-ribosylation of EF-2 and thereby caused its inactivation (8,24). The finding has once again made diphtheria toxin a model for the studies of the mechanism of action of bacterial toxins.

B. PURIFICATION AND PROPERTIES

Diphtheria toxin has been purified typically by the following procedure (10): The culture supernatant of a toxigenic diphtheria bacilli, grown under the optimum condition for toxin production, is fractionated by ammonium sulfate precipitation at 40–70% saturation. After dialysis it is chromatographed on DEAE-cellulose at pH 6.8 with a linear gradient of phosphate (Na) buffer from 0.005 to 0.2 M. The fraction with toxicity is then gel filtered through Sephadex G-100 twice. The toxic activity may conveniently be assayed by measuring the inhibition of protein synthesis (incorporation of radioactive leucine into cold trichloroacetic acid precipitate) in HeLa cells (10,25), or ADP-ribosylation of EF-2 after treatment with trypsin and thiols (26).

The purified toxin is highly lethal to guinea pigs, rabbits, dogs, and birds. The minimum lethal dose, MLD, for these species is 50–100 ng/Kg and as little as 10^8 molecules, or 10^{-12} g injected to the skin caused a visible lesion. Some animal species such as rats and mice, and cell lines derived from them, show considerable resistance to diphtheria toxin, with the MLD several orders of magnitude greater than that for susceptible species (3). The toxin is irreversibly inactivated by heating at 60° for 1 min, or by acidifying to pH below

5.5. Under these conditions, the protein is precipitated and cannot be redissolved with buffers at a neutral pH. Diphtheria toxoid for vaccination is prepared by formaldehyde treatment of the toxin solution in which toxicity is lost but immunogenicity is enhanced. Cross linking between tyrosine and lysine has been shown to occur within the molecule of the formaldehyde toxoid (27).

Early investigations of Gill et al. (9,28) and Collier et al. (10,26) in 1971 have demonstrated that most toxin preparations obtained by the above general procedure contain a mixture of intact protein of $M_r = 62,000$ and the "nicked" proteins that give rise to two polypeptides of $M_r = 24,000$ and $38,000$ on treatment with thiols and SDS (sodium dodecyl sulfate)-polyacrylamide gel electrophoresis. The two peptides are connected by a single disulfide bond and may be separated by gel filtration in the presence of thiols and urea or guanidine HCl. The smaller peptide, fragment A, is entirely responsible for the ADP-ribosyl transferase activity *in vitro*: the intact toxin or "nicked" toxin in the absence of thiols does not catalyze ADP-ribosylation of EF-2. The 38,000 M_r fragment B is enzymatically inactive but appears to contain major antigenic sites of the toxin and to be involved in the binding of the toxin to cell surfaces (Fig. 1). Fragment A is stable to heat at 100° for a few minutes and between pH 3 and 11, whereas fragment B precipitates out on standing (17). When diphtheria toxin is purified from a fast grown culture, processed rapidly in the cold, a preparation containing predominantly intact toxin may be obtained (25), indicating that the toxin is secreted as a single polypeptide chain from the bacteria. Fragments A and B may be generated by a mild trypsinolysis of the intact toxin followed by reduction of the disulfide linkage.

C. CHEMICAL STRUCTURE

The sequence of 193 amino acid residues in fragment A of diphtheria toxin was reported in 1976 (29). It shows that the fragment's only Cys residue is located at position 186 and 5 out of 7 Arg residues are located between positions 170 and 193, at its COOH terminus, with three clustered at positions 190, 192, and 193. The Cys-186 is linked to Cys-200 (Cys-7 of fragment B) of the intact toxin through a disulfide bond, thus forming a loop of 14 amino acid residues that is susceptible to cleavage by trypsinlike enzyme (Fig. 1). The information has provided an explanation for the earlier observation

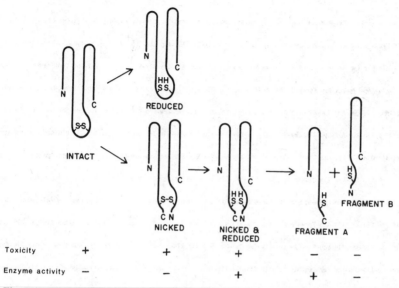

Figure 1. Schematic diagram of the A and B functional domains in diphtheria toxin: The polypeptide chain structure of the nicked and unnicked diphtheria toxins are depicted. N and C denote the NH$_2$ and COOH terminus, respectively, and –S–S–, the disulfide between Cys-186 and Cys-200. Proteolytic cleavages occur within the 14-residue segment between the two Cys residues, producing the enzymatically active Fragment A (the N-terminal segment) and Fragment B that function to bind to cell membranes (3).

that various preparations of diphtheria toxin contained enzymatically active fragments of slightly different chain lengths.

The structural analyses of most peptides from fragment B have been reported, but the complete sequence has not been available (30,31). In 1983, the nucleotide sequence of the DNA, deoxyribonucleic acid, segment containing the tox gene from corynebacteriophage beta (32) and of that carried by corynephage omega (33) have been determined. The predicted primary structure of the mature diphtheria toxin consists of a single polypeptide chain of 535 amino acid residues, with a calculated molecular weight of 58,342. The sequence of 193 residues from the NH$_2$-terminal Asn corresponds to that of fragment A and has been found to be identical to that determined with the protein except for the position of a Ser

```
                        10                          20                          30
Gly-Ala-Asp-Asp-Val-Val-Asp-Ser-Ser-Lys-Ser-Phe-Val-Met-Glu-Asn-Phe-Ser-Ser-Tyr-His-Gly-Thr-Lys-Pro-Gly-Tyr-Val-Asp-Ser-
                            40                          50                          60
Ile-Gln-Lys-Gly-Ile-Gln-Lys-Pro-Lys-Ser-Gly-Thr-Gln-Gly-Asn-Tyr-Asp-Asp-Asp-Trp-Lys⌐Gly⌐Phe-Tyr-Ser-Thr-Asp-Asn-Lys-Tyr-
                                                                                     ⌊Glu⌋
                            70                          ⌐‑‑‑ 80                       90
Asp-Ala-Ala-Gly-Tyr-Ser-Val-Asp-Asn-Glu-Asn-Pro-Leu-Ser-Gly-Lys-Ala-Gly⌊Gly⌋Val-Val-Lys-Val-Thr-Tyr-Pro-Gly-Leu-Thr-Lys-
                                                            ‑Asp⌐
                                                            ⌊‑‑‑⌋110
                        100                                                          120
Val-Leu-Ala-Leu-Lys-Val-Asp-Asn-Ala-Glu-Thr-Ile-Lys-Lys-Glu-Leu-Gly-Leu-Ser-Leu-Thr-Glu-Pro-Leu-Met-Glu-Gln-Val-Gly-Thr-
                        130                         140                         150
Glu-Glu-Phe-Ile-Lys-Arg-Phe-Gly-Asp-Gly-Ala-Ser-Arg-Val-Val-Leu-Ser-Leu-Pro-Phe-Ala-Glu-Gly-Ser-Ser-Ser-Val-Glu-Tyr-Ile-
                            160                         170                         180
Asn-Asn-Trp-Glu-Gln-Ala-Lys-Ala-Leu-Ser-Val⌐Glu⌐Leu-Glu-Ile-Asn-Phe-Glu-Thr-Arg-Gly-Lys-Arg-Gly-Gln-Asp-Ala-Met-Tyr-Glu-
                                                ⌊Lys⌋
                            190                         200                         210
Tyr-Met-Ala-Gln-Ala-Cys-Ala-Gly-Asn-Arg-Val-Arg-Arg⌐Ser-Val-Gly⌐Ser-Leu-Ser-Cys-Ile-Asn-Leu-Asp-Typ-Asp-Val-Ile-Arg-
                                                         ⌊Gly⌋
                        220                                               230                 240
Asp-Lys-Thr-Lys-Thr-Lys-Ile-Glu-Ser-Leu-Lys-Glu-His-Gly-Pro-Ile-Lys-Asn-Lys-Met-Ser-Glu-Ser-Pro-Asn-Lys-Thr-Val-Ser-Glu-
                        250                         260                         270
Glu-Lys-Ala-Lys-Gln-Tyr-Leu-Glu-Glu-Phe-His-Gln-Thr-Ala-Leu-Glu-His-Pro-Glu-Leu-Ser-Glu-Leu-Lys-Thr-Val-Thr-Gly-Thr-Asn-
                        280                         290                         300
Pro-Val-Phe-Ala-Gly-Ala-Asn-Tyr-Ala-Ala-Trp-Ala-Val-Asn-Val-Ala-Gln-Val-Ile-Asp-Ser-Glu-Thr-Ala-Asp-Asn-Leu-Glu-Lys-Thr-
                        310                         320                         330
Thr-Ala-Ala-Leu-Ser-Ile-Leu-Pro-Gly-Ile-Gly-Ser-Val-Met-Gly-Ile-Ala-Asp-Gly-Ala-Val-His-His-Asn-Thr-Glu-Glu-Ile-Val-Ala-
                        340                         350                         360
Gln-Ser-Ile-Ala-Leu-Ser-Ser-Leu-Met-Val-Ala-Gln-Ala-Ile-Pro-Leu-Val-Gly-Glu-Leu-Val-Asp-Ile-Gly-Phe-Ala-Ala-Tyr-Asn-Phe-
                        370                         380                         390
Val-Glu-Ser-Ile-Ile-Asn-Leu-Phe-Gln-Val-Val-His-Asn-Ser-Tyr-Asn-Arg⌐Pro⌐Ala-Tyr-Ser-Pro-Gly-His-Lys-Thr-Gln-Pro-Phe-Leu-
                                                                         ⌊Ser⌋                    ↑
                        400                         410                         420
His-Asp-Gly-Tyr-Ala-Val-Ser-Trp-Asn-Thr-Val-Glu-Asp-Ser-Ile-Ile-Arg-Thr-Gly-Phe-Gln-Gly-Glu-Ser-Gly-His-Asp-Ile-Lys-Ile-
                        430                         440                         450
Thr-Ala-Glu-Asn-Thr-Pro-Leu-Pro-Ile-Ala⌐Gly⌐Val-Leu-Leu-Pro-Thr-Ile-Pro-Gly-Lys-Leu-Asp-Val-Asn-Lys-Ser-Lys-Thr-His-Ile-
                                                ⌊Ser⌋
                        460                         470                         480
Ser-Val-Asn-Gly-Arg-Lys-Ile-Arg-Met-Arg-Cys-Arg-Ala-Ile-Asp-Gly-Asp-Val-Thr-Phe-Cys-Arg-Pro-Lys-Ser-Pro-Val-Tyr-Val-Gly-
                        490                         500                         510
Asn-Gly-Val-His-Ala-Asn-Leu-His-Val-Ala-Phe-His-Arg-Ser-Ser-Ser-Glu-Lys-Ile-His-Ser-Asn-Glu-Ile-Ser-Ser-Asp-Ser-Ile-Gly-
                        520                         530         535
Val-Leu-Gly-Tyr-Gln-Lys-Thr-Val-Asp-His-Thr-Lys-Val-Asn-Ser-Lys-Leu-Ser-Leu-Phe-Phe-Glu-Ile-Lys-Ser-
```

Figure 2. The primary structure of diphtheria toxin deduced from the nucleotide sequence of corynephage beta *tox* gene: The sequence of 193 residues in Fragment A (residue 1 to 193) has been determined on the protein and is in agreement with the structure shown above (32). Amino acid replacements coded by mutant genes are indicated under the wild type sequence, with a block at position 52 for CRM197, and with dotted blocks for CRM228. CRM45 contains the same sequence from residue 1 to 386 where the chain terminates, as indicated by an arrow.

residue. Reexamination of the peptide containing this Ser has confirmed the correctness of the structure deduced from the DNA sequence (Fig. 2).

The segment of 342 residues from Ser-194 to Ser-535 corresponds to fragment B. The data have provided the sequence information for regions where protein structural analysis have not been completed. Beside these regions, discrepancies have been found between this sequence and that obtained by analysis of the isolated fragment B, in the positions of 7 amino acid residues. Since the sequence analysis was performed on the toxin produced by a strain of *C. diphtheriae*, which carried a tox$^+$ gene different from that of phage beta, it was suggested that some or all of these discrepancies represented genetic differences between the two toxins.

The DNA fragment containing the structural gene for diphtheria toxin has also been found to contain the sequence that codes for a 25-residue leader peptide preceding Asn-1 of the toxin. This region is rich in hydrophobic amino acids and the sequence strongly resembled those of known signal peptides (32,34). The finding of a leader sequence for diphtheria toxin is consistent with the fact that the toxin is secreted from the bacteria. Recent observations on the expression in *E.coli* of a recombinant plasmid, which carries the sequence coding for the postulated tox regulatory region, signal peptide, and fragment A have also provided indication to the toxin's secretory mechanism. A protein of $M_r = 28,500$, which immunologically cross react with fragment A, was synthesized and transported to periplasmic compartment (35). On brief treatment with trypsin, the protein yielded a $M_r = 24,000$ polypeptide that comigrated with fragment A, and a peptide of $M_r = 45,000$, presumably the leader peptide. In another study in which restriction fragments from corynephage beta were expressed, ADP-ribosyl transferase activity was found in the periplasmic fraction of the clones (36). The region of fragment B is characterized by a stretch of hydrophobic sequence of 25 amino acid residues around position 350, which has been reported to resemble a portion of intrinsic membrane protein (30). The significance of the structural features in relation to recognition and transmembrane transfer of fragment A by fragment B remains unclear.

More recently, the nucleotide sequences of two mutant bacteriophage beta coding for CRM45 and CRM193 have been reported (37). The data show that a C to T point mutation has put ochre (TAA) stop signal after Thr-386, producing the shorter chain of CRM45 containing normal A fragment and nonfunctional B fragment. A single amino acid change has been found in the A region of CRM193 by these analyses. Fragment A from CRM193 is enzymatically inactive, indicating the importance of this amino acid, Gly-52, in the catalytic activity (Fig. 2, see the following discussion).

D. ENZYMOLOGY OF FRAGMENT A, A ADP-RIBOSYL TRANSFERASE

The free energy of hydrolysis for the bond between the ADP-ribose moiety and nicotinamide in NAD^+ is about -8.2 kcal/mol at pH 7 and 25°. It is thus conceivable that NAD^+ serves as a good ADP-ribosylating agent. However, the role of NAD^+ in the ADP-ribosylation reaction was not known until 1968 when diphtheria toxin was found to block protein synthesis by catalyzing the transfer of the ADP-ribose moiety from NAD^+ to EF-2 (see the preceding section).

It has now been well established that the ADP-ribosyl transferase activity of diphtheria toxin resides entirely with fragment A, residues 1–193 of the 535 residue polypeptide of the toxin. Fragment A may be obtained by a mild trypsin digestion of the intact toxin followed by reduction of the disulfide between residues 186 and 200 (26). Only the freed fragment A shows full enzymatic activity. This has also been supported by studies of the immunologically cross-reactive mutant proteins (CRMs) of diphtheria toxin. For example, CRM45, the $M_r = 45,000$ protein containing fragment A and a portion of fragment B, shows only 50% of the activity of fragment A, but is fully active after "nicking" and disulfide reduction (38). Apparently the portion of the molecule masked by the B fragment is directly involved in the catalysis.

Fragment A catalyzes ADP-ribosylation of EF-2 at the rate over 2000 mol EF-2 min^{-1} mol^{-1}, in the presence of physiological concentration of NAD^+ (0.1 mM) and EF-2 (2 μM). The Km value for NAD^+ has been estimated to be about 5 μM (16,25). As in any enzyme catalysis, the reaction is reversible, but only at a high concentration of nicotinamide because the equilibrium constant for this reaction is about 10^4 at pH 7. Fragment A also catalyzes hydrolysis of NAD^+ to yield ADP-ribose and nicotinamide. The rate of hydrolysis is, however, about 0.05 mol NAD^+ min^{-1} mol^{-1} at physiological pH and temperature, and is thus insignificant as compared with the ADP-ribosyl transfer reaction. Diphtheria toxin A fragment can only ADP-ribosylate eukaryotic EF-2. Its prokaryotic counterpart, EF-G in bacteria, or elongation factor G from mitochondria do not serve as the substrate. This stringent specificity makes fragment

A a convenient tool for the assay of EF-2 in the crude tissue extracts (17). The enzyme is also highly specific for NAD^+; $NADP^+$, NADH, and NADPH are not substrates.

Chemical modification studies have previously indicated that the only SH group in the molecule is not directly involved in the enzyme activity of fragment A. Nitration of one equivalent of Tyr has been shown to result in the loss of 75% of activity [discussed in ref. (29)]. Although the location of this Tyr in the molecule is unknown, it is noteworthy that 7 out of 10 Tyr residues are located between Tyr-20 and Tyr-85, indicating this area to be near the active site. NAD^+ has been shown to bind to fragment A with the K_d of about 8 μM, in the same range as the K_m in the catalysis of ADP-ribosylation. Binding apparently occurs at the site of one of the two Trp residues because it causes quenching of the Trp fluorescence of the enzyme (25). More recently, a covalent labeling of the A fragment with NAD^+ was observed when the enzyme was irradiated with ultraviolet light (39). The residue labled has now been identified as Glu-146 in fragment A, and the label is from the nicotinamide moiety of NAD^+ (40). The linking of nicotinamide is specific to Glu-146 and causes inactivation of both ADP-ribosyl transferase and NAD^+-glycohydrolase activities. It indicates that Glu-146 interact with the positive charge of the nicotinamide moiety in NAD^+ on binding, and must therefore be closed to the catalytic site of fragment A. The two Trp residues are at positions 50 and 153. The proximity of the latter to Glu-146 suggests that Trp-153 interacts with the nicotinamide moiety of NAD^+ to cause its fluorescence quenching. Trp-153 has also been implicated in the NAD^+ binding by chemical modification (41). On the other hand, recent sequence analysis of CRM197 has revealed a change of a single amino acid at position 52 in this inactive mutant protein, from Gly (in active fragment A) to Glu. It strongly indicates that this region of the molecule is involved in the activity of the enzyme (37). It is interesting that Gly-52 residue is in the Tyr rich area mentioned previously. It is apparent that Trp-50, Gly-52, Glu-146, and Trp-153 are all in proximity in the steric structure, constituting the active conformation of the enzyme.

The site of ADP-ribosyl transfer in EF-2 was shown to contain the sequence -Phe-Asp-Val-His-Asp-Val-Thr-Leu-His-Ala-Asp-Ala-Ile-"X"-Arg- in 1974 in which the residue X was linked to the ADP-ribose moiety (42–44). The identity of this amino acid was

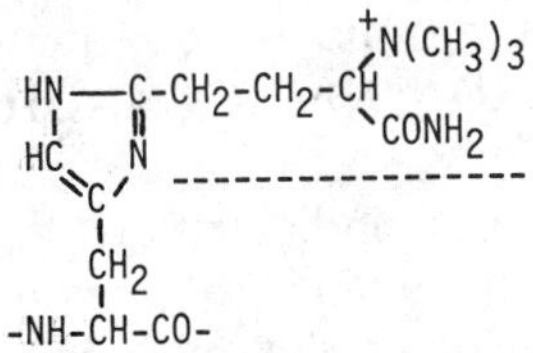

Figure 3. The structure of diphthamine: The imidazole group of histidine is modified posttranslationally. The moiety attached, indicated by dotted line, is derived from a methionine amide and 3 methyl groups (from 3 methionines).

elucidated by NMR analysis to be 2-[3-carboxyamido-3(trimethyl ammonio)propyl] histidine (Fig. 3), and was named diphthamide (45). That diphthamide was formed by post translational modification of the particular histidine was indicated by the studies of Chinese hamster ovary (CHO) cell mutants resistant to diphtheria toxin (46). In this study, EF-2 from a resistant cell was made sensitive to ADP-ribosylation by treatment with the extract of sensitive cells. More recently evidence has been obtained indicating that diphthamide is synthesized in yeast from 1 mol each of histidine and methionine and 3 methyl groups. The experiment also confirmed the correctness of the diphthamine structure (47). Further evidence for the post translational synthesis of diphthamide has been provided by the demonstration that the resistant EF-2 can be converted to sensitive EF-2 in CHO mutant cells by an addition of the S-adenosyl methionine-requiring methyl transferase (48).

E. FRAGMENT B IN THE MEMBRANE RECOGNITION AND TRANSPORT

Since only free fragment A is effective in inhibiting protein synthesis, intoxication by diphtheria toxin must involve the entry and release of fragment A into cytoplasm. The phenomena suggestive of this process was first observed in 1959 with HeLa cell culture (49), in which decline in amino acid uptake was found to occur after more than 90 min of lag period at 37°. The rate of decline was later found to follow the first-order kinetics (50), reaching half-maximal at $10^{-8}M$ of toxin. At $5 \times 10^{-7}M$ the entry site became saturated (the rate of amino acid uptake reached minimum), but the effect still followed a lag. Experiments with diphtheria antitoxin have suggested that the toxin initially interacts reversibly with cell membrane, then becomes inaccessible to antitoxin for the rest of the lag period during which fragment A is presumably "in transit" through the membrane. High pH or low salt in the medium were found to increase the time during which the cell could be rescued by antitoxin (51).

Fragment B is responsible for binding of diphtheria toxin and transfer of the A fragment through cell membrane. This has been evident because the isolated fragment A is not toxic to cells unless it is recombined with the B fragment. That the binding of the toxin to cell surfaces is through the B region of the intact toxin has been demonstrated by competitive inhibition of cell intoxication with CRM197 that contains a normal B region but a defective A region (50). Similarly, purified fragment B maintained in a borate buffer effectively prevented the effect of holotoxin (52). CRM45 is only partially toxic to cells because it contains only the NH_2-terminal third of the B fragment, and is ineffective in competing with intact toxin for the entry sites. There appear to be specific binding sites or entry sites for diphtheria toxin in the sensitive cells. The number of these sites, or receptors per HeLa cell has been estimated with ^{125}I-labeled toxin to be 4000. No other cells contain so many receptors, and resistant mouse cells contain undetectable amounts. From the CRM197 competition experiment, the dissociation constant between toxin and receptors has been calculated to be 10^{-8} M (53,54).

It was recognized early that fragment B was a hydrophobic protein. It formed a precipitate at neutral pH on standing in the cold, and could be purified from guanidine HCl solution by precipitation on dilution (9). Its hydrophobic character has been considered compatible with its role to bind and facilitate penetration through cell membranes. Studies with CRM45 have indicated that fragment B contains two functional regions, the 17,000 M_r COOH-terminal region involved in the cell surface receptor recognition and the 23,000 M_r NH_2-terminal region that interacts with the membrane bilayer (55,56). When most of the primary structure of fragment B was elucidated, the structural features relevant to its function became further apparent. Thus the 9000 M_r NH_2-terminal region has been found to contain a predicted secondary structure very similar to that found in the phospholipid-binding domains of human apolipoprotein A_2 (30). This region is followed by the 14,000 M_r segment containing a highly hydrophobic structure resembling that found in the intrinsic membrane protein. The COOH-terminal 17,000 M_r region, which also contains the intrachain disulfide, is hydrophilic in nature, but show no structural similarities to lipid-interacting proteins and presumably serve to interact only with cell surfaces.

F. MECHANISM OF ENTRY

The existence of resistant cells and a specific receptor for diptheria toxin make it unlikely that endocytosis is the major route of entry for the toxin. Besides, cytotoxic effect of the toxin is already apparent at the toxin concentration of 10^{-12} M, or only with about 20 molecule/cell. If these were to enter the cell via endocytosis, hardly any would be expected to escape lyosomal degradation (17). Experiments of Bonventre et al. indicated that less than one in 1 million molecules of diphtheria toxin escape intact from a phagosome of macrophage (57). Indeed as Gill has pointed out (16), "Topologically it (endocytosis) leaves toxin molecules in the same situation as they were before enclosure: one membrane away from their site of action." Despite these arguments, some investigators maintain that endocytosis is the route of entry for diphtheria toxin. This notion appears to be based on the observations that a low pH environment facilitates uptake of the toxin by cultured cells (51,58). Recently existence of an acidic nonlysosomal compartment in mammalian cells has been indicated (59). In the test tube, diphtheria toxin shows a drastic change in conformation when the pH is lowered to 4–4.5, to a more hydrophobic structure as measured by change in its fluorescence (60). A hydrophobic conformation would facilitate its interaction with the cell membrane. At 150 mM NaCl the change takes place between pH 5 and 5.5, suggesting that the toxin's insertion into the membrane is facilitated at this physiologically feasible condition. If endocytosis is followed by lowering of the pH within the endosomes, before their fusion with lysosomes, diphtheria toxin could penetrate the endosomes and escape into cytoplasm. Marnell et al. have observed that at 15°, diphtheria toxin enters into cytoplasm without being degraded by lysosomes, and suggested that such a condition exists in the cultured cells and that the toxin penetrates a prelysosomal membrane before reaching lysosome (61). With the use of a photoreactive glycolipid probe, Zalman and Wisniski have recently observed that both fragments A and B bind to the hydrocarbon domain of the lipid bilayer of an artificial liposome and Sendai virus (62). Low pH is shown to facilitate the binding. Permeability studies with the liposome indicate that nicked toxin forms pores of 24-Å diameters, much larger than those previously reported and large enough for fragment A to penetrate without unfolding. However, intact toxin has been found to poorly bind and form pores in lipid bilayers. Thus the *in vivo* significance of the experiments remains to be clarified.

The effect of disufide modification on the binding and penetration of diphtheria toxin through a cell membrane has recently been examined (63). Although the data are suggestive of the involvement of a disulfide bond (presumably that of the B region) in the toxin binding, the Cys residue involved has not been identified, and the results cannot be considered conclusive. For further discussion on the possible mechanisms of entry of bacterial toxins into mammalian cells, readers are referred to the review of Gill (16).

III. Pseudomonas Exotoxin A

A. THE EXOTOXINS OF PSEUDOMONAS AERUGINOSA

Pseudomonas aeruginosa is the gram-negative bacteria mainly responsible for the burn infection. Being gram negative, its pathogenicity had long been considered to be due to its lipopolysaccharide endotoxin until about 20 years ago. Many extracellular products of *P. aeruginosa*, including pigments, HCN, proteases, phopholipase, enterotoxin, exotoxin, and slime, are toxic to animals (64). However, none of these substances, except for the exotoxin A first described by Liu (65), show the toxicity that can account for the lethality due to Pseudomonas infection; the LD_{50} value in mice for the *P. aeruginosa* extracellular proteases is 75 μg as compared to 0.04 μg for diphtheria toxin in guinea pigs. *Pseudomonas aeruginosa* has been associated with diarrheal conditions, for which a heat-labile enterotoxin excreted in the medium has been considered responsible. However, the enterotoxic activity has not been isolated and characterized thus far.

B. PURIFICATION AND CHARACTERIZATION OF EXOTOXIN A.

The existence in the culture supernatant of *P. aeruginosa* of a heat-labile protein toxin lethal to mice was demonstrated in 1966. The toxin was found to be susceptible to proteases in the medium and its production inhibited by a variety of substances present in the common laboratory media. The production and isolation of the toxin has, therefore, been carried out with a nonproteolytic strain PA-103 that is also a good toxin producer when grown in a proper medium. The exotoxin from this strain, designated exotoxin A, may be purified by precipitation with zinc acetate and $(NH_4)_2SO_4$ followed by chromatography on DEAE-cellulose and Sephadex-200 (66). The purified toxin had a M_r of 50,000 as estimated from gel filtration and showed LD_{50} in mice of 0.1 μg. Passive immunization with antibody produced against exotoxin A in pony was found to prevent the animals infected with $P.$ $aeruginosa$ from death. The purified toxin also elicited various local and systemic conditions in animals that mimic symptoms of $P.$ $aeruginosa$ infection. These findings suggest that exotoxin A is likely the pathogen in the lethal infection by Pseudomonas. The M_r was later revised to 66,000 by the SDS-polyacrylamide gel electrophoresis of a highly purified exotoxin A (67).

A large-scale purification of exotoxin A from the culture supernatant of $P.$ $aeruginosa$ strain 103 with high efficiency and yield was reported in 1976 (67). In this procedure, the toxin was adsorbed onto DEAE-cellulose from the diluted culture supernatant, eluted and chromatographed successively on DEAE-cellulose and hydroxyapatite. From 50 L of culture supernatant, 135 mg of pure exotoxin A was obtained, with the overall yield of 30%.

The purified toxin showed cytotoxicity toward mouse L929 cells that could be blocked by the addition of antiserum. The median lethal dose for 20-g mice was found to be 0.1 μg in agreement with the previous finding. The protein had an isoelectric point of 5.1, M_r of 66,000, and amino acid composition different from that of diphtheria toxin. It was found to be a single chain polypeptide with Arg at the NH_2 terminus and contain 4 disulfide bonds. No further information on the structure of the toxin has been available.

C.   MECHANISM OF ACTION—ADP-RIBOSYLATION OF EF-2.

Pseudomonas exotoxin A had previously been shown to inhibit protein synthesis in the liver, kidney, and spleen of the intoxicated mice, and with cultured mammalian cells (68,69). In 1975, Iglewski and Kabat demonstrated that the toxin inhibited protein synthesis in reticulocyte lysates in the presence of NAD^+, and that EF-2 was ADP-ribosylated in the process (70). Furthermore, the site of ADP-ribosylation in EF-2 appeared identical to that obtained with dipth-eria toxin fragment A. On tryptic digestion of the modified EF-2 prepared by the two enzymes (fragment A and the Pseudomonas exotoxin A), labeled peptides with the same chromatographic mo-bility were obtained. Thus, although Pseudomonas exotoxin A and diphtheria toxin are structurally and serologically distinct, and have different specificities towards various cell types, their molecular mechanism of action appears to be identical. The exotoxin A has been shown to catalyze release of the ADP-ribose group from the ADP-ribosylated EF-2, prepared with diphtheria toxin fragment A, in the presence of nicotinamide (69,71).

Chung and Collier reported isolation of a protein of $M_r = 26,000$ with the ADP-ribosyl transferase activity, from the stationary cul-ture of the $P. aeruginosa$ strain PA-103 in 1977 (72). The purification procedure was similar to that employed by Liu et al. (66), starting with precipitation with zinc acetate. This protein catalyzed both hydrolysis of NAD^+ and ADP-ribosyl transfer with similar specific activity, kinetic constant, pH optima and sensitivities to inhibition by various nucleotides, to that found with diphtheria toxin fragment A. The findings suggested that a close similarity existed between the active site structures of the two toxins and that the 26,000 M_r active protein was a proteolytic fragment from the 66,000 M_r exo-toxin A of $P. aeruginosa$. The two enzymes were, however, clearly different in their amino acid compositions (no Cys residue was found in the Pseudomonas active protein), reactivities to antifragment A serum and thermal stabilities (69).

An interesting observation was reported in the following year by Leppla et al. (73). The intact exotoxin A was found to have little enzymatic activity unless it was exposed to both denaturing agents such as urea, guanidine HCl, or SDS and thiol reagents. Treatment only with thiols or urea did not activate the exotoxin A (to 100 times the original activity). Proteolysis was not required for the activation. It was suggested that reduction of disulfide(s) under denaturing con-dition was necessary in order to induce an active conformation for the catalysis. It is rather unlikely that this condition exists $in vivo$, and a conformational change due to proteolysis, with the formation of an active fragment like that reported by Chung and Collier (72), seems more plausible.

Recently a mutant of $P. aeruginosa$ producing a nontoxic, im-

munologically cross-reactive protein was obtained, after subjecting
a wild type strain to the nitrosoguanidine mutagenesis (74). The
CRM purified by immunoprecipitation with antiexotoxin A IgG
showed an identical M_r of 68,000 on SDS-gel electrophoresis. Availability of CRM shall facilitate the study of the structure–function
relationship in Pseudomonas exotoxin A.

IV.   Vibrio cholerae and E. coli Enterotoxins

A.   ENTEROTOXINS

Following the discovery of *Vibrio cholerae* as the pathogen for
cholera, Robert Koch in 1884 observed that the invading organisms
were localized in the gut of the patients without damaging the sur-
rounding tissues. This led him to suggest that the severe diarrhea
seen in the disease was due to a toxicosis. However, it was not till
1959 that supporting evidence for Koch's suggestion was obtained
(75,76). The main reasons for this delay appeared to be the difficulty
in finding a suitable animal model and the lack of reliable assays for
the diarrheagenic activity. In 1894 Metchinikoff reported that suckling rabbits could be infected orally with *V. cholerae* to produce
symptoms resembling that of human cholera, but reproducibility of
the procedure was poor. In 1953, De and Chatterje showed that
injection of a live vibrio suspension into ligated segments of the small
intestine of rabbits caused marked accumulation of fluid within the
segments. The fluid was found to resemble the "rice-water" stool
of cholera patients and the amount could be correlated with that
injected. The same result was obtained with a sterile culture filtrate
of *V. cholerae* in 1959 (75). Independently Dutta et al. (76) reported
that the similar filtrate of vibrio culture caused a diffused diarrhea
in the suckling rabbit model of Metchinikoff. By introducing the test
material directly into the stomach after irrigation to remove milk,
the original procedure was improved to attain repeatability and
quantitation. Availability of assay methods has since contributed
greatly to the study of diarrheal diseases. Enterotoxins, defined by
Craig as the agents responsible for the increased movement of water
and ions to lumen of the small intestine (77), have been isolated from
strains of *V. cholerae, E. Coli, B. cereus,* and *Cl. perfringens.* Protein toxins produced by *Staphylococcus, Salmonella,* and *Shigella*
also cause acute diarrhea in the animals, but their primary action

appears to be that of tissue damage, and thus are not considered as enterotoxins.

In 1969 Finkelstein and Lo Spalluto achieved the first purification of a diarrheagenic protein from the sterile culture filtrate of *V. cholerae* and called it choleragen (78). In the process, a protein that cross reacted with anticholera toxin serum but devoid of enterotoxicity was obtained, and was designated choleragenoid. Choleragen was subsequently shown to elicit fluid accumulation in the rabbit ileal loops as well as indurations in the skin in the dose-dependent manner. All effects were neutralized by the anticholeragen antibody (78). Availability of the pure toxin opened the way to its biochemical studies, and within a few years led to the discovery that cholera toxin caused an increase in the cAMP (cyclic adenosine monophosphate) production in the affected mammalian cells. Cholera toxin has been shown to consist of two types of subunits, A and B, in the molar ratio of 1:5. Subunit A stimulates adenylate cyclase activity in a membrane preparation, and subunit B binds to ganglioside G_{M1} on the cell surfaces. More recently stimulation of adenylate cyclase has been shown to be the result of ADP-ribosylation of a regulatory protein in the cell membrane, catalyzed by the A subunit of choleragen [for a review see ref. (79)].

Escherichia coli is a regular inhabitant of animal intestines. In 1967 Smith and Halls reported isolation of a strain of *E. coli* from piglets suffering acute diarrhea that gave a positive reaction in the rabbit ileal loop assay (80). Enterotoxigenic strains of *E. coli* have since been isolated from both porcine and human sources, and found to cause choleralike symptoms in animals. Two types of enterotoxin are produced by these bacteria, a heat-labile protein of $M_r = 86,000$ called LT, and a heat-stable toxin, ST. The former has been shown to have properties almost indistinguishable from, and the molecular mechanism of action identical to, that of choleragen. Thus investigations on *E. coli* LT have often been modeled after the cholera toxin research. Recent studies on the primary structures of the two toxins have even suggested their common evolutionary origin.

B. ASSAYS FOR ENTEROTOXINS

Enterotoxins are not lethal to experimental animals when administered through regular routes (i.v., i.p., or i.m. injections), and are not assayed by usual toxicological methods.

1. Enterotoxicity assay: One of the original methods of measuring diarrheagenic activity, and still being used, is the rabbit ileal loop test, in which the small intestine of adult rabbits are ligated into equal sections and injected with test solutions. After 18–24 h the animals are sacrificed, and the volume of fluid per unit length of ileal segments are determined (81). The procedure is cumbersome and produce only semiquantitative results. The use of infant mice instead of rabbits was first reported by Dean et al. in 1972 (82). In this assay, test samples are injected transabdominally into 2 to 4 day-old mice. The animals are killed after 1 h and the ratios of the weight of intestine to that of body are determined.

2. Vascular permeability assay: Enterotoxicity has been found to correlate with the ability to stimulate vascular permeability. The assay method developed by Craig in 1965 was based on this observation (83). The serially diluted test samples are intradermally injected to the marked spots on the back of rabbits, and after 18–24 h the size of indurations measured. The measurement is facilitated by i.v. injection of a blue dye (e.g., 5% Pontamine sky blue) that diffuses into the sites of induration within an hour and colors the areas. The size of the edema is directly proportional to the activity of test sample. The accuracy of assay is improved by premixing the test samples with known strengths of antitoxin.

3. ADP-ribosyl transferase assay: In the case of chorela toxin or *E. coli* enterotoxin LT, assay of their enzymatic activity as ADP-ribosyl transferase may conveniently be performed using [^{14}C] NAD^+ and polyarginine as an artificial acceptor (84). The toxins have to be activated by treatment with thiols, to release the active subunit as in the *in vitro* assay of diphtheria toxin, before testing.

4. Cell-biological assay: Guarrant et al. reported that enterotoxins of *V. cholerae* and *E. coli* caused a marked elongation of CHO cells in culture, mimicking the previously known effects of dibutyryl cyclic AMP and testosterone (85). The CHO cell suspensions are incubated with a various amount of test material in culture slides for 24 h, when morphological response reaches maximum. The percentage of cells elongated is determined by observation under a phase-contrast microscope. More recently the adhesivness of CHO cells has been found to increase by treatment with the toxin, and a simplified method of assay has been developed (86). In this method, CHO cells in confluency are incubated with test samples for 8 h, after which the cells floating in the medium are counted. The toxin reduces the number of floating cells.

C. PURIFICATION AND PROPERTIES

Practically all cholera toxin preparations used in research have been purified from the culture filtrate of *V. cholerae* strain Inaba 569B, grown under vigorous aeration in the TRY medium (87,88). Since the first purification of cholera toxin by Finkelstein et al. in 1969 (78), improved procedures have become available. One such procedure is that of Rappaport et al. (89), in which cholera toxin is first concentrated from the culture filtrate by absorption onto the sodium metaphosphate precipitate at pH 4.6. The concentrate is then adsorbed onto alumina at pH 5.6 and eluted at pH 8 with ammonium bicarbonate solution. With this procedure, choleragen (cholera toxin) with little or no contamination of choleragenoid (immunologically identical to choleragen but without enterotoxicity, a pentamer of B subunits, see in the following section) is obtained in high yield, 8 mg/L of culture filtrate. In Finkelstein's procedure, comparable yield was obtained except that over 60% of the product was as choleragenoid. The metaphosphate concentrate may be gel filtered on Bio-Gel P-100 to obtain essentially pure choleragen with 80% recovery (90). More recently Mekalanos et al. introduced a novel procedure to select hypertoxigenic mutants of Inaba strain 569B, in which filter paper impregnated with the ganglioside–serum albumin conjugates was used to bind the hyper *tox* mutants that produce two to three times the normal amount of choleragen. The toxin was concentrated by the metaphosphate precipitation, then purified by chromatography on phosphocellulose (91,92).

Purification of heat-labile enterotoxins, LT, of *E. coli* has met with more difficulties, and investigations from many laboratories throughout the world have produced different results, apparently due to the use of different strains, culture conditions, and stages of toxin production by the bacteria. LTs with M_r ranging from 23,000 to over 100,000 were reported, but all had activities 1000-fold less than that of cholera toxin in comparable assays. In 1979 Clements and Finkelstein (93) and independently Kunkel and Robertson (94) reported isolation of LTs with specific activities comparable to choleragen from two separate strains of *E. coli*. In the isolation of LT produced by a transformed *E. coli* K-12 bearing an LT gene from a porcine strain (93), the ability of agarose to specifically bind LTs was utilized. Concentrated and dialyzed culture filtrate was filtered through a column of Bio-gel A-5m and after an exhaustive washing, LT was eluted with 0.2 M D-galactose to yield an essentially pure form. The molecular weight of the toxin has been determined to be 91,440 by sedimentation equilibrium method. The isoionic point of LT is pH 8.0 as compared to 6.75 for choleragen and 7.75 for choleragenoid. During gel filtration of LT at pH 6.5 and room temperature, occurrence of a toxoid analogous to choleragenoid was noted and designated coligenoid. The pure LT showed a full activity when activated by trypsin treatment. The main feature in the purification of LT from a human strain was the two chromatographic steps using norleucin–Sepharose and hydroxyapatite (94). The M_r of 73,000 was determined by gel electrophoresis for the purified protein.

Analytical ultracentrifugation of choleragen and choleragenoid established their molecular weights to be 84,000 and 56,000, respectively (95). The frictional coefficients of 1.25 and 1.24 (for genoid) indicate that both are typical globular proteins. The solubility curve for cholera toxin in 2 M ammonium sulfate showed a sharp break at 1.7 mg of protein/mL, indicating its homogeneity (96). As mentioned previously, cholera toxin immunologically cross react with $E.\ coli$ LT. Both toxins have been found to dissociate into two subunits A and B under a denaturing condition. The corresponding subunits are immunologically very similar, but not identical. Their biological activities are identical, as discussed in the following section.

D. SUBUNIT STRUCTURES AND FUNCTIONAL DOMAINS

Various physicochemical studies under denaturing conditions have indicated that cholera toxin consists of two types of subunits, whereas choleragenoid consists of one [for a review see (79)]. Separation of the subunits of cholera toxin in a preparative scale was first achieved by gel filtration on Bio-Gel P-60 in acidic buffer containing guanidine HCl (97). The separation was independently achieved by gel filtration on Shephadex G-75 in 5% HCOOH, with a quantitative recovery indicating no other subunits to be present (98). The M_r of the two subunits as determined by SDS-gel electrophoresis in the presence of 8 M urea were 28,000 for A and 11,000 for B subunits. Choleragenoid (MW = 56,000) showed a single band of M_r = 11,000 corresponding to the B subunit in the electrophoresis, and the separated B subunit was found to form a soluble aggregate of MW = 56,000 after renaturation (98). Both choleragenoid and the renatured B subunit formed the precipitin bands against ganglioside G_{M1}, indicating their identity. Lai et al. calculated the molar ratio of subunits B to A in choleragen from the cysteine content in each subunit, and also by quantitation of the terminal dipeptide Ala–Asn from the B subunit, and arrived at the same value of 5.5. It was thus suggested that the choleragen preparation used contained a mixture of A_1B_5 and A_1B_6. With the use of a protein cross-linking reagent Gill observed 5 molecular species from choleragenoid on SDS-gel electrophoresis and 9 bands from the cross-linked choleragen. Based on this result he concluded that the subunit composition of choleragen was A_1B_5 (99).

The subunit composition of E. coli LT appears to vary with the strain and methods used for purification. LTs with high specific activities have been estimated to contain 1 A subunit and 4–6 B subunits per molecule (93,94). Beside variable subunit compositions, aggregation of LTA subunits with various proteins and tissue components in culture medium has been suggested as reasons for the wide range of M_r values as well as low specific activities reported for earlier preparations of E. coli LTs. The apparent M_r of subunits of E. coli LT have been estimated to be 30,000 for A and 11,000 for B, similar to that of choleragen (93,94). Together with the variable content of choleragenoid observed during preparations of cholera toxin, it is indicated that the bonds between A and B subunits in the toxins are sufficiently weak to be readily broken during manipulation. It also suggests that formation of the two subunits may be separately controlled.

On reduction and carboxymethylation, subunit A of choleragen dissociated into two polypeptides of M_r = 22,000 and 6000, designated A_1 and A_2, which could be separated by gel filtration in 5% HCOOH (Fig. 4). In washed membranes of pigeon erythrocytes, only A_1 polypeptide, among the isolated peptides of A_1, A_2, and B, was found to stimulate adenylate cyclase in the presence of NAD^+. The holotoxin and subunit A were found to show little activity unless these were treated with thiols in the cell-free test system (100,101). Both A_1 and A_2 polypeptides contain a single Cys residue, chemical modification of which has no effect on the A_1's activity. These find-

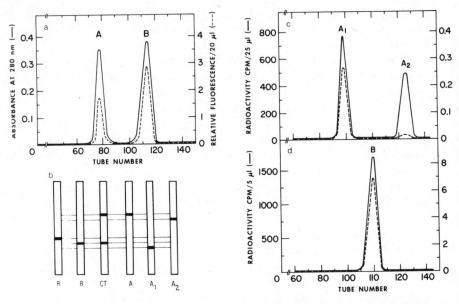

Figure 4. Subunit structure of cholera toxin: (*a*) Separation of subunits A and B by gel filtration on Shephadex G-75 in 5% HCOOH. (*b*) Electrophoretic patterns of CT (cholera toxin), subunits A, B, and *S*-carboxymethylated (*) A_1, A_2, and B in the presence of SDS and urea. (*c*) Separation of A_1 and A_2 by gel filtration. (*d*) Gel filtration of *S*-carboxymethylated B. The condition used in (*c*) and (*d*) is the same as that used in (*a*).

ings indicate that the active site for the adenylate cyclase stimulation is shielded in the A subunit in the native structure, and on reductive cleavage of the single disulfide bridge, becomes "expressed." The similar situation exists when the activity of A_1 is measured by ADP-ribosylation of polyarginine as an acceptor (84).

In contrast to the cholera toxin A subunit, the *E. coli* LTA subunit has been found to be a single polypeptide. The fact that LT preparations require proteolytic processing to attain full activity in bioassays (93,94) indicates that an active site region corresponding to choleragen A_1 polypeptide exists and is liberated into the cytoplasmic compartment to express its enzymatic activity. The presence of an "unnicked" A subunit in *E. coli* LT prompted a few laboratories to look for the occurrence of similar A subunit in cholera toxin. It was found that when purification of choleragen was carried out rapidly from a short culture, or the culture in the presence of protease inhibitors, the toxin containing the unnicked A subunit was obtained (102–104). More recently a protease very likely responsible for nicking the intact A subunit of choleragen has been purified from a strain of *Vibrio cholerae* (105).

None of the isolated subunits or polypeptides have been found to act on the intact cells to stimulate cAMP production, indicating the importance of a B subunit in the cell recognition and binding. Since only the freed A_1 polypeptide is active, and only in the broken cell systems requiring NAD^+, A_1 must enter the cell to exert its action. Unlike diphtheria toxin, cholera toxin appears to react ubiquitously with most mammalian cells, suggesting that there are no specific receptors for the toxin on the cell surfaces. In 1971, van Heyningen et al. reported that various activities of choleragen could be blocked by addition of chloroform–methanol extracts of rabbit tissues that contained varying amounts of gangliosides (106). Precipitation occurred at high concentrations of ganglioside and the toxin, and the supernatant was devoid of the activities. Interaction between cholera toxin and ganglioside was independently observed by Holmgren et al. (107, 108), who also identified ganglioside G_{M1} to have the specific affinity for choleragen. Only 0.1 ng of ganglioside G_{M1} was required to inactivate 5 ng of pure toxin. The high affinity of binding between choleragen and G_{M1} ganglioside suggests that the latter is the receptor substance on the cell surfaces for the toxin. It has been known that ganglioside G_{M1} is widely distributed among various tissues, with the highest concentration in the brain. The renatured subunit B, but not subunit A, formed a precipitin line on an agar diffusion plate against ganglioside G_{M1}, indicating that subunit B is involved in the toxin's cell binding. The finding is consistent with the observation that only the holotoxin is active on intact mammalian cells.

Thus a picture emerges that cholera toxin contains two functionally distinct subunits, one (subunit B) involved in the attachment of the toxin to the cell membrane, causing the other to dissociate and release A_1 into cytosol to cause activation of adenylate cyclase. *E. coli* LT acts on the cells in the identical manner.

E. STRUCTURE AND ADP-RIBOSYL TRANSFERASE ACTIVITY OF
SUBUNIT A

1. Structure: Cholera toxin subunit A consisted of A_1 and A_2 polypeptides linked to each other by a single disulfide bond. The primary structure of A_2 has been determined in 1980 (109) and that of A_1 has more recently been elucidated (110). The sequence of 192 amino acid residues in an A_1 polypeptide and that of 46 residues in A_2 have been found to be in complete agreement with those predicted from the nucleotide sequence of the tox gene coding for a cholera toxin A subunit (111). Amino acid analysis of unnicked subunit A has indicated the presence of two more Ser residues than that in A_1 and A_2 combined (104). In the course of sequence analysis of A_1 polypeptide, Xia et al. observed that 10% of the molecules contained Ser at A_1 from the subunit COOH terminus and suggested that proteolytic cleavage to form A_1 occurred predominantly at the carboxyl bond of Arg-192 but some (10%) after Ser-193 and/or Ser-194 (112). These observations are also in line with the finding that genes for A_1 and A_2 are connected and separated by the nucleotide sequence coding for Ser-Ser (111); the NH_2-terminal Met of A_2 follows Ser-194 (Fig. 5).

The amino acid sequence of *E. coli* LT has been deduced from the nucleotide sequence of the restriction fragment containing the LTA gene determined in 1982 (113). Comparison of the structures of LTA with that of choleragen A subunit now available indicated a great homology between the two (Fig. 5), especially in the region of A_1 where the homology is 81.3%. This would explain the functional similarity between the two toxin, and further suggest their common evolutionary origin.

2. Active site: It has been well established that the A_1 polypeptide of cholera toxin is a ADP-ribosyl transferase, and that this activity is primarily responsible for the activation of adenylate cyclase in the toxin treated cells. With polyarginine as an accepter, specific catalytic activity of A_1 has been determined as 0.22-mole ADPR (ADP-ribose) transferred per minute per mole at 23° at pH 6.6 (84). When isolated A_1 was subjected to a limited tryptic hydrolysis (substrate to enzyme ratio = 1000, 0°, at pH 7.0) and analyzed by SDS-gel electrophoresis, over 90% of the 22,000 M_r band corresponding to A_1 disappeared and was replaced by a 12,500 M_r band in 2 h. But, about 50% of the ADP-ribosylating activity remained, sug-

```
                            10                                      20                                  30
ASN-ASP-ASP-LYS-LEU-TYR-ARG-ALA-ASP-SER-ARG-PRO-ASP-GLU-ILE-LYS-GLN-SER-GLY-GLY-LEU-MET-PRO-ARG-GLY-GLN-SER-GLU-TYR-
-Gly-                                                                   -Arg-Phe-Arg-Ser-        -Asn- ■
                            40                                      50                                  60
PHE-ASP-ARG-GLY-THR-GLN-MET-ASN-ILE-ASN-LEU-TYR-ASP-HIS-ALA-ARG-GLY-THR-GLN-THR-GLY-PHE-VAL-ARG-HIS-ASP-ASP-GLY-TYR-VAL-
                                                                                             -Tyr-
                            70                                      80                                  90
SER-THR-SER-ILE-SER-LEU-ARG-SER-ALA-HIS-LEU-VAl-GLY-GLU-THR-ILE-LEU-SER-GLY-HIS-SER-THR-TYR-TYR-ILE-TYR-VAL-ILE-ALA-THR-
 ■          -Ala- -Gln-Ser-                                               -Tyr- -Leu-Thr- ■            -Ile-
                            100                                     110                                 120
ALA-PRO-ASN-MET-PHE-ASN-VAL-ASN-ASP-VAL-LEU-GLY-ALA-TYR-SER-PRO-HIS-PRO-ASP-GLU-GLN-GLU-VAL-SER-ALA-LEU-GLY-GLY-ILE-PRO-
 ■                                                     -Tyr-                         -Ile-Ser-Val-
                            130                                     140                                 150
TYR-SER-GLN-ILE-TYR-GLY-TRP-TYR-ARG-VAL-HIS-PHE-GLY-VAL-LEU-ASP-GLU-GLN-LEU-HIS-ARG-ASN-ARG-GLY-TYR-ARG-ASP-ARG-TYR-TYR-
                                          -Asn-                            -Arg-                -Glu-
                            160                                     170                                 180
SFR-ASN-LEU-ASP-ILE-ALA-PRO-ALA-ALA-ASP-GLY-TYR-GLY-LEU-ALA-GLY-PHE-PRO-PRO-GLU-HIS-ARG-ALA-TRP-ARG-GLU-GLU-PRO-TRP-ILE-
Arg-    -Glu-                            -Arg-                               -Asp- -Gln-
                            190                                     200                                 210
HIS-HIS-ALA-PRO-PRO-GLY-CYS-GLY-ASP-ALA-PRO-ARG-SER-SER-MET-SER-ASN-THR-CYS-ASP-GLU-LYS-THR-GLN-SER-LEU-GLY-VAL-LYS-PHE-
       -Gln-                              -Ser-Ser-  -Thr-Ile-Thr-Gly-Asp-    -Asn- -Glu-        -Asn-   -Ser-Thr-Ile-Tyr-
                            220                                     230                                 240
LEU-ASP-GLU-TYR-GLN-SER-LYS-VAL-LYS-ARG-GLN-ILE-PHE-SER-GLY-TYR-GLN-SER-ASP-ILE-ASP-THR-HIS-ASN-ARG-ILE-LYS-ASP-GLU-LEU
 -Arg-                                                  -Asp-      -Glu-Val- -Ile-Tyr-          -Arg-
```

Figure 5. The primary structure of cholera toxin A subunit: The structure is in agreement with that deduced from the nucleotide sequence of the coding DNA. The sequence of the corresponding subunit from E. coli LT, deduced from the gene sequence, shows a remarkable homology to the CT-A sequence. Only the amino acid changes found are indicated under the CT-A sequence. "-" indicate deletions. E. coli LT thus contains five fewer residues. Residues 1 to 192 corresponds to A_1 and residues 195 to 240, to A_2. Chorela toxin purified from the culture filtrates normally contains the two polypeptides linked through the disulfide between Cys-187 and Cys-199; Ser-Ser at positions 193 and 194 are absent.

127

gesting that the 12,500 M_r fragment of A_1 contained the partial activity and hence the active site (84). The fragment was found to contain the region Gly-47 to Arg-148. More recently, Lai et al. were able to obtain self ADP-ribosylated A_1 and showed that on incubation with polyarginine it transferred over 70% of ADP-ribose moiety onto the acceptor in 25 min. The peptide containing the ADP-ribose moiety in A_1 has been isolated and shown to contain residues 136–147 in the sequence. Arg-146 has been identified as the site of ADP-ribosylation (114). The secondary structure analysis based on the Chow–Fasman method has revealed that A_1 polypeptide contains a region with consecutive sheet–helix–sheet–helix configuration commonly found in NAD^+-binding proteins (e.g., lactic dehydrogenase), between residues 109 to 140 (115). This region is within the partially active fragment (residues 47–148) and in the proximity of Arg-146, and together with the region shielded by A_2 through disulfide at Cys-186, may constitute the active site structure. Examination of the sequence in the corresponding region of *E. coli* LTA revealed only 11 single amino acid substitutions in the 86 residue sequence, further confirming the functional significance of this area.

   3. In vivo site of ADP-ribosylation by choleragen A_1 polypeptide: Recent advance in the studies of membrane-mediated control of cell function, in which cholera toxin has played an important role, has provided evidence for the existence of a family of guanine nuceotide binding, regulatory proteins in cell membranes. These are N_s and N_i that, respectively, stimulate and inhibit cAMP production in response to a hormone action, and transducin that serves to couple photoexcitation of rhodopsin to stimulation of cGMP (cyclic guanosine monophosphate) degradation. The three proteins exhibit a similar subunit structure and amino acid compositions and the subunit of each contains a guanine nucleotide binding site (116). The present notion for the membrane-mediated control of cell function is as follows: The signal receptor (hormone receptor or rhodopsin) activate the coupling protein (N_s, N_i, or transducin) by promoting the exchange of GDP, guanosine dephosphate, to GTP, guanosine triphosphate, at the G-binding site. This exchange causes dissociation of α and β subunits in the coupling protein, and the GTP-bound α subunit stimulate or inhibit the effector enzyme (adenylate cyclase or cGMP phosphodiesterase). Finally GTPase activity of the coupling protein terminates the excited state of the system, readying it to receive a new signal.

A_1 polypeptide of cholera toxin is considered to stimulate adenylate cyclase by catalyzing ADP-ribosylation of N_s, upon which its GTPase activity is blocked and the system remains at the activated state (117). A_1 has also been found to ADP-ribosylate transducin in the vertebrate retina (118). More recently, a peptide containing the ADP-ribosyl moiety has been isolated from the cholera toxin modified transducin and its sequence determined to be Ser-Arg-Val-Lys where the ADP-ribose group is attached to the guanido group of Arg (119).

F. THE B SUBUNIT AND THE GANGLIOSIDE G_{M1} BINDING

As stated in the preceding section, isolated subunit B forms a pentamer on renaturation at a neutral pH and in all respect behaves like choleragenoid. Rabbit antibody prepared against cholera toxin has been found to react strongly with B subunits but not with the A subunit, suggesting that the pentamer of B forms the outer surface of the holotoxin molecule. Similarly ganglioside G_{M1} forms a precipitin line against 5B in a immunodiffusion plate but not with the A subunit. This observation has led to the notion that the B subunit is responsible for the toxin's binding to cell surfaces. The complete primary structure of subunit B was established in 1977 (120,121). It contains only 3 Arg residues in the sequence of 102 amino acid residues. By a controlled chemical modification of one or two Arg in the molecule, Duffy and Lai have found that binding of B subunits to ganglioside G_{M1} is completely blocked when Arg-35 is modified with cyclohexanedione (122). This region of the molecule is thus indicated to be in contact with the cell surfaces during the intoxication process (Fig. 6). At present no receptor substance other than ganglioside G_{M1} has been identified for cholera toxin. In view of the fact that intestinal cells contain little ganglioside G_{M1}, and yet are the major site for the cholera toxin action, existence of a receptor substance other than G_{M1} ganglioside is likely.

G. DEVELOPMENTS IN THE IMMUNOLOGY OF ENTEROTOXINS

It has been recognized that B subunits of cholera toxin and of *E. coli* LT are mainly responsible for the toxin's immunogenicity. Choleragenoid or separated B subunits generate antibodies that neutral-

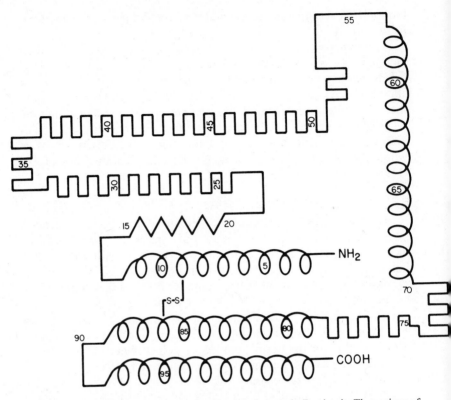

Figure 6. Predicted secondary structure of cholera toxin B subunit: The regions of α helix and β-pleated sheets are indicated by 000 and VVV, respectively. Turns are indicated by 360° bends with 4 corners. The rest of the structure is in random coil. Arg-35 has been implicated in the ganglioside G_{M1} binding.

ize enterotoxicity of holotoxins in bioassays (123). Glutalaldehyde detoxified choleragen has not been an effective vaccine probably because it mainly elicits the circulating antibodies that cannot reach the intestinal surface to inactivate the toxin. Oral doses, and to a somewhat less extent i.m. injection, of purified B subunits have recently been reported to produce IgA (secretory) antitoxin in the intestine of volunteers (124), suggesting its use as a potential vaccine. Choleragenoid itself competes with holotoxin for cell surface binding and thereby protects cells from the toxin's action. This observation has led to the use of a purified B subunit for the treatment of cholera, but apparently with little success. Probably most of the B subunits taken orally get destroyed in the stomach and does not reach the small intestine to compete with the toxin *in situ*.

It was noted that most nonreplicating antigens were relatively ineffective at eliciting the mucosal antibody (IgA) formation, and the B subunit of cholera toxin was a notable exception (125,126). Recently, cholera toxin B subunit was conjugated with horseradish peroxidase for immunization in mice and was found to cause a 33- to 120-fold increase in the IgA antibody production in comparison to the use of a mixture of the two (126). The work suggests a potential use of the B subunit as a carrier protein to stimulate mucosal immunity.

The other development has been the use of genetic engineering in producing bacterial mutants that can serve as live vaccines. Immunity can be generated by promoting mucosal antibody formation in the gut against enterotoxin or against the colonization pili (127). *Virus cholerae* and *E. coli* mutants that carry the genes defective in A subunit or that code only for B subunit have been cloned in *E. coli* K-12, but this *E. coli* strain poorly colonizes the intestine and is thus not suitable as a live vaccine. Recently a bacterial plasmid containing a LT gene defective in A subunit and the gene of conjugal transfer has been constructed and used to transform *E. coli* K-12 C600. This plasmid was found to be transferred with high frequency to colonizing porcine strains of *E. coli* on mating. The transconjugants produced high titers of the B subunits and no active LT (127), indicating their potential use as live vaccines for porcine diarrhea.

V. Pertussis Toxin

A. EXOTOXINS OF *BORDETELLA PERTUSSIS*

Bordetella pertussis is a gram-negative bacteria tht has been recognized as the pathogen of whooping cough. The organism produces a variety of toxic substances into culture medium, including the pertussis toxin, dermonecrotic toxin, tracheal cytotoxin, and soluble adenylate cyclase (128–131). Although none of these has been documented to contribute to the pathogenesis of whooping cough, the purtussis toxin appears to play the major role; the detoxified toxin has recently been shown to elicit immunity in mice that effectively protect the animals against intracerebral challenge of viable *Bordettela pertussis* (132). Pertussis toxin has a variety of biological activities and has been known under different names. It induces increase in insulin production and is thus called islet activating protein (Iap), increased sensitivity to histamine (histamine-sensitizing factor), increased leukocytosis (leukocytosis-promoting factor, LPF), and agglutination of erythrocytes (hemoagglutinin). That a single protein is responsible for all the biological effects has become evident in recent years when highly purified preparations have become available. A unifying name of pertussigen has been proposed for this protein (133).

In 1979 Katada and Ui reported that injection of Iap (pertussigen) into rats caused an increased response of pancreatic islets to agents that stimulate adenylate cyclase, and attenuated its response to agents that inhibit the cyclase activity (134). Similar effects have subsequently been observed with a variety of tissues *in vitro* and cultured cells (135,136). The apparent lack of tissue specificity was reminiscent of cholera toxin action. In 1982, the same authors demonstrated that pertussigen caused enhancement of adenylate cyclase activity in the membrane preparation from rat C6 glioma cells (137). Furthermore, the effect was seen only in the presence of NAD^+ and ATP, and was accompanied by ADP-ribosylation of a $M_r = 41,000$ membrane protein. Pertussis toxin has thus become the latest addition to the list of bacterial toxins that exert their action by catalyzing the ADP-ribosylation of a cellular component.

B. PURIFICATION AND PROPERTIES

Pertussin toxin was obtained in a crystalline form for the first time in 1981 (133), in which the sterile culture filtrate was passed through an antipertussigen Sepharose column and the toxin eluted with the buffer containing 2.5 M KSCN. After gel filtration through Sepharose CL6B in a buffer containing 5% ethanol, the hemagglutinin peak was pooled and concentrated to 0.8 mg/mL by ultrafiltration. Crystals were formed in this solution on standing in the cold for 2 to 3 weeks. Crystalline pertussigen was found to induce hypersensitivity in mice at a dose of 0.5 ng, promote leukocytosis with 10 to 40 ng, increased insulin production with 2 ng, increased production of IgE and IgG1 with 0.1 ng, and increased vascular permeability of striated muscle with 0.5 ng. The LD_{50} in mice (i.p.) was about 430 ng.

More recently, purification of pertussigen through two affinity columns, Affi-Gel blue and fetuin–Sepharose 4B, has been reported (138). Affi-Gel blue has an affinity toward NAD^+-binding proteins, and fetuin is a sialoglycoprotein. A very high overall yield of 76% was attained in this procedure. The purified toxin exhibited a single band in gel electrophoresis, but showed 4 bands in the presence of SDS, indicating a multiple subunit structure. It was suggested that pertussigen contained 5 subunits with M_r ranging from 13,200 to 26,100, with 2 copies of the second smallest subunit present in the holotoxin. A similar subunit composition was suggested by Katada et al. independently (139). A highly efficient purification has also been achieved by hydroxyapatite chromatography (140).

C. ADP-RIBOSYL TRANSFERASE ACTIVITY OF PERTUSSIGEN

Tamura et al. reported a dissociation of the largest subunit of pertussigen from the rest of the molecule in 8 M urea (139). This subunit catalyzed ADP-ribosylation of a M_r = 41,000 protein in the membrane in the absence of other cellular components, and was named A protomer. The A protomer alone was not active towards intact cells. The rest of the pertussigen molecule is composed of an aggregate of 5 smaller subunits that appears to function like B subunits in cholera toxin or E. coli LT (139). The molecular mechanism of action similar to that of cholera toxin has been postulated for pertussis toxin [reviewed by Ui et al. (141)]. Sekura et al. observed that although no interchain disulfide bonds appeared to be present, treatment of pertussis toxin with thiols was required to cause ADP-ribosylation of the membrane protein (140). In this case, an active conformation is brought about by the cleavage of intrachain disulfide within the A protomer. Ui et al. found that the native pertussigen was as effective as an A protomer with membranes from some tissues, whereas liberation of an A protomer was necessary to act on that from other tissues (141), suggesting the existence of an "activating" mechanism for pertussigen in some cell types.

Evidence has now been accumulating that pertussis toxin and cholera toxin catalyze ADP-ribosylation of different sites in the adenylate cyclase system. Thus pertussigen treated membrane of 3T3 cells lose its responsiveness to inhibition of adenylate cyclase by GTP while the cyclase activity in the choleragen treated membrane is inhibited by GTP in a concentration dependent manner. It indicated that pertussigen caused ADP-ribosylation of N_i, the regulatory protein coupled to the inhibitory receptor, resulting in their uncoupling and becoming unresponding (142). ADP-ribosylation occurs at N_s with cholera toxin, leading to stimulation of adenylate cyclase but no effect on the GTP inhibitory site.

More recently, the site of ADP-ribosylation in transducin by pertussis toxin has been identified (143). The peptide containing the ADP-ribose moiety has been isolated and its structure determined as Glu-Asn-Leu-Lys-Asn-Gly-Leu-Phe, and Asn residue in the middle shown to link to ADP-ribose. It is to be recalled that cholera toxin catalyzed ADP-ribosylation of transducin occurs at the Arg residue, in the sequence, Ser-Arg-Val-Lys. The molecular basis for the difference in the target (or acceptor) specificities of the two toxins will become apparent when the structure of pertussigen is elucidated.

VI. Concluding Remarks

Biochemical studies of bacterial protein toxins have brought to light some very interesting features of their structure–function relationship. Most of the large molecular weight protein toxins have been found to contain two functional domains. One of this, domain A contains the active site and is responsible for the toxin's primary action within the cytoplasmic compartment. The other, domain B, is responsible for the toxin's binding to cell surfaces and facilitate penetration of domain A through the membrane into cytoplasm. The two domains are either synthesized as one polypeptide chain (e.g., diphtheria toxin and Pseudomonas exotoxin A) and connected by a single disulfide bridge, or formed separately as subunits (e.g., cholera toxin, E. coli LT, and pertussis toxin) and bound to each other via a noncovalent force. In the case of cholera toxin, a portion of subunit A, A_2, is linked to the active subunit A_1 through a single disulfide. A_2 polypeptide is considered to serve as anchor for A_1 polypeptide to the pentamer of B subunits, and probably remains on the cell surface with B subunits after A_1 penetrates the membrane. Based on this consideration, Gill suggests that the B domain of cholera toxin is $1A_2 \cdot 5B$ (16).

All bacterial toxins discussed in this chapter are proenzymes with latent ADP-ribosyl transferase activity. Upon or during penetration through the membrane, a change in the conformation of domain A takes place that makes it a fully active ADP-ribosyl transferase. Fragment A of diphtheria toxin and Pseudomonas exotoxin cause the ADP-ribosylation of EF-2 and inhibit the protein synthesis on the ribosomes. Subunit A_1 of cholera toxin and the corresponding subunits in *E. coli* LT and pertussis toxin cause ADP-ribosylation of membrane proteins that are involved in the regulation of membrane-mediated cell responses to stimuli of hormones or light (transducin). The ADP-ribosyl transferases of cholera toxin and *E. coli* LT are structurally related and essentially identical in their activities. Their target of ADP-ribosylation is different from that of pertussis toxin; the former act on the stimulatory component of the adenylate cyclase system and the latter, the inhibitory component. The two toxins therefore serve as probes of membrane-mediated cell functions in a complementary manner.

The mechanism by which domain B facilitates entry of domain A remains to be fully elucidated.

References

1. Bonventre, P.F., in *Microbial Toxins*, Vol. 1, S. Kadis, T.C. Montie, and S.J. Ajil, Eds., Academic, New York, 1970 pp. 29–66.

2. Kabir, S., Rosenstreich, D.L., and Mergenhagen, S.E., in *Bacterial Toxins and Cell Membranes*, J. Jeljaszewicz and T. Wadstroem, Eds., Academic, New York, 1978 pp. 59–87.

3. Lai, C.Y., in *Survey of Contemporary Toxicology*, Vol. I., A. T. Tu, Ed., Wiley, New York, 1980 pp. 245–284.

4. A.W. Bernheimer, Ed., *Mechanisms in Bacterial Toxinology* Wiley, New York, 1976.

5. J. Jeljaszewicz and T. Wadstroem, Eds., *Bacterial Toxins and Cell Membranes*, Academic, New York 1978.

6. O. Hayaishi and K. Ueda, Eds., *ADP-Ribosylation Reactions*, Acdemic, New York 1982.

7. Collier, R.J. *J. Mol. Biol.*, **25**, 83–98 (1967).

8. Honjo, T., Nishizuka, Y., and Hayaishi, O. *J. Biol. Chem.*, **243**, 3553–3555 (1968).

9. Gill, D.M. and Pappenheimer, A.M., Jr. *J. Biol. Chem.*, **246**, 1492–1495 (1971).

10. Collier, R.J. and Kandel, J., *J. Biol. Chem.*, **246**, 1496–1503 (1971).

11. Iglewski, B.H. and Kabat, D., *Proc. Natl. Acad. Sci. USA*, **72**, 2284–2288 (1975).

12. Cassel, D. and Pfeuffer, T. *Proc. Natl. Acad. Sci. USA*, **75**, 2669–2673 (1978).

13. Gill, D.M. and Meren, R. *Proc. Natl. Acad. Sci. USA*, **75**, 3050–3054 (1978).

14. Moss, J., Garrison, S., Oppenheimer, N.J., and Richardson, S.H., *J. Biol. Chem.*, **254**, 6270–6272 (1979).

15. Katada, T. and Ui, M. *Proc. Natl. Acad. Sci. USA*, **79**, 3129–3133 (1982).

16. Gill, D.M. in *Bacterial Toxins and Cell Membranes*, J. Jeljaszewicz and T. Wadstroem, Eds., Academic, New York, 1978 pp. 291–332.

17. Pappenheimer, A.M., Jr., *Ann. Rev. Biochem.*, **46**, 69–94 (1977).

18. Freeman, V.J., *J. Bacteriol.*, **61**, 675–688 (1951).

19. Groman, N.B., *J. Bacteriol.*, **66**, 407–414 (1953).

20. Barksdale, L. and Pappenheimer, A.M., Jr., *J. Bacteriol.*, **67**, 220–232 (1954).

21. Hirai, T., Uchida, T., Shinmen, Y., Yoneda, M., *Biken J.*, **9**, 19–31 (1966).

22. Collier, R.J. and Pappenheimer, A.M., Jr., *J. Exp. Med.*, **120**, 1019–1039 (1964).

23. Goor, R.S. and Pappenheimer, A.M., Jr., *J. Exp. Med.*, **126**, 899–912 (1967).

24. Gill, D.M., Pappenheimer, A.M., Jr., Brown, R., and Kurnick, J.T. *J. Exp. Med.*, **129**, 1–21 (1969).

25. Kendal, J., Collier, R.J., and Chung, D.W., *J. Biol. Chem.*, **249**, 2088–2097 (1974).

26. Drazin, R., Kendal, J., and Collier, R.J., *J. Biol. Chem.*, **246**, 1504–1510 (1971).

27. Blass, E., Bizzini, B., and Raynaud, M., *Bull. Soc. Chim. Fr.*, **10**, 3957–65 (1967).

28. Gill, D.M. and Dinius, L.L., *J. Biol. Chem.*, **246**, 1485–1491 (1971).

29. DeLange, R.J., Drazin, R.E., and Collier, R.J., *Proc. Natl. Acad. Sci. USA*, **73**, 69–72 (1976).

30. Lambotte, P., Falmagne, P., Capiau, C., Zanen, J., Ruysschaert, J.-M., and Dirkx, J., *J. Cell. Biol.*, **87**, 837–840 (1980).

31. Falmagne, P., Capiau, C., Zanen, J., Kayser, G., and Ruysschaert, J.-M., *Toxicon* **20**, 243–246 (1982).

32. Greenfield, L., Bjorn, M.J., Horn, G., Fong, D., Buck, G.A. Collier, R. J., and Kaplan, D.A., *Proc. Natl. Acad. Sci. USA*, **80**, 6853–6857 (1983).

33. Ratti, G., Rappuoli, R., and Giannini, G., *Nuc. Acid Res.*, **11**, 6589–6595 (1983).

34. Michaelis, S. and Beckwith, J., *Ann. Rev. Microbiol.*, **36**, 435–465 (1982).

35. Leong, D., Coleman, K.D., and Murphy, J.R., *J. Biol. Chem.*, **258**, 15016–15020 (1983).

36. Leong, D., Coleman, K.D., and Murphy, J.R., *Science*, **220**, 515–517 (1983).

37. Giannini, G., Rappuoli, R. and Ratti, G., *Nuc. Acid Res.*, **12**, 4063–4069 (1984).

38. Murphy, J.B. Pappenheimer, A.M., Jr., and Tayart de Borms, S., *Proc. Natl. Acad. Sci. USA*, **71**, 11–15 (1974).

39. Carroll, S.F., Lory, S. and Collier, R.J. *J. Biol. Chem.*, **255**, 12020–12024 (1980).

40. Carroll, S.F. and Collier, R.J., *Proc. Natl. Acad. Sci. USA*, **81**, 3307–3311 (1984).

41. Michel, A. and Dirkx, J., *Biochim. Biophys. Acta*, **491**, 286–295 (1977).

42. Robinson, E.A., Henrikson, D., and Maxwell, E.S., *J. Biol. Chem.*, **249**, 5088–5093 (1974).

43. van Ness, B.G., Howard, J.B., and Bodley, J.W., *J. Biol. Chem.*, **253**, 8687–8690 (1978).

44. Browen, B.A. and Bodley, J.W., *FEBS Lett.*, **103**, 253–255 (1979).

45. van Ness, B.G., Howard, J.B., and Bodley, J.W., *J. Biol. Chem.*, **255**, 10710–10716 (1980).

46. Moehring, J.M., Moehring, T.J., and Danley, D.E., *Proc. Natl. Acad. Sci. USA*, **77**, 1010–1014 (1980).

47. Dunlop, P.C. and Bodley, J.W., *J. Biol. Chem.*, **258**, 4754–4758 (1983).

48. Moehring, T.J., Danley, D.E., and Moehring, J.M., *Mol. Cell. Biol.*, **4**, 642–650 (1984).

49. Strauss, N. and Hendel, E.D., *J. Exp. Med.*, **109**, 145–163 (1959).

50. Uchida, T., Pappenheimer, A.M., Jr., and Harper, A.A., *J. Biol. Chem.*, **248**, 3845–3850 (1973).

51. Duncan, J.L. and Groman, N.B., *J. Bacteriol.*, **98**, 963–969 (1969).

52. Zanen, J., Muyldermans, G., and Beugnier, N., *FEBS Lett.*, **66**, 261–263 (1976).

53. Ittelson, T.R. and Gill, D.M., *Nature*, **242**, 330–332 (1973).

54. Bopuet, P. and Pappenheimer, A.M. Jr., *J. Biol. Chem.*, **251**, 5770–5778 (1976).

55. Uchida, T., Pappenheimer, A.M., Jr., and Harper, A.A., *Science*, **175**, 901–903 (1972).

56. Boquet, P., Silverman, M.S., Pappenheimer, A.M., Jr., and Vernon, W.B., *Proc. Natl. Acad. Sci. USA*, **73**, 4449–4453 (1976).

57. Bonventre, P.F., Saelinger, C.B., Ivins, B., Woscinski, C., and Amorini, M., *Infect. Immun.* **11**, 675–684 (1975).

58. Sandvig, K. and Olsnes. S., *Cell Biol.*, **87**, 828–832 (1980).

59. van Renswoude, J., Bridges, K.R., Harford, J.B., and Klausner, R.D., *Proc. Natl. Acad. Sci. USA*, **79**, 6186–6190 (1982).

60. Blewitt, M.G., Chao, J.M., McKeever, B., Sarma, R., and London, E., *Biochem. Biophys. Res. Commun.*, **120**, 286–290 (1984).

61. Marnell, M.H., Shia, S.-P., Stookey, M., and Draper, R.K., *Infect. Immun.*, **44**, 145–150 (1984).

62. Zalman, L.S. and Wisnieski, B.J., *Proc. Natl. Acad. Sci. USA*, **81**, 3341–33454 (1984).

63. Wright, H.T., Marston, A.W., and Goldstein, D.J., *J. Biol. Chem.*, **259**, 1649–1654 (1984).

64. Liu, P.V., *J. Infect. Dis.*, **130**, S94–99 (1974).

65. Liu, P.V., *J. Infect. Dis.*, **116**, 112–116 (1966).

66. Liu, P.V. and Hsieh, H., *J. Infect. Dis.*, **128**, 520–528 (1973).
67. Leppla, S.H., *Infect. Immun.*, **14**, 1077–1086 (1976).
68. Pavlovskis, O.R. and Gordon, F.B., *J. Infect. Dis.*, **125**, 631–636 (1972).
69. Pavlovskis, O.R. and Shackelford, A.H., *Infect. Immun.*, **9**, 540–546 (1974).
70. Iglewski, B.H. and Kabat, D., *Proc. Natl. Acad. Sci. USA*, **72**, 2284–2288 (1975).
71. Iglewski, B.H., Liu, P.V., and Kabat, D., *Infect. Immun.*, **15**, 138–144 (1977).
72. Chung, D.W. and Collier, R.J., *Infect. Immun.*, **16**, 832–841 (1977).
73. Leppla, S.H., Martin, O.C., and Muehl, L.A., *Biochem. Biophys. Res. Commun.*, **81**, 532–538 (1978).
74. Crys, S.J., Jr., Friedman, R.L., and Iglewski, B.H., *Proc. Natl. Acad. Sci USA*, **77**, 7199–7203.
75. De, S.N., *Nature*, **183**, 1533–1535 (1959).
76. Dutta, N.K., Panse, N.V., and Kulkarni, D.R., *J. Bact.*, **78**, 594–600 (1959).
77. Craig, J.P., Symposium on Cholera, US–Japan Corp. Med. Sci. Program, pp. 131–135 (1972).
78. Finkelstein, R.A. and Lo Spalluto, J.J., *J. Exp. Med.*, **130**, 185–202 (1969).
79. Lai, C.Y., *CRC Crit. Rev. Biochem.*, **9**, 171–206 (1980).
80. Smith, H.W. and Halls, S., *J. Pathol. Bacteriol*, **93**, 531–543 (1967).
81. Burrows, W. and Musteikis, G.M., *J. Infect. Dis.*, **116**, S183 (1966).
82. Dean, A.G., Ching, Y.C., Williams, R.G., and Harden, L.B., *J. Infect. Dis.*, **125**, S407–S411 (1972).
83. Craig, J.P., *Nature*, **207**, 614–616 (1965).
84. Lai, C.Y., Cancedda, F., and Duffy, L.K., *Biochem. Biophys. Res. Commun.*, **102**, 1021–1027 (1981).
85. Gurrant, R.L., Brunton, L.L., Schnaitman, T.C., Rebhun, L.I., and Gilman, A.G., *Infect. Immun.*, **10**, 320–325 (1974).
86. Nozawa, R., Yokota, T., and Kuwahara, S., *13th Joint Conference on Cholera*, US–Japan Corp. Med. Sci. Program, 299–303 (1977).
87. Finkelstein, R.A., Norris, H.T., and Dutta, N.K., *J. Infect. Dis.*, **114**, S203–210 (1964).
88. Richardson, S.H., *J. Bacteriol.*, **100**, 27–31 (1969).
89. Rappaport, R.S., Rubin, B.A., and Tint, H., *Infect. Immun.*, **9**, 294–303 (1974).
90. Lai, C.Y., *J. Biol. Chem.*, **252**, 7249–7256 (1977).
91. Mekalanos, J.J., Collier, R.J., and Romig, W.R., *Infect. Immun.*, **20**, 552–559 (1978).
92. Mekalanos, J.J., Collier, R.J., and Romig, W.R., *Proc. Natl. Acad. Sci. USA*, **75**, 941–944 (1978).
93. Clements, J.D. and Finkelstein, R.A., *Infect. Immun.*, **24**, 760–769 (1979).
94. Kunkel, S.L. and Robertson, D.C., *Infect. Immun.*, **25**, 586–596 (1979).

95. Lo Spalluto, J.J. and Finkelstein, R.A., *Biochim. Biophs. Acta.*, **257**, 158–162 (1972).

96. Finkelstein, R.A. and Lo Spalluto, J.J., *Science*, **175**, 529–530 (1972).

97. Kurosky, A., Markel, D.E., Touchstone, B., and Peterson, J.W., *J. Infect. Dis.*, **133**, S14–21 (1976).

98. Lai, C.Y., Mendez, E., and Chang, D., *J. Infect. Dis.*, **133**, S23–30 (1976).

99. Gill, D.M., *Biochemistry*, **15**, 1242–1247 (1976).

100. Wodnar–Filipowicz, A. and Lai, C.Y. *Arch. Biochem. Biophys.*, **176**, 475–482 (1976).

101. Gill, D.M. and King, C.A., *J. Biol. Chem.*, **250**, 6424–6429 (1975).

102. Gill, D.M. and Rappaport, R.S., *J. Infect. Dis.*, **139**, 674–681 (1979).

103. Mekalanos, J.J., Collier, R.J., and Romig, W.R., *J. Biol. Chem.*, **254**, 5855–5861 (1979).

104. Duffy, L.K., Peterson, J.W., and Kurosky, A., *FEBS Lett.*, **126**, 157–161 (1981).

105. Booth, B.A., Boesman-Finkelstein, M., and Finkelstein, R.A., *Infect. Immun.*, **45**, 558–560 (1984).

106. van Heynigen, W.E., Carpenter, C.C.J., Jr., Pierce, N.F., and Greenough, W.B., III, *J. Infect. Dis.*, **124**, S415–418 (1971).

107. Holmgren, J., Lonroth, I., and Svennerholm, L., *Infect. Immun.*, **8**, 208–212 (1973).

108. Holmgren, J., Lonnroth, I., and Svennerholm, L. *Scand. J. Infect. Dis.*, **5**, 77–82 (1973).

109. Duffy, L.K., Peterson, J.W., and Kurosky, A., *J. Biiol. Chem.*, **256**, 12252–12256 (1981).

110. Lai, C.Y., Duffy, L.K., Xia, Q.C., Chang, D., and Kurosky, A., *J. Biol. Chem.*, submitted for publication.

111. Mekalanos, J.J., Swartz, D.J., Pearson, G.D., Harford, N., Gyrone, F., and Wilde, M., *Nature*, **306**, 551–557 (1983).

112. Xia, Q.C., Chang, D., Blacher, R., and Lai, C.Y., *Arch. Biochem. Biophs.*, **234**, 363–370 (1984).

113. Spicer, E.K. and Noble, J.A., *J. Biol. Chem.*, **257**, 5716 (1982).

114. Lai, C.Y., Xia, Q.C., and Salotra, P.T., *Biochem. Biophys. Res. Commun.*, **116**, 341–348 (1983).

115. Duffy, L.K., Kurosky, A., and Lai, C.Y., *Arch. Biochem. Biophs.*, submitted for publication.

116. Manning, D.R. and Gilman, A.G., *J. Biol. Chem.*, **258**, 7059–7063 (1983).

117. Cassel, D. and Selinger, Z., *Proc. Natl. Acad. Sci. USA*, **74**, 3307–3311 (1977).

118. Abood, M.E., Hurley, J.B., Pappone, M.-C., Bourne, H.R., and Stryer, L., *J. Biol. Chem.*, **257**, 10540–10543 (1982).

119. van Dop, C., Tsubokawa, M., Bourne, H.R., and Ramachandran, J., *J. Biol. Chem.*, **259**, 696–698 (1984).

120. Lai, C.Y., *J. Biol. Chem.*, **252**, 7249–7265 (1977).

121. Kurosky, A., Markel, D.E., and Peterson, J.W., *J. Biol. Chem.*, **252**, 7257–7264 (1977).

122. Duffy, L.K. and Lai, C.Y., *Biochem. Biophys. Res. Commun.*, **91**, 1005 (1979).

123. Finkelstein, R.A., *Crit. Rev. Microbiol.*, **2**, 553 (1973).

124. Svennerholm, A., Sack, D.A., Bardhan, P.K., Huda, S., Jertborn, M., and Holmgren, J., *Proceedings of the 16th Joint Conference, US–Japan Coop. Med. Sci. Program* 310–316 (1981).

125. Svennerholm, A.M., Lange, S., and Holmgren, J., *Infect. Immun.*, **21**, 1–17 (1978).

126. McKenzie, S.J. and Halsey, J.F., *J. Immunol.*, **133**, 1818–24 (1984).

127. Chen, T.-M., Mazaitis, A.J., and Maas, W.K., *Infect. Immun.*, **47**, 1, in press.

128. Hewlett, E.L., Cronin, M.J., Moss, J., Anderson, H., Myers, G.A., and Pearson, R.D., *Adv. Cyclic Nuc. Prot. Phos. Res.*, **17**, 173–182 (1984).

129. Cowell, J.L. Hewlett, E.L., and Manclark, C.R., *Infect. Immun.*, **25**, 896–901 (1979).

130. Goldman, W.E., Klapper, D.G., and Baseman, J.B., *Infect. Immun.*, **36**, 782–794 (1982).

131. Olansky, L., Myers, G.A., Pohl, S.L., and Hewlett, E.L., *Proc. Natl. Acad. Sci. USA*, **80**, 6547–6551 (1983).

132. Munoz, J.J., Arai, H., Bergman, R.K., and Sadowski, P.L., *Infect. Immun.*, **33**, 820–826 (1981).

133. Arai, H. and Munoz J.J., *Infect. Immun.*, **31**, 495–499 (1981).

134. Katada, T. and Ui, M., *J. Biol. Chem.*, **254**, 469–479 (1979).

135. Katada, T. and Ui, M., *J. Biol. Chem.*, **255**, 9580–9588 (1980).

136. Katada, T. and Ui, M., *J. Biol. Chem.*, **256**, 8310–8317 (1981).

137. Katada, T. and Ui, M., *Proc. Natl. Acad. Sci. USA*, **79**, 3129–3133 (1982).

138. Sekura, R.D., Fish, F., Manclark, C.R., Meade, B., and Zhang, Y.-L., *J. Biol. Chem.*, **258**, 14647–14651 (1983).

139. Tamura, M., Nogimori, K., Murai, S., Yajima, M., Ito, K., Katada, T., Ui, M., and Ishii, S., *Biochem.*, **21**, 5516–5522 (1982).

140. Sato, Y. Cowell, J.L., Sato, H. Burstyn, D.G., and Manclark, C.R., *Infect. Immun.*, **41**, 313–320 (1983).

141. Ui, M., Katada, T., Murayama, T., Kurose, H., Yajima, M., Tamura, M., Nakamura, T., and Nogimori, K., *Adv. Cyclic Nuc. Prot. Phos. Res.*, **17**, 145–151 (1984).

142. Murayama, T., Katada, T., and Ui, M., *Arch. Biochem. Biophys.*, **221**, 381–390 (1983).

143. van Dop, C. Yamanaka, G., Steinberg, F, Sekura, R.D., Manclark, C.R., Stryer, L., and Bourne, H.R., *J. Biol. Chem.* **259**, 23–26 (1984).

SUCCINYL-CoA SYNTHETASE STRUCTURE-FUNCTION RELATIONSHIPS AND OTHER CONSIDERATIONS

By JONATHAN S. NISHIMURA, *Department of Biochemistry, University of Texas Health Science Center, San Antonio, Texas 78284*

CONTENTS

I. Introduction

Succinyl-CoA synthetase (Succinate:CoA ligase [GDP* forming] EC 6.2.1.4 and [ADP forming] EC 6.2.1.5), also referred to as succinic thiokinase (4) and the P (phosphorylating) enzyme (39,43), catalyzes the reaction shown in Eq. (1).

$$\text{succinate } + \text{ CoA } + \text{ NTP } \xleftrightarrow{\text{M}^{2+}} \text{ succinyl-CoA } + \text{ NDP } + \text{ Pi}\quad(1)$$

where NTP and NDP are purine ribonucleoside triphosphate and diphosphate, respectively, and M^{2+} is a divalent metal ion with Mg^{2+} and Mn^{2+} being the most active. The subject has been reviewed previously (11,35,63). More recently, a review concerning the use of ^{31}P NMR in a mechanistic study of the enzyme has appeared (99).

Succinyl-CoA synthetase has had an interesting and, at times, controversial history. A variety of ideas concerning its mechanism of action have been expressed, including proposal and exclusion of such intermediates as thiophosphoryl-CoA (33,34,91) and a high-energy nonphosphorylated form of the enzyme (6,20,58,97). A succinyl phosphate intermediate was considered on the basis of ^{18}O

---

* Abbreviations used: ADP, adenosine diphosphate; ATP, adenosine triphosphate; EPR, electron paramagnetic resonance; GDP, guanosine diphosphate; GTP, guanosine triphosphate; ITP, inosine triphosphate; NADH, reduced nicotinamide adenine dinucleotide; PRR, proton relaxation rate.

exchange experiments (26), but was ruled out (43). Later, direct evidence for enzyme-bound succinyl phosphate was obtained (65), followed by other experimental results that were consistent with the concept (30,40,62,101).

Taken together with evidence that a phosphorylated form of the enzyme fulfilled the kinetic requirement as an intermediate (13), Eq. (2a)–(2c) describe the covalent steps in the overall reaction.

$$ATP + enzyme \xleftrightarrow{M^{2+}} P\text{-enzyme} + ADP \qquad (2a)$$

$$P\text{-enzyme} + succinate \xleftrightarrow{M^{2+}} enzyme \cdot succinyl\text{-P} \qquad (2b)$$

$$enzyme \cdot succinyl\text{-P} + CoA \xleftrightarrow{M^{2+}} enzyme + succinyl\text{-CoA} + P_i$$

$$(2c)$$

Steady state kinetic studies (24,55,56) suggest that the pig heart enzyme and *E. coli* enzyme mechanisms are more complex than that represented by Eqs. (2a)–(2c). This has been discussed in some detail by Bridger (11). The mechanism appears to involve both sequential and random components with NTP binding first and NDP released last in the direction of succinyl-CoA formation.

This review will be concerned primarily with developments that have come to light since the last comprehensive review (11). These data will be presented in the context of older information. A preponderance of effort has been associated with the study of the enzyme from *E. coli* and pig heart. However, significant complementary work has been done with enzymes from other sources.

II. Studies of General Interest

Most of the published work on the succinyl-CoA synthetase reaction has had to do with the focused mechanistic and structural studies of highly purified enzymes. Other studies, of more general interest, have been carried out for the most part with less purified systems.

A. LARGE AND SMALL FORMS OF SUCCINYL-CoA SYNTHETASE

Weitzmann (103) has drawn an interesting correlation between the size of succinyl-CoA synthetase and the nature of the bacterium from which it is isolated. Thus, in general, large forms of the enzyme are found in gram-negative bacteria and small forms in gram-positive bacteria and in eukaryotic organisms. Large and small refer to relative elution volumes from a Sephadex G-200 column and correspond to M_r values of ~140,000 and 70,000. It is not clear what selective advantage, if any, is realized by a particular organism in having one form of the enzyme or the other. Classical allosteric kinetics have not been observed in extensive studies of *E. coli* succinyl-CoA synthetase that would be categorized as a large form. On the other hand, it is of interest that a large allosteric form of citrate synthase seems to be common to most gram-negative organisms and a small form of the enzyme is found in most gram-positive bacteria. The large citrate synthase has been shown to be allosterically inhibited by NADH (103).

B. NUCLEOSIDE TRIPHOSPHATE (DIPHOSPHATE) SPECIFITY OF VARIOUS SUCCINYL-CoA SYNTHETASES

At one time it was believed that bacterial and plant succinyl-CoA synthetases were specific for ATP (ADP) and that the animal enzyme could only utilize GTP (GDP) or ITP (IDP) as substrates. This still appears to be the case for the plant enzyme. However, Burnham (16) demonstrated that the enzyme from the photosynthetic bacterium *Rhodopseudomonas spheroides* could utilize ATP, GTP, or ITP almost equally well. Ability to use these purine nucleoside triphosphates as substrates has also been observed in the *E. coli* enzyme (46,59) and in other bacterial enzymes (104). A correlation between relative K_m values for GDP and ADP in the classification of bacteria has been suggested (104). However, the number of organisms studied was not extensive.

It does appear that plant succinyl-CoA synthetases exhibit a strong preference for ATP and ADP as substrates (17,44,61,80,105). Succinyl-CoA synthetase that has been isolated from *Saccharomyces cerevisiae* in a partially purified state is specific for ATP (89).

Hansford (37) demonstrated the first animal ATP-dependent succinyl-CoA synthetase in the blowfly. Subsequently, Hamilton and Ottaway (36) reported that the enzyme from pigeon breast muscle was specific for ATP. Other tissues in both pigeon and chicken contained synthetase activities that utilized both ATP and GTP as substrates. It is of interest that Severin et al. (90) had reported experiments in which the phosphorylation of the pigeon breast muscle enzyme by [γ-^{32}P]ATP was described. However, an earlier report from the same laboratory described the purification and properties of a GTP-specific enzyme from pigeon breast muscle (54). Using antibody against commercially prepared pig heart succinyl-CoA synthetase, Steiner and Smith (92) have reported phosphorylation of rat brain enzyme by both [γ-^{32}P]GTP and [γ-^{32}P]ATP. On the basis of immunotitration experiments, they suggest that there are ATP-specific and GTP-specific enzymes and that these enzymes are distributed differently in the brain.

Ottaway (77) has suggested that a GTP-specific enzyme may be desirable in various tissues, for example, heart muscle, in the utilization of ketone bodies. Succinyl-CoA is also a substrate for 3-keto acid CoA transferase (EC 2.8.3.5). The K_m of the latter for succinyl-CoA is 4.2 mM, approximately 10 times higher than that of succinyl-CoA synthetase for succinyl-CoA (11). Ottaway suggests that, if the latter enzyme were ATP(ADP)-dependent, steady state levels of succinyl-CoA would be too low, because of unfavorable ATP/ADP ratios [see reverse of Eq. (1)], to provide an effective substrate concentration for 3-keto acid CoA transferase. On the other hand, if the GTP/GDP ratio were higher than that of ATP/ADP, one might expect the steady state levels of succinyl-CoA to be higher in the presence of a GTP(GDP)-specific enzyme. At the present time, there is no basis upon which to compare the two ratios in mitochondrial matrix, where both 3-keto acid CoA transferase and succinyl-CoA synthetase are located.

C. EQUILIBRIUM CONSTANTS FOR THE REACTION

Lynn and Guynn have made a careful study of the equilibrium constants of the reaction using the pig heart enzyme (51). The observed equilibrium constant (K_{obs}), expressed as follows, was measured at physiological ionic strength (0.25 M) and physiological ranges of pH and free [Mg^{2+}]:

$$K_{obs} = \frac{[\Sigma \text{ succinyl-CoA}][\Sigma \text{ GDP}][\Sigma P_i]}{[\Sigma \text{ CoA}][\Sigma \text{ GTP}][\Sigma \text{ succinate}]}$$

where Σ and brackets indicate total concentrations. At pH 7.0 and

38°C, K_{obs} values of 3.63 (free $[Mg^{2+}]$ = 0), 0.84 (free $[Mg^{2+}]$ = 10^{-3} M) and 0.55 (free $[Mg^{2+}]$ = 10^{-2} M) were determined. The corresponding values of K_{obs} at pH 7.0 and 25°C were 6.37 (free $[Mg^{2+}]$ = 0), 2.09 (free $[Mg^{2+}]$ = 10^{-3} M), and 1.31 (free $[Mg^{2+}]$ = 10^{-2} M). Thus, it is quite clear that free $[Mg^{2+}]$ has a pronounced effect on K_{obs}. Values of K_{obs} were relatively insensitive to changes in $[H^+]$ above pH 7. However, significant differences in K_{obs} were observed with changes in $[H^+]$ below pH 7. Earlier, Kaufman and Alivisatos had obtained a value of 0.27 for K_{obs} at 20°C and pH 7.4 (44).

Calculation of K (51) was based on the equation

$$K = \frac{[\text{succinyl-CoA}^{1-}][\text{GDP}^{3-}][\text{HPO}_4^{2-}]}{[\text{CoA}][\text{GTP}^{4-}][\text{succinate}^{2-}]}$$

Values of 3.02 ± 0.07 and 1.29 ± 0.05 were obtained for K at 25 and 38°C, respectively.

D. INTERMOLECULAR ASSOCIATIONS OF PIG HEART SUCCINYL-CoA SYNTHETASE

Evidence was presented recently for a physical interaction between pig heart succinyl-CoA synthetase and the α-ketoglutarate dehydrogenase complex of pig heart (83). This association was observed through the application of polyethylene glycol precipitation, ultracentrifugation, and gel chromatography. Of seven proteins tested, only succinyl-CoA synthetase was found to associate with the complex. Srere (93a) has also found that significant quantities of succinyl-CoA synthetase are associated with inner mitochondrial membrane. These observations support the prediction that citric acid cycle enzymes are present in the mitochondrial membrane as a complex [see ref. (93)]. The association of such a coupled complex with membranes is also of interest, particularly in light of the report that enzymes of fatty acid β-oxidation are associated with inner mitochondrial membrane (94). Since succinyl-CoA is known to be an inhibitor of the α-ketoglutarate dehydrogenase complex [see ref. (83)], physical interaction of succinyl-CoA synthetase with the latter might present possibilities for a sophisticated mechanism of control at this point in the citric acid cycle.

III. Properties of Well-Studied Succinyl-CoA Synthetases

A. THE *E. coli* ENZYME

A wide variety of experimental approaches to understanding the succinyl-CoA synthetase reaction has been made possible by the availability of the *E. coli* enzyme in 100- to 200-mg quantities. Purification of such quantities has been facilitated by the fact that the enzyme is derepressed in succinate medium (47) and by the use of a key acetone fractionation step performed at high salt concentration (29,97). Highly purified preparations of *E. coli* succinyl-CoA synthetase have been described by a number of laboratories (14,29,30,50,87).

1. Molecular Weight and Subunit Structure

Molecular weight values of 160,000 (29), 141,000 (87), 146,000 (50), and 136,000 (48) have been estimated for the *E. coli* enzyme. Polyacrylamide gel electrophoresis in sodium dodecylsulfate revealed the presence of two molecular species, one (the α subunit) corresponding to a M_r of 29,600 $\pm$ 500 and the other (the β subunit) with a M_r of 38,700 $\pm$ 300 (10). When the assumption was made that each polypeptide chain bound the same amount of Coomassie Blue dye per unit weight, it was deduced that there were equimolar amounts of each subunit. Addition of the molecular weights led to the conclusion that the native protein consisted of an $(\alpha\beta)_2$ subunit structure. Cross-linking studies with dimethylsuberimidate and *o*-phenylenedimaleimide confirmed the $(\alpha\beta)_2$ structure and gave an indication of an intimate association between α and β subunits (96). No evidence was obtained for cross-linked homologous pairs (α_2 or β_2).

2. Amino Acid Composition and Extinction Coefficient

Amino acid analyses of the enzyme (50) and of the enzyme as well as the subunits (11) have been reported. The subunits do not appear to be similar to each other. The presence of both methionine and serine as *N*-terminal amino acid residues of the native enzyme (11) is also consistent with the presence of dissimilar chains. There are three tryptophan residues in each β subunit and none in the α subunit (84). The low tryptophan content of the enzyme would account for the values obtained for $E_{1\ cm}^{0.1\%}$ at 280 nm of 0.511 (87) and 0.49 (48).

3. Substrate Specificity

The substrate specificities with regard to succinate, NTP, and CoA have been discussed most recently by Bridger (11). Of the divalent metal ions tested as cofactors, Mg^{2+}, Mn^{2+}, and Co^{2+}, in decreasing order of effectiveness, were active (29). Slight activity was observed with Zn^{2+} and Ca^{2+}.

4. Divalent Metal Ion Binding

Mn(II) ion binding has been studied by PRR, proton relaxation rate, measurements and EPR, electron paramagnetic resonance (49,18). Enzymes of high specific activity contained 3.5 ± 0.7 metal ion binding sites with indistinguishable dissociation constants (K_D = 6.9 × 10^{-4} M). Enzymes of lower specific activity, but of comparable purity, contained 1.6 strong binding sites (K_D = 6.6 × 10^{-4} M) and 2.0 weak binding sites (K_D = 4.0 × 10^{-3} M). It was found that dephosphorylated enzyme has a slightly lower binding affinity for the Mn(II) ion (K_D = 1.4 × 10^{-3} M) than the phosphorylated enzyme (K_D = 7.74 × 10^{-4} M). Mn·ADP was bound at one or two sites (18). Mn·ATP did not bind to the phosphorylated enzyme, which is in agreement with the findings of Bowman and Nishimura with Mg·ATP binding (8). This observation was consistent with the findings of Moffet et al. (57) who measured binding of ADP in the presence of magnesium ion. Progression decreases in PRR enhancement were observed with each successive addition of substrate, indicating that the bound manganese ion is associated with the active site region (18).

5. Stoichiometry of Phosphorylation of the Enzyme and Half-Sites Reactivity

In a number of cases, incubation of succinyl-CoA synthetase with $[\gamma^{32}P]ATP$ has resulted in the incorporation of approximately 1 mol of P/mol of enzyme (50,57,87). Since there are at least two possible phosphorylation sites, these data have been interpreted as evidence for half-sites reactivity in the enzyme (57). The higher (1.38 mol of P/mol of enzyme) incorporation of $[^{32}P]P_i$ in the presence of succinyl-CoA (87) has been attributed to errors in the determination of the specific radioactivity of $[^{32}P]P_i$ (11). An enigma that remains unsolved is that homogeneous preparations of the enzyme have been shown to express different specific enzyme activities (11). Futhermore, incorporation of more than 1 mol of P/mol of enzyme has been

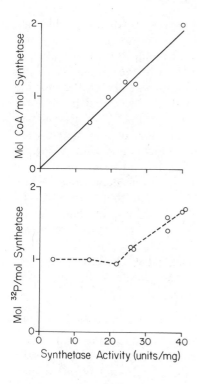

Figure 1. Relationship of specific enzyme activity to phosphorylation capacity of *E. coli* succinyl-CoA synthetase. The data were taken from Bowman and Nishimura (8).

observed. For example, a stoichiometry of approximately 2 mol of P/mol of enzyme has been reported (31). It was suggested that the method of protein determination was in error (11). However, in the protein method employed, the value for $E^{0.1\%}_{1\,cm}$ at 280 nm of 0.511 (87) was used as reference. In addition, Bowman and Nishimura (8) showed that homogeneous preparations of relatively low specific activity (up to 22 unit/mg) incorporated 1 mol of P from ATP/mol of enzyme. However, as specific activity increased, proportionately greater phosphorylation (up to approximately 1.7 mol of P/mol of enzyme) was observed. In contrast, binding of up to 2 mol of CoA/ mol enzyme was proportional to enzyme activity. These points are illustrated in Fig. 1. In this context, it is of interest that Lam et al. (49) have reported that enzyme of relatively high specific activity (29–34 μmol min^{-1} mg^{-1}) exhibited four high affinity Mn(II) binding sites, whereas enzyme of relatively low specific activity (18–27 μmol min^{-1} mg^{-1}) contained two high affinity sites for Mn(II) and two low affinity sites (see Section III.A.4). The reasons for these differences in phosphorylation capacity and metal ion binding are not immediately clear. However, with regard to the former, it should be mentioned that Bowman and Nishimura incubated the enzyme under dephosphorylating conditions and removed the dephosphorylating substrates prior to introduction of [γ-^{32}P]ATP (8). Ramaley et al. (87) also mentioned that certain preparations of enzyme that were treated with succinate and CoA prior to incubation with ATP gave phosphorylation stoichiometries as high as 1.6 to 1.8 mol of P/mol of enzyme. Moffet et al. (57) have reported that dephosphorylation of the enzyme with ADP resulted in a protein that was more susceptible to proteolysis and lost activity more rapidly than enzyme that was not dephosphorylated. Thus, there can be little doubt that the enzyme, as isolated, is phosphorylated, the extent of which is probably variable. It would seem reasonable, therefore, that less equivocal phosphorylation stoichiometries would be obtained after prior dephosphorylation of the enzyme.

6. Unfolding of the Enzyme in Acidic Urea and Refolding

In a major contribution toward developing means to study structure–function relationships in succinyl-CoA synthetase, Pearson and Bridger (81) isolated α and β subunits by gel filtration on a Sephadex G-150 column that had been equilibrated with 6-M urea–5% acetic acid–0.1-mM EDTA–0.5-mM dithiothreitol. The subunits could be refolded at approximately pH 7.4 to yield an active enzyme. ATP was required for optimal refolding. Thus, it is possible that phosphorylation of the α subunit is necessary for correct intrachain and interchain relationships to be established as requisites for proper refolding. The reconstituted enzyme behaved like the native enzyme on a gel-filtration column.

The isolated α subunit was capable of autophosphorylation by ATP (82). However, phosphorylation was greatly accelerated in the presence of the β subunit. This observation provided additional evidence that the active site might be composed of components from both subunits.

A study of extracts of *E. coli* by Wolodko et al. (106) revealed that neither α or β subunits accumulated in excess. These investigators then obtained evidence for a heat-stable substance of low molecular weight (~2000) that precludes folding of the α and β chains to give reconstituted enzyme. The effect is irreversible, but proteolysis seems not to be involved. The factor has no effect on folded protein. It also appears that the factor is present in cells from mid-log, late-log, and stationary phase cultures. As yet, nothing is known of the chemistry of the factor or how it interacts with the subunits.

7. Tryptophan Fluorescence and the Effect of Tryptophan Modification on Catalytic and Structural Properties of the Enzyme

There are three tryptophan residues per $\alpha\beta$ dimer of the enzyme and all three are located in the β subunit (84). Succinyl-CoA synthetase shows a fluorescence emission maximum at 335 nm. Quenching of the fluorescence by acrylamide was enhanced significantly by ATP and by CoA and, to a lesser extent, by succinate. It would appear from these data that at least one of the three tryptophan residues can act as a reporter group for events that occur at the active site. Further, the fact that both ATP and CoA have similar effects on quenching adds support to models that place the active site at the points of contact between subunits.

Chemical modification of succinyl-CoA synthetase with N-bromosuccinimide has resulted in the loss of one tryptophan residue at complete loss of enzymatic activity (76,110). The modified β subunit was refolded with unmodified α subunit to yield a hybrid molecule that was inactive but was capable of phosphorylation. Thus, it would seem that at least one tryptophan residue of the enzyme is located in close proximity to the active site region.

8. Sulfhydryl Groups of the Enzyme

There are some 16–18 titratable sulfhydryl groups per molecule of the enzyme (72). Given the fact that there are also approximately 24 half-cystine residues (11), it seems reasonable to conclude that there must be 3–4 disulfide bonds or 6–8 inaccessible sulfhydryl groups (even in the presence of sodium dodecyl sulfate, 6 M urea, or 6 M guanidine hydrochloride) in the enzyme. Titration of four sulfhydryl groups by N-ethylmaleimide (52), iodoacetamide (52), or methyl methanethiosulfonate (64) led to total loss of enzymatic activity. Modification of sulfhydryl groups can result in apparent changes in quaternary structure that may be attributable to dissociation, perhaps to $\alpha\beta$ dimers, and aggregation (72). However, this occurs only when the modifying group adds charge to the protein, for example, p-mercuribenzoate or 5,5'-dithiobis(2-nitrobenzoate) (72). On the other hand, alkylation by iodoacetamide (72) or methanethiolation by methyl methanethiosulfonate (64) results in no apparent change in quaternary structure. Perhaps introduction of charged groups favors dissociation due to charge–charge interactions. The changes brought about by p-mercuribenzoate and 5,5'-dithiobis(2-nitrobenzoate) can be completely reversed by incubation of the modified enzyme with dithiothreitol (64). However, methanethiolation is difficult to reverse, requiring incubation with the reagent tributylphosphine for several days (64). In addition, the fact that N-hexylmaleimide and N-butylmaleimide were more effective inhibitors of the enzyme than N-ethylmaleimide (53) gives reason to speculate that at least one type of important sulfhydryl group may be located in a more hydrophobic environment than the rest.

9. Crystallization of the Enzyme

Wolodko et al. (108) have reported the crystallization of tetragonal prisms of succinyl-CoA synthetase from solutions of ammonium sulfate and mixtures of sodium and potassium phosphates. In phosphate medium, the crystals have a space group of P_222 with unit cell dimensions $a = b = 94$ Å, $c = 248$ Å and, evidently, 0.5 $(\alpha\beta)_2$ tetramer/asymmetric unit. In ammonium sulfate medium, the crystals have a space group of $P4_122(P4_322)$ and unit cell dimensions $a = b = 99$ Å and $c = 399$ Å. the crystals contain one tetramer/asymmetric unit. The latter result was deemed important because it is consistent with an asymmetric arrangement of $\alpha\beta$ dimers in the phosphorylated tetramer and thus supports the alternating site hypothesis (see Section IV.B).

B. THE PIG HEART ENZYME

Cha and his co-workers were responsible for devising a large-scale purification procedure of the enzyme from pig heart (19,21,23). Improvements of the procedure have been described (15,60). The overall yield of enzyme is relatively poor. However, sufficient quantities of homogeneous enzyme can be obtained with which to do mechanistic studies and certain structural work.

1. Molecular Weight and Subunit Structure

Molecular weight values of 70,000 (21), 75,000 (15), and 80,000 (60) have been estimated for the pig heart enzyme. Brownie and Bridger (15) reported the presence of one α subunit and one β subunit of M_r values 34,500 and 42,500, respectively. Values of 33,000 and 47,000 for the subunits were obtained by Murakami and Nishimura (60). Almost identical values have been obtained for the molecular weight of the subunits from the rat liver enzyme (3).

2. Tryptophan Content and Extinction Coefficient of the Enzyme

The subunits of the pig heart enzyme have been separated preparatively and the amino acid compositions of each determined (75). As was reported for the *E. coli* enzyme (11), the amino acid compositions do not suggest homology between the chains. Of interest is the finding that only one tryptophan residue is present and it is located in the β subunit. The low tryptophan content of the enzyme is consistent with the similarly low value of $E_{1\,cm}^{0.1\%}$ at 280 nm of 0.35 (60).

3. Substrate Specificity

This subject has been discussed by Bridger (11).

4. Stoichiometry of Phosphorylation of the Enzyme

Brownie and Bridger (15) have shown that upon incubation of the enzyme with GTP, a maximum of 1 mol of phosphoryl group was incorporated per mole synthetase. Cha et al. (22) had reported an incorporation of at least 2 mol of P/mol of enzyme. However, as has been pointed out (11), the latter calculation was made on the basis of specific enzyme activity and it is likely that, while overall catalytic activity has been lost, the enzyme is still able to undergo phosphorylation by GTP. This result is analogous to that referred to earlier concerning the *E. coli* enzyme (see Section III.A.5).

5. Fluorescence Studies of the Enzyme

Fluorimetric experiments similar to those performed with the *E. coli* enzyme (84) have been carried out with the pig heart enzyme (75). Thus, quenching of tryptophan fluorescence by acrylamide was enhanced by addition of GTP or CoA. However, in contrast to the result observed with the *E. coli* enzyme, succinate protected against acrylamide quenching. These data suggest that the tryptophan residue is close to the active site. It is also suggestive that the conservation of this single tryptophan is important to the activity of the enzyme. As was mentioned earlier, modification of one of the three tryptophan residues in the *E. coli* αβ dimer did not alter the ability of the modified β subunit to refold with unmodified α subunit (see Section III.A.7). The refolded product was inactive with respect to catalysis of the overall reaction, but was capable of phosphorylation by ATP.

6. Sulfhydryl Groups and Multiple Interconvertible Forms of the Enzyme

Five forms of the enzyme with pI values of 6.4, 6.2, 6.0, 5.9, and 5.8 were detected by the technique of isoelectric focusing (1). The first four appeared to be interconvertible if incubated appropriately with CoA or with GTP. Thus, it would appear that the differences are not related to variations in amino acid sequence. The pI-6.2 species was converted to the pI-6.0 species by incubation with mercaptoethanol or dithiothreitol (2). The latter enzyme form contained about three sulfhydryl groups per mole more than the pI-6.2 species. Thus, it would appear that disulfide bonding may provide the basis for at least some of the interconvertibility among enzyme forms. Murakami and Nishimura (60) have noted a tendency toward loss of enzyme activity that appears to be related nonproportionately to disulfide bond formation. Baccanari and Cha (2) have also observed pI-5.3 and pI-5.2 forms that were produced by disulfide bonding with CoA during incubation of the enzyme with CoA disulfide. Amino acid analysis of the enzyme gave 14 half-cystine residues per mole (75). The highest number of sulfhydryl groups measured was 12/mol of enzyme (60). It seems likely that, at most, there is only one disulfide bond in the native enzyme protein.

In analogy to the *E. coli* enzyme, an active site sulfhydryl group appears to react with the CoA affinity analog CoA disulfide-*S,S*-dioxide (see Section IV.A.5) and may be involved in the cross linking of α and β subunits by the thiol-reactive reagent *p*-phenylenedimaleimide (60).

7. Can the Subunits Be Refolded into an Active Enzyme Species?

Although the α and β subunits of the enzyme have been separated on Sephadex G-150 columns (75), no reconstitution of enzymatic activity has been achieved in attempts to refold subunits. Some activity has been regained from preparations that were made 6 M with respect to urea and then after 30 minutes at 0°C diluted under refolding conditions that were similar to those described by Pearson and Bridger (81). However, no activity was regained when the incubations in urea were longer prior to adding diluent (69).

IV. Molecular Mechanisms

A. BINDING SITES AND ACTIVE SITE RESIDUES

Since it is highly likely in the case of the succinyl-CoA synthetases that catalytic residues are located in both α and β subunits and quite possible that binding sites for substrates may consist of amino acid residues from either subunit, it is helpful to approach this study in a simplified manner. However, it is difficult to think, for example, of the binding site of the γ phosphoryl group of NTP and the P_i binding site in the same terms. The reason is that there are intervening reactions by which these groups are interconverted: phosphorylation by NTP and reaction with succinate to yield succinyl phosphate in one direction and reaction of P_i with succinyl-CoA to yield succinyl phosphate and phosphorylation of the enzyme in the reverse direction. Analogous considerations would apply to the relationships between the succinate binding site and that for the succinyl group of succinyl-CoA. On the other hand, it is probably useful to think in terms of NTP and NDP as sharing the binding site for NDP and of succinyl-CoA and CoA as sharing the CoA binding site.

1. Phosphorylation Site

It is well established that in the phosphorylation step [Eq. (2a) and reverse of Eq. (2c)] a histidine residue of the α subunit is phosphorylated (10,15) in the N^τ (N-3) position of the imidazole ring (41). This is envisioned as a nucleophilic attack by the N^τ atom on the terminal phosphoryl group of ATP or on the phosphoryl group of succinyl phosphate. Since the tautomeric form of the imidazole ring containing the lone pair of electrons on the N^π (N-1) position is favored, it has been suggested that the tautomer in which the lone pair is associated with N^τ is stabilized by an anionic group of the protein (99). A tryptic dodecapeptide containing the histidine phosphorylation site of the *E. coli* enzyme has been sequenced (102). This dodecapeptide remains the only portion of the enzyme containing an active site residue that has been sequenced.

Recently, the phosphoryl enzyme was studied by ^{31}P NMR (98–100). A downfield shift and widening of the linewidth of the His*P* (phosphohistidine) resonance was observed in the presence of CoA (98,99). Addition of the substrate analog 2,2′-difluorosuccinate to phosphorylated enzyme in the presence of saturating CoA caused an upfield shift and a reversal of the broadening of the linewidth. These data suggest that binding of CoA brings about an increase of the mobility of the His*P* group, whereas succinate has the reverse effect. The observations of Vogel, et al. (100) of the ^{31}P NMR of phosphorylated proteins suggest that phosphorylated residues that are catalytic intermediates are immobilized, whereas phosphoserine residues at regulatory sites, as in the case of glycogen phosphorylase, have mobility. Vogel and Bridger (98,99) propose that in the presence of CoA the phosphoryl histidine residue has two conformations, one that reacts with ADP to form ATP and another that reacts with succinate to form succinyl phosphate. In the presence of CoA and difluorosuccinate, the latter conformation is favored. This hypothesis is illustrated in Fig. 2. Since the optimal effect is observed in the presence of saturating CoA, both CoA binding sites are occupied and it is not known if the CoA effect is mediated at the same site. In any case, the CoA effect may be a manifestation of the phenomenon referred to as substrate synergism in which substrates enhance rates of partial reactions (13). In this context, it is of interest that the binding of CoA to the *E. coli* enzyme brought about an enhanced quenching of tryptophan fluorescence of the enzyme by acrylamide (84). This observation is consistent with a change in conformation of the protein upon binding of CoA. It was also observed that binding of CoA induced changes in the flexibility of the protein in regions labeled with dansyl chloride (85).

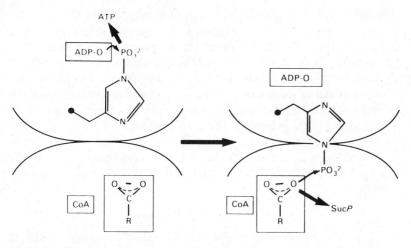

Figure 2. Schematic representations of the active site of succinyl-CoA synthetase at the point of contact between the two subunit types (98,99). In this representation, the His P residue is visualized as being sufficiently mobile due to binding of CoA to allow phosphoryl transfer either to ADP on the α subunit (left) or to succinate on the β subunit (right).

2. The NTP (NDP) Binding Site

Since phosphorylation occurs on the α subunit and the isolated α subunit is capable of autophosphorylation (82), it would seem reasonable to suggest that the NTP (NDP) binding site is also located on the α subunit. However, phosphorylation of the subunit by ATP is significantly stimulated by addition of the β subunit. Thus, it is possible that part of the β subunit is involved in the NTP (NDP) binding site. On the other hand, physical association with the β subunit may impart to the α subunit a conformational stability that is conducive to optimal binding of NTP with respect to the histidine phosphorylation site. The use of GDP-dialdehyde and ADP-dialdehyde as affinity labels of the rat liver (3) and *E. coli* enzymes (70), respectively, resulted in cross linking of α and β subunits. It is tempting to speculate that the NTP (NDP) binding site bridges the gap between α and β subunits. However, the high reactivity of these affinity labels is such that extreme caution must be exercised in the interpretation of such results. ADP-2'-semialdehyde (88) also inactivated the enzyme but resulted in no cross linking (70). It was not determined if modification of the enzyme occurred in the α or β subunit or both. The finding that the *E. coli* enzyme is inactivated by the affinity label 5'-fluorosulfonylbenzoyladenosine (86) did not provide evidence as to which subunit was involved either, since the compound did not remain attached to the protein. Here it is proposed that inactivation was brought about by formation of a disulfide bond involving thiol groups close to or at the active site. This is believed to be mediated by a thiosulfonate intermediate between one of the thiol groups and 5'-sulfonylbenzoyladenosine portion of the affinity analog.

3. The Phosphate Binding Site

There is no firm evidence as to the subunit location of the P_i binding site. Permanganate ion, which has been described as an analog of P_i (5) appears to inactivate the *E. coli* enzyme rapidly by bringing about oxidation of sulfhydryl groups. However, attempts to show that P_i or arsenate, an analog of P_i, could protect the enzyme against permanganate were not successful (74). On the other hand, protection was afforded by ATP-Mg^{2+}, succinate, and the CoA analog desulfo-CoA.

4. The Succinyl and Succinate Binding Sites

The observation that preferential labeling of the β subunit by [^{14}C]succinate occurs in the presence of ATP and CoA would appear to implicate the β subunit in the binding of the succinyl group of succinyl-CoA (6,82). This labeling is covalent but does not affect the enzymatic activity of the *E. coli* enzyme. Thus, it would appear that a nucleophile of the β subunit, for example, the ε-amino group of a lysyl residue, which is in close proximity to the succinyl binding site attacks the succinyl group, resulting in succinylation of the nucleophile. On the other hand, no evidence has been presented to define the subunit location of the succinate binding site.

5. The CoA Binding Site: Sulfhydryl Groups and a Possible Second Important Histidine Residue

The finding that one sulfhydryl group of the β subunit in both the *E. coli* and pig heart enzymes would react with the CoA affinity

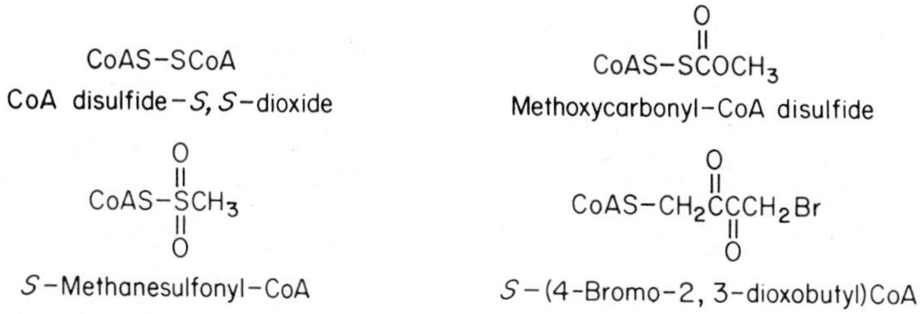

CoAS–SCoA

CoA disulfide–S,S–dioxide

$$\overset{\overset{\displaystyle O}{\|}}{CoAS-SCOCH_3}$$

Methoxycarbonyl–CoA disulfide

$$\overset{\overset{\displaystyle O}{\|}}{\underset{\underset{\displaystyle O}{\|}}{CoAS-SCH_3}}$$

S–Methanesulfonyl–CoA

$$\overset{\overset{\displaystyle O}{\|}}{\underset{\underset{\displaystyle O}{\|}}{CoAS-CH_2CCCH_2Br}}$$

S–(4-Bromo–2, 3-dioxobutyl)CoA

Figure 3.   Affinity analogs of CoA that react with active site sulfhydryl groups.

analog CoA disulfide-S,S-dioxide (73) to form an enzyme-bound di-sulfide-linked CoA provided the first firm evidence that a sulfhydryl group was located close to or at the active site (27). The inhibition could be completely reversed by incubation of the modified enzyme with dithiothreitol. A related analog, S-methane sulfonyl-CoA has been shown to inactivate the pig heart enzyme (79). The inhibitory effects of this reagent were reversed by dithiothreitol. In addition, methoxycarbonyl-CoA disulfide inactivated malate thiokinase, an enzyme that also catalyzes succinyl-CoA synthesis by a similar mechanism (38). The enzyme was protected against inactivation by addition of low concentrations of succinyl-CoA. It would appear that these reagents have reacted with important thiol groups of the enzymes. Yet another affinity analog of CoA, S-(4-bromo-2,3-dioxo-butyl)CoA (78), has been shown to be a potent inhibitor of the $E.$ $coli$ synthetase (71) and probably reacts with the same sulfhydryl group that is modified by CoA disulfide-S,S-dioxide. The structures of the CoA analogs are shown in Fig. 3.

Differential labeling of the $E.$ $coli$ enzyme with ethoxyformic anhydride led to the conclusion that a second histidine residue may be involved in the catalytic mechanism of succinyl-CoA synthetase (28). Blockage with ^3H-labeled anhydride of approximately 1 mol of histidine residue per $\alpha\beta$ dimer resulted in a dimunition of reactivity in the region of pH 7 with CoA disulfide-S,S-dioxide and with the sulfhydryl reagent methylmethanethiosulfonate, but did not appear to interfere with phosphorylation of the enzyme. Thus, the histidine phosphorylation site appeared not to be affected by the modification. It was proposed that the histidine residue side chain in the imida-

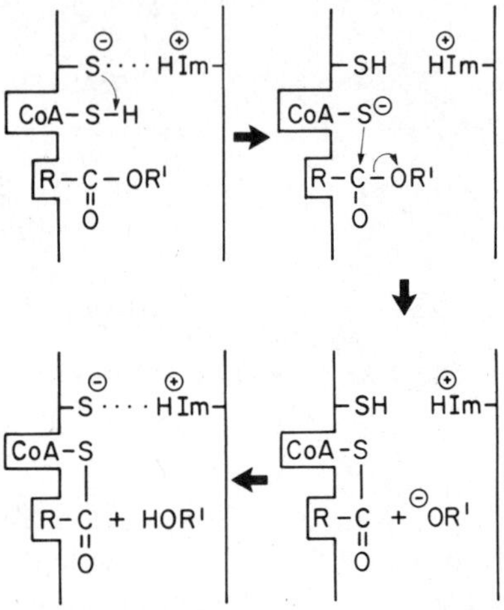

Figure 4. Proposed mechanism for the roles of important thiol and imidazole groups in the catalysis of the succinyl-CoA synthetase reaction, Upper left, ion pair between thiolate anion and imidazolium cation; upper right, dissociation of CoA thiol group, promoting nucleophilic attack on the carbonyl carbon atom on enzyme-bound succinyl phosphate, RCO–OR′; lower right, succinyl-CoA is produced and inorganic phosphate, –OR′, is released; lower left, ion pair association is reestablished.

zolium form stabilizes the anionic form of the reactive thiol group. Roles for both the reactive thiol group and the imidazolium group are proposed in Fig. 4 (28). Similar positioning of thiol groups in relation to the CoA binding site has been suggested for the enzymes phosphotransacetylase (EC 2.3.1.8) and carnitine acetyltransferase (EC 2.3.1.7) (28). The latter enzyme is known to have a histidine residue close to its CoA binding site (25). Unfortunately, the [³H]ethoxyformyl group did not remain covalently bonded to succinyl-CoA synthetase protein during sodium dodecylsulfate-polyacrylamide gel electrophoresis under nonreducing conditions. It was, therefore, not possible to identify which subunit was modified.

B. THE ALTERNATING SITES COOPERATIVITY MODEL

An interesting concept that has been proposed for *E. coli* succinyl-CoA synthetase by Bild et al. (7) is that of an alternating sites cooperativity mechanism. There would appear to be good evidence that such a mechanism applies in the case of beef heart mitochondrial ATPase (32,42). The observation of apparent half-sites phosphorylation of succinyl-CoA synthetase discussed earlier made this enzyme a logical system to test for this mechanism. In the case of the *E. coli* enzyme, synthesis of succinyl-CoA [see Eq. (2a)–(2c)] would be envisioned to occur at one active site. Binding of ATP at the second site would facilitate formation and release of succinyl-CoA at the first site. ATP that is bound to the second site may have simultaneously undergone reaction to phosphorylate that site. Another round of succinyl-CoA synthesis would proceed at this site, and so forth. To test this possibility, Bild et al. (7) performed ^{18}O exchange experiments in which ATP, the enzyme, Mg^{2+}, succinate, and $[^{18}O]P_i$ were incubated in the presence of hydroxylamine, as a succinyl-CoA trap, and a pyruvate kinase–lactate dehydrogenase ADP trap. Under these conditions, $[^{18}O]P_i \leftrightarrow$ succinate exchange would be a measure of interaction of $[^{18}O]P_i$ with enzyme-bound succinyl-CoA before release. According to the hypothesis, one would predict that ^{18}O exchange per molecule of succinyl-CoA released would be higher at low ATP concentrations than it would be at higher ATP. The rationale for this expectation is that higher ATP would favor release of succinyl-CoA, lowering the residence time of the latter on the enzyme and reducing the amount of ^{18}O exchanged. This is what was observed. In addition, the corresponding experiment was done with the dimeric pig heart enzyme at different GTP concentrations. In this case, no effect of GTP concentration on the extent of exchange was observed. In anticipation of the possibility that the succinyl-CoA trap would not completely eliminate released succinyl-CoA, the experiments were run at relatively high (96 μM) to low (0.3 μM) CoA concentrations. No significant differences were seen in ^{18}O exchanged per mole product released. It was concluded from this that exchange arising as a consequence of rebinding of succinyl-CoA that had escaped reaction with hydroxylamine was negligible. Vogel and Bridger (99) confirmed and extended these observations. They observed that the ^{18}O exchanged per mole of product formed was dependent on protein concentration. In view of the evidence that the tetrameric form of the enzyme seems to dissociate upon dilution (48), this observation appeared to strengthen the argument that the phenomenon was displayed by the tetramer. In addition, Vogel and Bridger (99) reported that the pig heart enzyme behaved like the *E. coli* enzyme at higher protein concentrations. This suggests that an aggregated form of the pig heart enzyme is capable of mediating cooperative effects. In an elaborate experiment, Wolodko et al. showed that the *E. coli* enzyme has a capacity for an alternating sites mechanism (109). Nishimura and Mitchell (66) observed that under the conditions used by Bild et al. (7), significant amounts of medium succinyl-CoA were maintained, despite the presence of the hydroxylamine trap. Thus, with an initial CoA concentration of 190 μM, apparent steady state succinyl-CoA concentrations of 13, 25, 58, and 74 μM were observed at initial ATP concentrations of 3.6, 20, 50, and 150 μM, respectively. The magnitude of the succinyl-CoA concentrations is not difficult to appreciate if one considers that the enzyme concentration (0.4 μM) is quite high and the presence of the ADP trap ensures that the reaction will be driven toward succinyl-CoA. Even at 6 μM CoA, approximately 5 μM succinyl-CoA was found under steady state condition (69). Succinate $\leftrightarrow$ succinyl-CoA exchange reactions were run at the succinyl-CoA concentrations found at the ATP concentrations described previously. One would expect that this exchange rate would parallel that of [^{18}O]P$_i$ $\leftrightarrow$ succinate exchange. In the absence of the ADP trap, the exchange reaction was stimulated by ATP, as expected. However, in the presence of the ADP trap, there was significant inhibition, which was greater at the higher ATP concentrations. On the basis of these experiments, an alternative explanation of the data of Bild et al. (7) was proposed. Thus, the rebinding of untrapped succinyl-CoA to the enzyme is a major factor in ^{18}O exchange of P$_i$ with succinate. In the presence of ATP, the ADP trap inhibits the exchange. The reason for this is that at these low concentrations of succinyl-CoA, ADP is required for binding of succinyl-CoA to the enzyme. The ADP that is inadvertently present in ATP solutions is converted to ATP by the ADP trap, shutting down succinyl-CoA binding. Since inhibition is greater at high ATP levels, it appears that ATP competes with succinyl-CoA for binding to the enzyme.

From this discussion, it is clear that more than one interpretation can be made of the data of Bild et al. (7). However, other data supporting the concept of alternating sites cooperativity with respect to *E. coli* succinyl-CoA synthetase have emerged. Some of this has been cited earlier. In addition, corroborating evidence has evolved

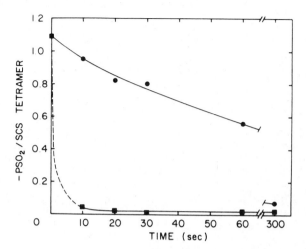

Figure 5. Effect of ATP on the rate of discharge of thiophosphate from thiophosphorylated succinyl-CoA synthetase by succinate and CoA. Thiophosphorylated succinyl-CoA synthetase was prepared by incubation of the enzyme with ATPγS followed by removal of the nucleotide. At zero time, thiophosphorylated succinyl-CoA synthetase was mixed with 10 mM succinate, 0.5-mM CoA, and, when present, 1 mM ATP. ■—■, with ATP; ●—●, without ATP. Details are described in ref. (107).

from two approaches, ^{31}P NMR and the use of [^{35}S]thiophosphoryl enzyme. Thus, Vogel and Bridger observed an NMR signal corresponding to succinyl phosphate only when ATP was added to a reaction solution containing succinate and phosphoryl enzyme (98,99). Wolodko et al. then made the important observation that ATPγS could serve as a substrate of the *E. coli* enzyme, albeit with a k_{cat} that was 6000 times lower than that observed with ATP (107). They also observed a relatively slow thiophosphorylation of the enzyme by ATPγS that was greatly stimulated by succinyl-CoA. The [^{35}C]thiophosphorylated enzyme was isolated and its reactivity with substrates tested, as shown in Fig. 5. A slow release of label was observed upon addition of succinate *plus* CoA [analogous to Eqs. (2b) and (2c)]. However, when ATP was added to the latter reaction, release of label was greatly stimulated. This observation was consistent with other site binding of ATP in affecting events at the thiophosphorylated site. A model for the phosphorylated enzyme based

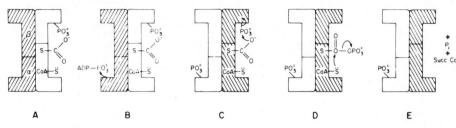

Figure 6.   Model for alternating sites catalytic cooperativity in the action of succinyl-CoA synthetase (107). The *shaded* and *unshaded* portions of the molecule represent αβ dimers with different conformations. *A*, E-PO₃·succinate·CoA complex, not in the optimum configuration for catalysis of transfer of the phosphoryl group from the enzyme to succinate. *B*, ATP interacts with the neighboring active site. Either following the phosphorylation of the adjacent site, or in concert with phosphorylation, the two halves of the molecule undergo reciprocal change in conformation. *C*, production of an active site on the right that promotes phosphoryl transfer, the step of the reaction that is believed to be rate limiting (98). *D*, displacement of phosphate by CoA; and *E*, release of the products from the enzyme.

on these findings is shown in Fig. 6. Nishimura and Mitchell (67) confirmed the findings of Wolodko et al. (107) and also observed that succinyl-CoA stimulated reaction of thiophosphoryl enzyme with ADP to form the corresponding analog of ATP. In this case, it did not appear that displacement of the label involved phosphorylation at the site to which succinyl-CoA was bound. Overall, the findings using ^{31}P NMR and thiophosphoryl enzyme appeared to be consistent with an alternating sites model for succinyl-CoA synthetase from *E. coli*. It was then of interest to examine the kinetics of reactivity of the thiophosphorylated pig heart enzyme, since this enzyme contains only one active site.

GTPγS proved to be a fairly efficient substrate of pig heart succinyl-CoA synthetase (68), in contrast to ATPγS with the *E. coli* enzyme. The k_{cat} for GTPγS was only 0.23 s^{-1}, compared with 65 s^{-1} for GTP as substrate. However, the K_m for GTPγS was 3 μM compared with 48 μM for GTP. Thus, k_{cat}/K_m for GTPγS was 7.7 × 10^4 M^{-1} s^{-1}, in contrast to 135 × 10^4 M^{-1} s^{-1} for GTP. By comparison, the corresponding values in the *E. coli* system were 71 for ATPγS and 5 × 10^6 M^{-1} s^{-1} for ATP (107). Reaction of GTPγS with the pig heart enzyme in the presence of Mg^{2+}, but in the ab-

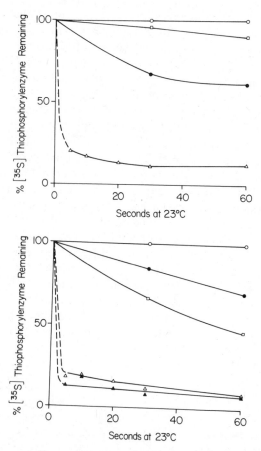

Figure 7. Release of [³⁵S]thiophosphoryl groups from [³⁵S]thiophosphoryl pig heart succinyl-CoA synthetase by GTP (top) and by CoA plus succinate (bottom). Top: Incubation solutions contained 100 mM Tris-HCl (pH 7.4), 100 mM KCl, 12 mM MgCl$_2$, and [³⁵S]thiophosphorylenzyme (1 mg/mL), in a final volume of 50 μL. Additions were as follows: –O–, none; □, 1- mM succinyl-CoA; ●, 0.1-mM GDP; △, 1-mM succinyl-CoA, and 0.1 mM GDP. Bottom: Incubation solutions were as described above, with the following additions: O, GDP trap; ●, 1 mM GTP; □, GDP trap, 50 mM succinate, and 0.1 mM CoA; ▲, 50 mM succinate, 0.1 mM CoA, and 1 mM GTP; △, 50 mM succinate, 0.1 mM CoA, 1 mM GTP, and GDP trap. The GDP trap consisted of 1 mM phosphoenolpyruvate, 0.4 mM NADH, 29 unit/mL of pyruvate kinase, and 63 unit/mL of lactate dehydrogenase. Following incubation, each reaction was quenched with 0.3 mL phenol. Washing and counting of the phenol layer was done as described (9,67).

sence of other substrates, resulted in the incorporation of approximately 0.8-mol thiophosphoryl group per mole enzyme. The isolated thiophosphoryl enzyme displayed some surprising reactions. As shown in Fig. 7, GTP stimulated reaction of [^{35}S]thiophosphorylated enzyme with succinate plus CoA and succinyl-CoA stimulated release of label when ADP was added, results that were virtually identical to those observed with the *E. coli* enzyme (67,107). These experiments were performed at 1-mg of enzyme protein/mL. Although ample evidence is at hand to indicate that the enzyme is an αβ dimer at this concentration, Vogel and Bridger (99) did report that at relatively high protein concentrations the enzyme exhibited properties similar to those exhibited by the *E. coli* enzyme in ^{18}O exchange studies described earlier. Therefore, the same experiments were conducted at protein concentrations of 0.1 and 0.01 mg/mL. Essentially the same results observed at 1 mg/mL were obtained. While this outcome does not necessarily rule out a cooperative alternating sites mechanism for the *E. coli* enzyme, the data for both the *E. coli* and pig heart enzymes can be rationalized by a comparatively simple displacement mechanism at a single site. This is explained in Fig. 8 for the phosphorylated enzyme. The upper path represents nucleoside triphosphate-stimulated succinyl-CoA syn-

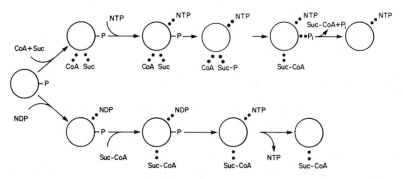

Figure 8. Model for same-site stimulation of succinyl-CoA and NTP synthesis by NTP and succinyl-CoA, respectively. Starting with E–P, the upper path shows binding of succinate *plus* CoA. Binding of NTP at this point stimulates succinyl-CoA formation via succinyl phosphate. Phosphorylation of the enzyme (not shown) could occur as a result of NTP binding. The lower path shows succinyl-CoA stimulation of reaction of NDP with E–P. In this case, no phosphorylation of the enzyme occurs.

thesis and release, and the lower path the succinyl-CoA stimulated synthesis and release of nucleoside triphosphate. The observation made by Vogel and Bridger (98,99) that ATP was needed for release of succinyl phosphate from phosphorylated enzyme and succinate can be rationalized by the upper path. The fact that nonhydrolyzable analogs were ineffective in mimicking the effect of ATP is not inconsistent with this hypothesis. Phosphorylation of the enzyme may provide the driving force for dissociation of the ternary complex (enzyme·ATP·succinyl phosphate) in the absence of CoA. It is of interest that, as Surendranathan and Hersh have reported for the malate thiokinase system (95), ATP randomly displaces one of two tightly bound succinyl-CoA molecules from the enzymatically active $(\alpha\beta)_2$ species, resulting in the incorporation of 1 mol of $P/(\alpha\beta)_2$.

V. Concluding Comments

It has been said that succinyl-CoA synthetase is an underrated enzyme (12). While other enzymes have received broad attention, succinyl-CoA synthetase has gone relatively unnoticed. A number of fundamental enzymological problems are approachable through study of this enzyme. The powerful techniques of ^{31}P NMR and fluorimetry have been established as excellent methods of monitoring molecular changes that come about as a result of binding of substrates. Thus, the presence of phosphohistidine (α subunit) and tryptophan (β subunit) presents the investigator with built-in reporter groups for these methods. Fluorimetry may be well suited for the scrutiny of fine changes in structure that occur during unfolding and refolding of the enzyme protein. X-ray crystallography of the enzyme gives promise of providing data that will complement that arising from other techniques and furnishing for the first time a correlation between relationships in fine structure and models that have been proposed for succinyl-CoA synthetase. It would also be desirable to know the amino acid sequence of the E. coli enzyme. Given the highly refined techniques of molecular biology, determination of the sequence of each polypeptide chain is quite feasible and should be undertaken. Thus, a highly informative future for succinyl-CoA synthetase can be anticipated. I hope that the approaches that have recently been instituted in the study of this enzyme activity will have borne much fruit by the time this subject is again reviewed.

Acknowledgments

The author thanks Professors A. Battle de Albertoni, W. A. Bridger, J. H. Ottaway, and P. D. J. Weitzman for reprints of their publications and for sharing unpublished research data. The author also thanks Ms. Pamela Lenow for typing this manuscript.

Work of the author that was cited in this review was supported in part by grants from the National Institutes of Health and the Robert A. Welch Foundation.

REFERENCES

1. Baccanari, D.P. and Cha, S. *J. Biol. Chem.*, **248**, 15–24 (1973).

2. Baccanari, D.P. and Cha, S., *Biochim. Biophys. Acta*, **334**, 226–234 (1974).

3. Ball, D.J. and Nishimura, J.S., *J. Biol. Chem.*, **225**, 10805–10812 (1980).

4. Beinert, H., Green, D.E., Hele, P., Hoffman-Ostenhof, O., Lynen, F., Ochoa, S., Popjak, G., and Ruyssen, R., *Science*, **124**, 614–616 (1956).

5. Benisek, W.F., *J. Biol. Chem.*, **246**, 3151–3159 (1971).

6. Benson, R.W., Robinson, J.L., and Boyer, P.D., *Biochemistry*, **8**, 2496–2502 (1969).

7. Bild, G.S., Janson, C.A., and Boyer, P.D. *J. Biol. Chem.*, **255**, 8109–8115 (1980).

8. Bowman, C.M. and Nishimura, J.S., *J. Biol. Chem.*, **250**, 5609–5613 (1975).

9. Boyer, P.D. and Bieber, L.L., *Methods Enzymol.*, **10**, 768–773 (1967).

10. Bridger, W.A., *Biochem. Biophys. Res. Commun.*, **42**, 948–954 (1971).

11. Bridger, W.A., in *The Enzymes*, Vol. X, P.D. Boyer, Ed. Academic Press, New York, 1974, pp. 581–606.

12. Bridger, W.A., *Can. J. Biochem.*, **59**, 1–8 (1981).

13. Bridger, W.A., Millen, W.A., and Boyer, P.D., *Biochemistry*, **7**, 3608–3616 (1968).

14. Bridger, W.A., Ramaley, R.F., and Boyer, P.D., *Methods Enzymol.*, **13**, 70–75 (1969).

15. Brownie, E.R. and Bridger, W.A., *Can. J. Biochem.*, **50**, 719–724 (1972).

16. Burnham, B.F., *Acta Chem. Scand.*, **17**, S123–S128 (1963).

17. Bush, L.P., *Plant Physiol.*, **44**, 347–350 (1969).

18. Buttlaire, D.H., Cohn, M., and Bridger, W. A., *J. Biol. Chem.*, **252**, 1957–1964 (1977).

19. Cha, S., *Methods Enzymol.*, **13**, 62–69 (1969).

20. Cha, S., Cha, C.-J.M., and Parks, R.E., Jr., *J. Biol Chem.*, **240**, PC3700–3702 (1965).

21. Cha, S., Cha, C.-J.M., and Parks, R.E., Jr., *J. Biol. Chem.*, **242**, 2577–2581 (1967).

22. Cha, S., Cha, C.-J.M., and Parks, R.E., Jr., *J. Biol. Chem.*, **242**, 2582–2592 (1967).

23. Cha, S. and Parks, R.E., Jr., *J. Biol. Chem.*, **239**, 1961–1967 (1964).

24. Cha, S. and Parks, R.E., Jr., *J. Biol. Chem.*, **239**, 1968–1977 (1964).

25. Chase, J.F.A. and Tubbs, P.K., *Biochem. J.*, **116**, 713–720 (1970).

26. Cohn, M., in *Phosphorus Metabolism*, Vol. I, W.D. McElroy and B. Glass, Eds., Johns Hopkins Press, Baltimore, 1951, pp. 374–376.

27. Collier, G.E. and Nishimura, J.S., *J. Biol. Chem.*, **253**, 4938–4943 (1978).

28. Collier, G.E. and Nishimura, J.S., *J. Biol. Chem.*, **254**, 10925–10930. (1979).

29. Gibson, J., Upper, C.D., and Gunsalus, I.C., *J. Biol. Chem.*, **242**, 2474–2477 (1967).

30. Grinnell, F.L. and Nishimura, J.S., *Biochemistry* **8**, 562–568 (1969).

31. Grinnell, F.L. and Nishimura, J.S., *Biochemistry*, **8**, 568–574 (1969).

32. Grubmeyer, C. and Penefsky, H.S., *J. Biol. Chem.*, **256**, 3728–3734 (1981).

33. Grunau, J.A., Knight, E., Hart, E.S., and Gunsalus, I.C., *J. Biol. Chem.*, **242**, 3531–3537 (1967).

34. Gunsalus, I.C. and Smith, R.A., in *Proceedings of the International Symposium on Enzyme Chemistry, Tokyo and Kyoto, 1957*, K. Ichihara, Ed., Academic Press, New York, 1958, pp. 77–86.

35. Hager, L.P., in *The Enzymes*, Vol. VI, Boyer, P.D., Ed., Academic Press, New York, 1962, pp. 387–399.

36. Hamilton, M.L. and Ottaway, J.H., *FEBS Lett.*, **123**, 252–254 (1981).

37. Hansford, R.G., *FEBS Lett.*, **31**, 317–32 (1973).

38. Hersh, L.B. and Surendranathan, K.K., *J. Biol. Chem.*, **257**, 11633–11638 (1982).

39. Hift, H., Ouellet, L., Littlefield, J.W., and Sanadi, D.R., *J. Biol. Chem.*, **204**, 565–575 (1953).

40. Hildebrand, J.G. and Spector, L.B., *J. Biol. Chem.*, **244**, 2606–2613 (1969).

41. Hultquist, D.E., Moyer, R.W., and Boyer, P.D., *Biochemistry*, **5**, 322–331 (1966).

42. Hutton, R.L. and Boyer, P.D., *J. Biol. Chem.*, **254**, 9990–9993 (1979).

43. Kaufman, S., *J. Biol. Chem.*, **216**, 153–164 (1955).

44. Kaufman, S. and Alivisatos, S.G., *J. Biol. Chem.*, **216**, 141–152 (1955).

45. Kaufman, S., Gilvarg, C., Cori, O., and Ochoa, S., *J. Biol. Chem.*, **203**, 869–888 (1953).

46. Kelly, C.J. and Cha, S., *Arch. Biochem. Biophys.*, **178**, 208–217 (1977).

47. Knight, E., Jr., Ph.D. thesis, University of Illinois,1961.

48. Krebs, A. and Bridger, W.A., *Can. J. Biochem.*, **52**, 594–598 (1974).

49. Lam, Y.-F., Bridger, W.A., and Kotowycz, G., *Biochemistry*, **15**, 4742–4748 (1976).

50. Leitzmann, C., Wu, J.-Y., and Boyer, P.D., *Biochemistry*, **9**, 2338–2346 (1970).

51. Lynn, R. and Guynn, R.W., *J. Biol. Chem.*, **253**, 2546–2553 (1978).

52. Matula, J.M., Ph.D. thesis, University of Texas Health Science Center at San Antonio, 1979.

53. Matula, J.M. and Nishimura, J.S., *Int. J. Biochem.*, **9**, 213–215 (1978).

54. Meshkova, N.P. and Matveeva, L.N., *Biokhimiya*, **35**, 374–381 (1970).

55. Moffet, F.J. and Bridger, W.A., *J. Biol. Chem.*, **245**, 2758–2762 (1970).

56. Moffet, F.J. and Bridger, W.A., *Can. J. Biochem.*, **51**, 44–55 (1972).

57. Moffet, F.J., Wang, T., and Bridger, W.A., *J. Biol. Chem.*, **247**, 8139–8144 (1972).

58. Moyer, R.W., Ramaley, R.F., Butler, L.C., and Boyer, P.D., *J. Biol. Chem.*, **242**, 4299–4309 (1967).

59. Murakami, K., Mitchell, T., and Nishimura, J.S., *J. Biol. Chem.*, **247**, 6247–6252 (1972).

60. Murakami, Y. and Nishimura, J.S., *Biochim. Biophys. Acta*, **336**, 252–263 (1974).

61. Nandi, D.L. and Waygood, E.R., *Can. J. Biochem. Physiol.*, **43**, 1605–1614 (1965).

62. Nishimura, J.S., *Biochemistry*, **6**, 1094–1099 (1967).

63. Nishimura, J.S. and Grinnell, F., *Advan. Enzymol.*, **36**, 183–202 (1972).

64. Nishimura, J.S., Kenyon, G.L., and Smith, D.J., *Arch. Biochem. Biophys.*, **170**, 461–467 (1975).

65. Nishimura, J.S. and Meister, A., *Biochemistry*, **4**, 1457–1462 (1965).

66. Nishimura, J.S. and Mitchell, T., *J. Biol. Chem.*, **259**, 2144–2148 (1984).

67. Nishimura, J.S. and Mitchell, T., *J. Biol. Chem.*, **259**, 9642–9645 (1984).

68. Nishimura, J.S. and Mitchell, T., *J. Biol. Chem.* **260**, 2077–2079 (1985).

69. Nishimura, J.S. and Mitchell, T., unpublished work.

70. Nishimura, J.S., Mitchell, T., Collier, G.E., Matula, J.M., and Ball, D.J., *Eur. J. Biochem.*, **136**, 83–87 (1983).

71. Nishimura, J.S., Mitchell, T., Collier, G.E., and Prasad, A.R.S., *Biochem. Int.*, **1**, 339–345 (1980).

72. Nishimura, J.S., Mitchell, T., and Grinnell, F., *J. Biol. Chem.*, **248**, 743–748 (1973).

73. Nishimura, J.S., Mitchell, T., Hill, K.A., and Collier, G.E., *J. Biol. Chem.*, **257**, 14896–14902 (1982).

74. Nishimura, J.S., Mitchell, T., and Matula, J.M., *Biochem. Biophys. Res. Commun.*, **69**, 1057–1064 (1976).

75. Nishimura, J.S., Ybarra, J., Mitchell, T., and Horowitz, P., in preparation.

76. Nishimura, J.S., Ybarra, J., and Prasad, A.R.S., *Fed. Proc.*, **43**, 1545 (1984).

77. Ottaway, J.H., *Int. J. Biochem.*, **13**, 401–410 (1981).

78. Owens, M.S. and Barden, R.E., *Arch. Biochem. Biophys.*, **187**, 299–306 (1978).

79. Owens, M.S., Clements, P.R., Anderson, D.A., and Barden, R.E., *FEBS Lett.*, **124**, 151–154 (1981).

80. Palmer, J.M. and Wedding, R.T., *Biochim. Biophys. Acta*, **113**, 167–174 (1966).

81. Pearson, P.H. and Bridger, W.A., *J. Biol. Chem.*, **250**, 4451–4455 (1975).

82. Pearson, P.H. and Bridger, W.A., *J. Biol. Chem.*, **250**, 8524–8529 (1975).

83. Porpaczy, Z., Sumegi, B., and Alkonyi, I., *Biochim. Biophys. Acta*, **749**, 172–179 (1983).

84. Prasad, AAR.S., Horowitz, P.M., and Nishimura, J.S., *Biochemistry*, **22**, 4272–4275 (1983).

85. Prasad, A.R.S., Nishimura, J.S., and Horowitz, P.M., *Biochemistry*, **22**, 5142–5147 (1982).

86. Prasad, A.R.S., Ybarra, J., and Nishimura, J.S., *Biochem., J.*, **215**, 513–518 (1983).

87. Ramaley, R.F., Bridger, W.A., Moyer, R.W., and Boyer, P.D., *J. Biol. Chem.*, **242**, 4287–4298 (1967).

88. Rosenthal, L.P., Hogenkamp, H.P.C., and Bodley, J.W., *Carbohydr. Res.*, **111**, 85–91 (1982).

89. Schwartz, H., Steitz, H.-O., and Radler, F., *Antonie van Leeuwenhoek J.*, **49**, 69–78 (1983).

90. Severin, S.E., Meshkova, N.P., Matveeva, L.N., and Mikeladze, D.G., *Dok. Akad. Nauk SSSR* [Sou. Phys. Dokl.], **227**, 1010–1013 (1976).

91. Smith, R.A., Frank, I.R., and Gunsalus, I.C., *Fed. Proc.*, **16**, 251 (1957).

92. Steiner, A.W. and Smith, R.A., *J. Neurochem.*, **37**, 582–593 (1981).

93. Srere, P.A., *Trends Biochem. Sci.*, **6**, 4–6 (1981).

93a. Srere, P.A., personal communication (1984).

94. Sumegi, B. and Srere, P., *J. Biol. Chem.*, **259**, 8748–8752 (1984).

95. Surendranathan, K.K. and Hersh, L.B., *J. Biol. Chem.*, **258**, 3794–3798 (1983).

96. Teherani, J.A. and Nishimura, J.S., *J. Biol. Chem.*, **250**, 3883–3890 (1975).

97. Upper, C.D., Ph.D. thesis, University of Illinois, 1964.

98. Vogel, H.J. and Bridger, W.A., *J. Biol. Chem.*, **257**, 4834–4842 (1982).

99. Vogel, H.J. and Bridger, W.A., *Biochem. Soc. Trans.*, **11**, 315–323 (1983).

100. Vogel, H.J., Bridger, W.A., and Sykes, B.D., *Biochemistry*, **16**, 1126–1132 (1982).

101. Walsh, C.T., Jr., Hildebrand, J.G., and Spector, L.B., *J. Biol. Chem.*, **245**, 5699–5708 (1970).

102. Wang, T., Jurasek, L., and Bridger, W.A., *Biochemistry*, **11**, 2067–2070 (1972).

103. Weitzman, P.D., *Soc. Appl. Bacteriol. Symp. Ser.*, **8**, 107–125 (1980).

104. Weitzman, P.D. and Jaskowska-Hodges, H., *FEBS Lett.*, **143**, 237–240 (1982).

105. Wider, E.A. and Tigier, H.A., *Enzymologia*, **41**, 217–231 (1971).

106. Wolodko, W.T., Brownie, E.R., and Bridger, W.A., *J. Bacteriol.*, **143,** 231–237 (1980).
107. Wolodko, W.T., Brownie, E.R., O'Connor, M.D., and Bridger, W.A., *J. Biol. Chem.*, **258,** 14116–14119 (1983).
108. Wolodko, W.T., James, M.N.G., and Bridger, W.A., *J. Biol. Chem.*, **259,** 5316–5320 (1984).
109. Wolodko, W.T., O'Connor, M.D., and Bridger, W.A., *Proc. Natl. Acad. Sci, USA,* **78,** 2140–2144 (1981).
110. Ybarra, J., Prasad, A.R.S., and Nishimura, J.S., submitted.

β-LEUCINE AND THE β-KETO PATHWAY OF LEUCINE METABOLISM

J. MICHAEL POSTON, *Laboratory of Biochemistry, National Heart, Lung, and Blood Institute, National Institutes of Health, Bethesda, Maryland 20205*

CONTENTS

I. Introduction

The first study of the metabolism of leucine was that reported by Proust in 1819 (1). He was investigating the principle to which various types of cheese owe their flavor. In addition to the production of carbon dioxide, ammonia, and acetic acid, he noted the presence of an "oxide caséeux." Examination of this oxide caséeux showed it to be organic and he noted its solubility in various solvents and its precipitability by salts of heavy metals. It is now quite clear that what he isolated was leucine, although it was not a pure preparation.

173

Henri Braconnot (2,3) hydrolyzed muscle fiber and wool in 1820 and isolated a white material that he named leucine (from the Greek λευκόσ, white).

The excellent review of Vickery and Schmidt (4) describes the work of Proust and Braconnot as well as that of von Gorup-Basenez who first found valine in extracts of pancreas in 1856 (5) and the work of Ehrlich (6) who isolated isoleucine from sugar beet molasses. In addition, the review contains detailed accounts of the discoveries of all the other amino acids in proteins (as recognized in 1931).

Many of the early studies were descriptive or dealt primarily with the isolation of the amino acid from crude material. It was not until 1906 that Embden et al. (7) [quoted by Meister (8)] recognized that leucine and isovaleric acid gave rise to strong ketone body formation. The later studies of Dakin (9), Ringer et al. (10), and Cohen (11) on ketone body formation led to extensive examination of the fate of the carbon atoms of the leucine skeleton. These latter studies were greatly facilitated by the introduction of isotopic tracer analysis and are discussed in the treatise by Meister (8).

Rose (12) and his students demonstrated that leucine was essential for maintenance for nitrogen balance in rats and later showed that leucine was essential for nitrogen balance in humans as well (13–15).

In 1954, Menkes et al. (16) published the details of four infant siblings who shared a syndrome characterized by rapidly progressive cerebral dysfunction and the passage of urine with an odor strikingly similar to that of maple syrup. Westall et al. (17) were the first to describe the elevated levels of the branched-chain amino acids in the plasma and urine of children with maple syrup urine disease, or MSUD as the syndrome has come to be known. They also showed that there were high levels of the corresponding branched-chain keto acids in the body fluids and later showed that the blockage in MSUD is at the ketodecarboxylase step. These findings from MSUD patients, coupled with the earlier studies of isotopic distribution in products of leucine metabolism have led to the establishment of the metabolic pathway for the branched-chain amino acids shown in Fig. 1. Several reviews of the work associated with the metabolic pathway have been published (8,18–20), in particular, the clinical aspects of MSUD and its related syndromes, isovalericacidemia are re-

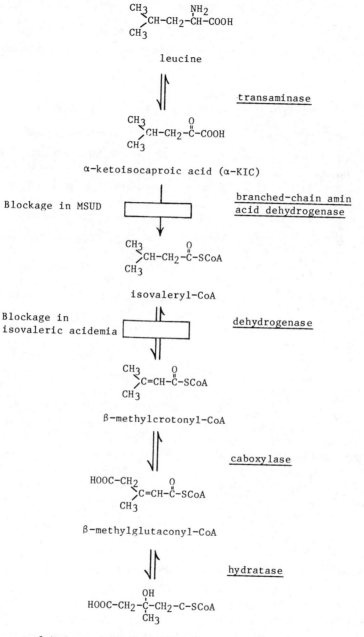

Figure 1. The catabolic pathway of leucine metabolism.

viewed by Tanaka and Rosenberg (21) and the disorders of propionate and methylmalonate metabolism are reviewed by Rosenberg (22).

II.   The β-Keto Pathway of Leucine Metabolism

A.   DISCOVERY OF AN UNUSUAL METABOLITE OF LEUCINE

A strain of *Clostridium sporogenes* was isolated from mud from the bank of the River Doule in the Amazon basin of Ecuador (23,24). The mud had been enriched with L-leucine and, although the organism grew minimally in single amino acid fermentations, it grew luxuriently in rich media. When the cultures of the strain of *C. sporogenes* were exposed to radiolabeled L-leucine, a variety of radioactive metabolites were recovered. One of these was isobutyrate. This finding was puzzling, since it seemed clear from the metabolic pathway in Fig. 1 that isobutyrate was an unlikely primary metabolite and the labeling was sufficiently strong to preclude secondary recycling of the metabolites. When the spent medium was treated with 2,4-dinitrophenylhydrazine, a hydrazone of methyl isopropyl ketone was obtained. Similar results were obtained from incubations with cell-free extracts of the bacterium.

Because lysine metabolism had been shown to involve migrations of the amino groups, it was reasoned that perhaps the amino group of leucine was being moved from the α carbon to the β carbon. If the amino group was then removed, the resulting keto acid would react with 2,4-dinitrophenylhydrazine to yield the hydrazone. But, because β-keto acids are unstable in strong acid, the end product of the reaction would be the derivative of the product of decarboxylation, methyl isopropyl ketone. Moreover, it was recognized that the β-keto acid could give rise to isobutyrate by a cleavage reaction. This reasoning allowed a metabolic scheme to be proposed (Fig. 2).

B.   ORIGINS OF β-LEUCINE

In the early 1950s there was a general search for antimetabolites and other materials that might serve as useful drugs for the treatment of infections. In Yugoslavia, Dvornik (25) synthesized β-leucine as a possible antimetabolite. When his synthetic material was tested on several gram-negative and gram-positive organisms, no antime-

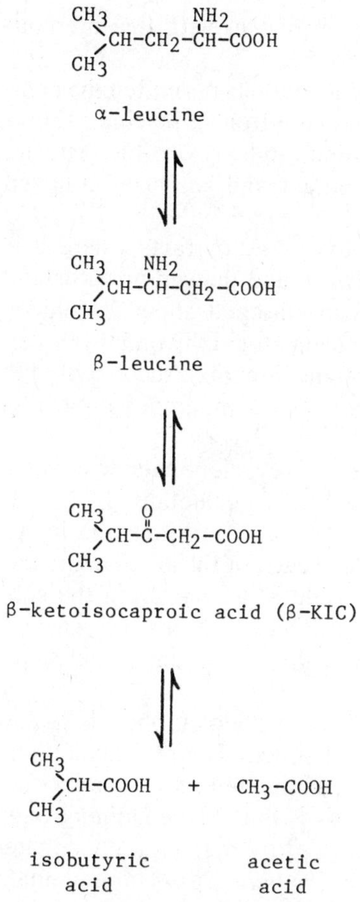

CH$_3$ NH$_2$
>CH–CH$_2$–CH–COOH
CH$_3$

α-leucine

⇅

CH$_3$ NH$_2$
>CH–CH–CH$_2$–COOH
CH$_3$

β-leucine

⇅

CH$_3$ O
>CH–C–CH$_2$–COOH
CH$_3$

β-ketoisocaproic acid (β-KIC)

⇅

CH$_3$
>CH–COOH   +   CH$_3$–COOH
CH$_3$

isobutyric   acetic
  acid   acid

Figure 2.   The proposed novel pathway of leucine metabolism.

tabolic activity was observed. Indeed, there was marked stimulation of growth of the organisms. As a result, the compound was abandoned as a therapeutic agent.

C.   INTERRELATION OF LEUCINE AND β-LEUCINE

When radiolabeled leucine was incubated with crude cell-free extracts of *C. sporogenes* in the presence of a pool of synthetic DL-β-leucine, it was possible to isolate radioactive β-leucine (24). This offered support for the scheme shown in Fig. 2, since the trapping of label in the pool of β-leucine was consistent with the interconversion of the two isomers of leucine.

Incubation of DL-β-leucine with crude extracts permitted the demonstration of the net production of leucine from β-leucine. Therefore, the mutation reaction was presumed to be reversible. The activity was named leucine 2,3-aminomutase and has been assigned the number, EC 5.4.3.7 (26).

The enzyme has been studied in the crude extracts where it is stable for several days and can be frozen and thawed or withstand 60°C for 10 min. Although it has been enriched about 20-fold by ammonium sulfate precepitation and chromatography on DEAE-cellulose or with hydrophobic interaction medium (42), the activity becomes unstable upon partial purification and undergoes rapid loss of activity.

Enzyme is assayed in the presence of CoA, nicotinamide adenine dinucleotide NAD^+, and flavin adrenine dinucleotide, FAD. Although it is doubtful that these cofactors play a direct role in the reaction catalyzed by the aminomutase, each of these cofactors has been found to provide an increased yield of leucine from the substrate, β-leucine (24). Possibly they are necessary for the regeneration of coenzyme B_{12}. Clarification of this uncertainty must await study of purified enzyme.

In addition to the cofactors discussed, pyridoxal phosphate and either glutamate or α-ketoglutarate are added. The pyridoxal phosphate gives some stimulation of the reaction and may actually play an intimate role in the mutation. The α–β shift of the amino groups of both lysine and ornithine are catalyzed by pyridoxal phosphate-dependent enzymes of *Clostridium sticklandii* (28). The glutamate or α-ketoglutarate addition serves to provide a donor or acceptor of amino groups that may interact with a transaminase in the metabolism of β-leucine.

D. INVOLVEMENT OF ADENOSYLCOBALAMIN

The migration of some of the amino groups in the catabolism of lysine and ornithine has been shown to be catalyzed by enzymes that use adenosylcobalamin (coenzyme B_{12}, AdoCbl) as a cofactor (28). In order to test whether or not the mutation of leucine to β-leucine was similarly catalyzed, crude extracts of *C. sporogenes* were incubated with purified porcine intrinsic factor. (Mammals can-

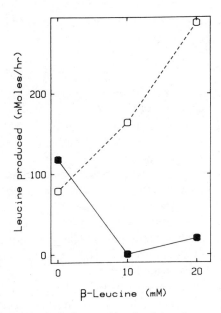

Figure 3. The effect of intrinsic factor of leucine 2,3-aminomutase The conditions of incubation were those of Table II in which the enzyme was a crude extract of *C. sporogenes*. Symbols: O—O, no addition; ●—●, intrinsic factor added.

not absorb cobalamins from the lumen of the intestine in significant quantities unless they are first bound to a mucoprotein, intrinsic factor, which is secreted by the gastric mucosa. By incubating the extracts with intrinsic factor, free cobalamins are effectively removed from the reaction mixture due to the extremely tight binding between the cobalamin and the intrinsic factor. Thus, any inhibition of the reaction that may result is consistent with an involvement of cobalamin in the catalysis.) Intrinsic factor was found to cause striking inhibition of the production of leucine from β-leucine as shown in Fig. 3. Addition of AdoCbl restored the activity. Thus, it was clear that the enzymic activity involved a B_{12} cofactor and it may be inferred that the binding of the coenzyme to the enzyme is less tight than that between the coenzyme and intrinsic factor. This is the first instance of an α–β shift involving B_{12}. Although such shifts at more distal sites on the carbon chain of amino acids have been shown to be B_{12} dependent, previously described α–β shifts have not implicated cobalamins as cofactors.

E. OCCURENCE OF LEUCINE 2,3-AMINOMUTASE

The enzyme that causes the mutation of leucine has been detected in a wide variety of organisms. Initially found in *C. sporogenes* (24), it has been found as well in *C. lentoputrescens* (24), in the livers of rat, sheep, and rhesus and African green monkeys (24). It has been demonstrated in human hair roots and leukocytes (29) and in human liver.* It has been observed in plants such as potato tubers and spinach (30), navy beans (*Phaseolus vulgaris*), and annual rye grass (*Lolium* spp.) (31). In addition, leucine 2,3-aminomutase has been demonstrated in the yeast *Candida utilis* (32). In each of these instances, the enzyme has been shown to have a requirement for coenzyme B_{12} (AdoCbl). An interesting exception to this trend, however, has been reported by a Scottish laboratory; Freer et al. (33) have described the stereochemistry of the mutase reaction catalyzed by cell-free extracts of tissue cultures of the plant *Andographis paniculata*. They found that the activity mediates the equilibrium between (2*S*)-α-leucine and (3*R*)-β-leucine and that the activity does not depend on coenzyme B_{12}. AdoCbl did not appear to stimulate the reaction nor was there inhibition by intrinsic factor.

F. DISTRIBUTION OF LEUCINE 2,3-AMINOMUTASE IN THE RAT

Examination of various organs of the rat revealed (34) that the activity of the enzyme is present throughout the body of the animal, as shown in Table I. [The gas chromatographic analysis used was that of Gehrke and Leimer (35)]. The greatest specific activity is that in muscle, skeletal muscle being somewhat higher than cardiac muscle. The distribution in the subcellular organelles of the liver showed that the greatest activity was found in the cytosol. This is particularly surprising in view of the virtually total localization of the pathway of Fig. 1 in mitochondria. There is some leucine 2,3-aminomutase activity in the mitochondria, but the specific activity of the cytosol is 100 times that found in the mitochondria.

---

* From a 12-year old child with leukemia who died of sepsis (27).

TABLE I
Distribution of Leucine 2,3-Aminomutase in the
Rat[a,b]

Tissue	Leucine formed (nmol/h (mg protein)
Brain	204
Heart	421
Kidney	76
Liver	100
Nuclear fraction	29
Mitochondrial fraction	2
Microsomal fraction	11
Cytosol	290
Skeletal muscle	525
Small intestine	62

[a] Each reaction mixture (0.25 mL) contained 100 mM triethanolamine/HCl buffer, pH 8.5, 5 × 10^{-7} M FAD, 0.5 mM CoA, 0.5 mM NAD$^+$, 0.5 mM pyridoxal phosphate, 20 mM glutamate, and adenosylcobalamin was 5.8 × 10^{-8} M. After incubating in the dark in open tubes for 60 min, the reactions were stopped by the addition of 0.025 mL 6 N HCl and 0.0125 mL of 10% (w/v) NaWO$_4$ and centrifuged; 0.05 mL of the protein-free supernatant solution was taken for gas chromatographic analysis by the method of Gehrke and Leimer (35).
[b] Taken from ref. (34).

G.   PRODUCTION OF β-KETOISOCAPROIC ACID (β-KIC)

When crude extracts of *C. sporogenes* were incubated with L-α-[U^{-14}C] leucine in the presence of β-ketoisocaproic acid, β-KIC, the keto-acid pool was labeled and radioactive 2,4-dinitrophenylhydrazine of methyl isopropyl ketone could be prepared from the incubations (24). This finding was consistent with the idea that leucine could be catabolized by the series of reactions shown in Fig. 2. Liver extracts were also capable of this type of conversion. Similarly, when radioactive β-leucine was incubated with crude extracts of rat liver, radiolabel appeared in the non-CO$_2$ anionic fractions. Thus it seemed clear that the pathway was probably constituted as outlined in Fig. 2.

H.   CLEAVAGE OF β-KETOISOCAPROIC ACID

It was reasoned that the cleavage of the β-KIC should be similiar to the cleavage of acetoacetic acid (β-KIC may be considered as 4,4-dimethylacetoacetic acid). Incubation of extracts with β-KIC, CoA, and a variety of additional biochemicals did not yield any evidence of a thiolytic cleavage of the β-KIC and it seemed probable that there should be an activation of the β-KIC, most likely production of a thioester. There was no evidence however, for an ATP-dependent esterification. Accordingly, effort was devoted to finding some other type of activation.

When extracts of liver were incubated with free β-KIC, succinyl-SCoA, and magnesium ion, there was a detectable change in the optical density of the reaction mixture when monitored at 303 nm. This wavelength is characteristic of the magnesium–enol complex of β-keto thioesters (37). The activity detectable in liver extracts was not especially strong, but when extracts from the kidney were examined, it was found that the organ had a vigorous production of β-KIC thiol ester. This distribution of activity was reminiscent of that of 3-keto acid CoA-transferase (EC 2.8.3.5) and when commercial preparations of pure porcine-heart enzyme were tested, it was clear that the β-KIC was a satisfactory substrate for the enzyme (36,37). Purification of the transferase activity showed it to be identical to the commercial enzyme.

I.   THIOLASE

The CoA thioester of β-KIC was prepared (38) and incubated in the presence of liver extract supplemented with magnesium ion and free CoA. It was possible to demonstrate (39) a CoA-dependent loss of absorption at 303 nm. This loss was consistent with a thiolytic cleavage of the β-KIC-SCoA. The enzymic activity was purified to near electrophoretic homogeneity. It was shown to be active against both β-KIC-SCoA and acetoacetyl-SCoA and appeared to be a member of the thiolase I class of general thiolase enzymes (40).

As a result of the findings concerning the activation and cleavage of the β-ketoisocaproic acid it was possible to write the course of metabolism of leucine by way of this new pathway as

leu ⇄ β-leu ⇄ β-KIC ⇄ β-KIC-SCoA ⇄ isobutyryl-SCoA + acetyl-SCoA

Table   II

Formation of Leucine and β-Leucine by Human Leukocytes[a]

Substrate	Intrinsic factor	No cobalamin added		Adenosylcobalamine added	
		Leucine found[c]	β-Leucine found[c]	Leucine found[c]	β-Leucine found[c]
β-Leucine	−	233		320	
β-Leucine	+	166		246	
L-Valine	−	70	0	211	0
L-Valine	+	0	254	473	0

[a] Taken from ref. (29).

[b] Reaction conditions were those indicated in Table I. When added intrinsic factor was 0.2 mg/mL and adenosylcobalamin was 5.8×10^{-8} M. protein was 5.7 mg/mL.

[c] Units are nmol/mL reaction.

III.   Involvement of β-Leucine Metabolism and Other Metabolic Pathways

A.   INTERCONVERSION OF LEUCINE AND VALINE

It has long been assumed that there is no interconversion of the branched-chain amino acids. In light of the production of isobutyryl-SCoA from leucine by way of the β-leucine or β-keto pathway and the production of the same compound from the catabolism of valine (29), it seemed reasonable to see whether mammals could carry out such an interconversion. When rat livers were incubated with radiolabeled isobutyryl-SCoA, labeled α-leucine was produced. Similar incubations carried out with human leukocytes, showed that radiolabeled leucine was produced in a substrate-dependent fashion (29). The incubation of human leukocytes with L-valine showed that the production of leucine from that substrate was stimulated by the addition of AdoCbl and that the complete inhibition by intrinsic factor could be overcome by the addition of more coenzyme B_{12} (Table II). Such results are consistent with the interpretation that mammals, specifically man, can convert valine into leucine. The interconversion is shown in Fig. 4. These findings were supported and extended by the work of Monticello and Costilow (41) who showed that $C.$ $sporogenes$ can incorporate tritium from [4,5-^{3}H]leucine into valine

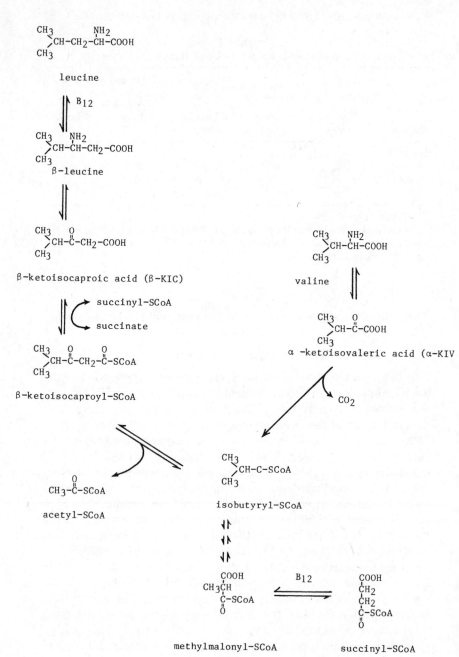

Figure 4. The interrelation of the pathways of leucine, β-leucine, and valine.

184

and that this organism can convert [^{14}C]leucine into valine and [^{14}C]valine into leucine.

The observations of these two laboratories establishes the interconversion of leucine and valine. While it is not difficult to understand that valine can give rise to leucine by the general reversal of the pathway through β-leucine, how the reverse conversion is accomplished remains obscure. There is little evidence, other than that of Monticello and Costilow (41), to support a carboxylation of isobutyryl-SCoA. Such a reaction is required if the biosynthesis of valine can proceed via the same pathway that is used for its catabolism.

B. ISO-FATTY ACIDS AS AN ALTERNATE SOURCE OF THE LEUCINE CARBON CHAIN

Another source of isobutyryl-SCoA and, therefore, another source of leucine and β-leucine is the branched-chain fatty acids (29). When isoacids such as isocaproate, isomyristate, or isostearate were incubated with rat liver extracts, there was a B_{12}-dependent oxidation of the iso-fatty acid yielding β-ketoisocaproyl-SCoA which, of course, could then be converted to β-leucine. When intrinsic factor was added to an experiment using isostearate as substrate, the production of leucine was abolished and β-leucine accumulated. These results suggest that one result of the catabolism of iso-fatty acids may be the supplementation of the leucine pool (29).

IV. Clinical Aspects

A. THE β-LEUCINE PATHWAY IN COBALAMIN DEFICIENCY STATES

The observation that adenosylcobalamin is a cofactor for leucine 2,3-aminomutase prompted an examination of serum from patients suffering from pernicious anemia. In this disease, there are a variety of abnormal conditions (among others, a lack of gastric secretion of intrinsic factor, achlorhydria, and, frequently, the presence of circulating antibodies to intrinsic factor) that contribute to the individual's inability to absorb cobalamins from the intestine. This inability to make use of dietary cobalamin manifests itself in the classic megaloblastic, macrocytic anemia of cobalamin deficiency. A common finding in patients with severe or prolonged pernicious anemia is a

TABLE   III
Levels of Branched-Chain Amino Acids in Serum[a]

Serum	β-Leucine[b]	Leucine[b]	Isoleucine[b]	Valine[b]
Normal (n = 37)	4.8 ± 3.1	81.2 ± 24.1	58.7 ± 22.9	136.5 ± 24.8
Pernicious anemia (n = 17)	24.7 ± 12.4	52.0 ± 18.2	65.2 ± 16.1	127.1 ± 24.1

[a] Taken from ref. (28).
[b] Units are μM.

characteristic neurologic dysfunction of the spinal cord. This "sub-acute combined degeneration" of the spinal cord, which may sometimes precede the onset of hematologic signs, has a gradual onset with paresthesia and weakness progressing to leg stiffness and abnormal propioception, often expressing itself in the patient's broad-based gait. The principal degeneration is in the peripheral nerves and in the myelinated tracts of the posterior column of the spinal cord. Some degeneration may be seen in cortical neurons as well.

Since the discovery of vitamin B_{12} and its wide availability as an injectable therapy for pernicious anemia, the incidence of untreated pernicious anemia is relatively rare. By good fortune, serum was provided by Professor A. V. Hoffbrand of the Royal Postgraduate Medical School.* His patients all had pernicious anemia. Seventeen sera were examined and found to have a generally elevated level of β-leucine as shown in Table III (29).

The finding that there was an accumulation of β-leucine when leucine 2,3-aminomutase was blocked by the deficiency of AdoCbl, came as a great surprise, for it had been assumed that the β-leucine pathway was catabolic in nature. Instead, the finding prompted the notion that the pathway must be biosynthetic. Although Rose and his students have shown leucine to be essential for man (12–15) that determination was based on nitrogen balance studies. Such studies showed that leucine was necessary to maintain nitrogen balance, but

* These sera were originally drawn for Dr. S. Harvey Mudd of the National of Mental Health, Bethesda, Maryland, who provided aliquots of them for the estimation of the branched-chain amino acid content.

they did not address the question of whether or not man was capable of limited biosynthesis of leucine. It now seems clear than man can synthesize some leucine in the β-keto pathway, although not sufficent to satisfy the needs of the body.

The interaction of iso-fatty acid metabolism with leucine metabolism raises the interesting probability that the leucine metabolism may somehow be involved with maintenance of normal nerve tissue. Frenkel (42) showed that acids with odd numbers of carbon atoms are biosynthesized from propionate in nerve tissue. The effects of vitamin B_{12}-deficiency in fruit bats has shown that there is a marked increase in 15:1 fatty acids in cholineglycerophosphatide; no evidence was presented concerning iso-fatty acids (43). Since methyl malonate and propionate accumulate to varying degrees in pernicious anemia (because of the varying inactivity of the AdoCbl-dependent enzyme, methylmalonyl-SCoA mutase), it is possible to speculate whether the findings of Frenkel and those involving β-leucine point the way to a possible understanding of the neurological involvement of pernicious anemia. It may be that the increased levels of branched-chain fatty acids promote the decreased synthesis of normal fatty acids and interfere with the integrity or renewal of the myelin. Alternatively, it might also be that myelin with increased levels of iso- and odd-carbon-fatty acids may result in defects in neuronal function.

B.   INCIDENCE OF β-LEUCINE IN OTHER DISEASE STATES

Inborn errors in metabolism have intrigued researchers ever since Garrod discussed them with reference to specific syndromes in his famous Croonian lectures of 1908 (44) and his monograph of 1909 (45). Study of these errors has often provided the key that opened up the understanding of a metabolic pathway. The observation of the dysfunction of the metabolism of the branched-chain amino acids in maple syrup urine disease provided the clue that led to the establishment of what had been a postulated pathway of leucine metabolism.

Little has been observed that would relate the metabolism by way of β-leucine and the β-keto pathway of leucine metabolism to any disease state other than pernicious anemia. In that ailment, of course, the proximal cause is simply lack of a cofactor. No evidence has been uncovered that can firmly link an error in the metabolism of β-leucine to any syndrome. There are some observations, however, that may eventually have some bearing on this matter. An inborn error that would seem to be an obvious candidate for demonstrating a relationship with β-leucine would be maple syrup urine disease. When serum from genuine MSUD patients has been examined, no β-leucine has been found (27,29). It is not at all certain whether this represents an inoperative β-keto pathway or a consequence of the shutdown of leucine 2,3-aminomutase by the very high levels of circulating leucine seen in that disease. It will be necessary to examine tissues or tissue cultures of MSUD patients to establish whether or not there is, in fact, a relationship.

References

1. Proust, *Ann. Chim. Phys.*, **10**, [2] 29–49 (1819) [quoted in ref. (4)].

2. Braconnot, H., *Ann. Chim. Phys.*, **13**, [2] (1820) [quoted in ref. (4)].

3. Braconnot, H., *Ann. Chim. Phys.*, **36**, [2] (1827) [quoted in ref. (4)].

4. Vickery, H.B. and Schmidt, C.L.A., *Chem. Rev.*, **9**, 169–318 (1931).

5. von Gorup-Besanez, E., *Ann.*, **98**, 1–43 (1856) [quoted in ref. (4)].

6. Ehrlich, F., *Ber.*, **37**, 1809–1840 (1904) [quoted in ref. (8)].

7. Embden, G., Salomon, H., and Schmidt, F., *Beitr. Chem. Physiol. Pathol.*, **8**, 129 (1906) [quoted in ref. (8)].

8. Meister, A., in *Biochemistry of the Amino Acids*, Vol. II, 2nd ed., Academic, New York 1965, pp. 729–757 and 1051–1053.

9. Dakin, H.A., *J. Biol. Chem.*, **14**, 321–333 (1913).

10. Ringer, A.I., Frankel, E.M., and Jonas, L., *J. Biol. Chem.*, **14**, 525–538 (1913).

11. Cohen, P.P., *J. Biol. Chem.*, **119**, 333–346 (1937).

12. Rose, W.C., *Physiol. Revs.*, **18**, 109–136 (1938).

13. Rose, W.C., *Fed. Proc.*, **8**, 546–552 (1949).

14. Rose, W.C., Eades, C.H., Jr., and Coon, M.J., *J. Biol. Chem.*, **216**, 225–234 (1955).

15. Rose, W.C., Wixom, R.L., Lockhart, H.B., and Lambert, G.F., *J. Biol. Chem.*, **217**, 987–1004 (1955).

16. Menkes, J.H., Hurst, P.L., Craig, J.M., *Pediatrics*, **14**, 462–466 (1954).

17. Westall, R.G., Dancis, J., Miller, S., *Am. J. Dis. Child*, **94**, 571–572 (1957).

18. Walzer, M. and Williamson, J.R., Eds., *Metabolism and Clinical Implications of Branched Chain Amino and Ketoacids* Proceedings of the International Symposium held at Kiawah Island Conference Center, Charleston, S.C., Nov. 15–16, 1980) *Developments in Biochemistry*, Vol. 18, Elsevier/North-Holland, New York, 1981).

19. Adibi, S.A., *Metabolism*, **25**, 1287–1302 (1976).

20. Adibi, S.A., *J. Lab. Clin. Med.* **95**, 475–484 (1980).

21. Tanaka, K. and Rosenberg, L.E., in *The Metabolic Basis of Inherited Disease*, J.B. Stanbury, J.B. Wyngaarden, D.S. Fredrickson, J.L. Goldstein, and M.S. Brown, eds., 5th ed., McGraw-Hill, New York, 1983, pp. 440–473.

22. Rosenberg, L.E., in *The Metabolic Basis of Inherited Disease*, J.B. Stanbury, J.B. Wyngaarden, D.S. Fredrickson, J.L. Goldstein, and M.S. Brown, eds. 5th ed., McGraw-Hill, New York, 1983 pp. 474–497.

23. Stadtman, T.C., *J. Biol. Chem.*, **238**, 2766–2773 (1963).

24. Poston, J.M., *J. Biol. Chem.*, **251**, 1859–1863 (1976).

25. Dvornik, D., *Arh. Kem.*, **26**, 211–214 (1954).

26. International Union of Biochemistry, *Eur. J. Biochem.*, **125**, 1–13 (1982).

27. Poston, J.M., unpublished observations.

28. Baker, J.J. and Stadtman, T.C., in B_{12}, *Volume 2: Biochemistry and Medicine*, David Dolphin, Ed., Wiley, New York, 1982, pp. 203–232.

29. Poston, J.M., *J. Biol. Chem.*, **255**, 10067–10072 (1980).

30. Poston, J.M., *Phytochemistry*, **17**, 401–402 (1978).

31. Poston, J.M., *Science*, **195**, 301–302 (1977).

32. Poston, J.M. and Hemmings, B.A. *J. Bact.*, **140**, 1013–1016 (1977).

33. Freer, I., Pedrocchi-Fantoni, G., Picken, D.J., and Overton, K.H., *J. Chem. Soc. Chem. Commun.*, **1981**, 80–82 (1981).

34. Poston, J.M., *Biochem. Biophys. Res. Commun.*, **96**, 838–843 (1980).

35. Gehrke, C.W. and Leimer, K., *J. Chromatogr.*, **57**, 219–238 (1971).

36. Poston, J.M., Abstracts of the 186th National Meeting of the American Chemical Society, Washington, D.C., Aug. 28–Sept. 2, 1983, BIOL 143 (1983).

37. Stern, J.R., Coon, M.J., del Campillo, A., and Schneider, M.C., *J. Biol. Chem.*, **221**, 15–31 (1956).

38. Vagelos, P.R. and Alberts, A.W., *Anal. Biochem.*, **1**, 8–16 (1960).

39. Poston, J.M., *Fed. Proc.* **42**, 1978 (1983).

40. Aragon, J.J. and Lowenstein, J.M., *J. Biol. Chem.*, **258**, 4725–4733 (1983).

41. Monticello, D.J. and Costilow, R.N., *J. Bact.*, **152**, 946–949 (1982).

42. Frenkel, E.P., *J. Clin. Invest.* **52**, 1237–1245 (1973).

43. van der Westhuyzen, J., Cantrill, R.C., Fernandes-Costa, F., and Metz, J., *J. Nutr*, **113**, 531–537 (1983).

44. Garrod, A.E., *Lancet* 2:1, 73, 143, 214 (1908).

45. Garrod, A.E., *Inborn Errors of Metabolism*, Oxford University Press, London (1909), modified edition (1923).

ENZYMOLOGY AND PHYSIOLOGY OF RETICULOCYTE LIPOXYGENASE: COMPARISON WITH OTHER LIPOXYGENASES

By TANKRED SCHEWE, SAMUEL M. RAPOPORT, and HARTMUT KÜHN, *Institute of Biochemistry, Humboldt University, DDR-1040 Berlin, GDR*

CONTENTS

I. Introduction

Lipoxygenases (formerly called lipoxidases in plants) have been known for a long time. The first lipoxygenase was crystallized from soybeans in 1947 by Theorell (1). Later on lipoxygenases were detected in other plants, particularly in legumes, cereals, Solanaceae, and some fruits. Their biological role remained unclear. The lipoxygenases were believed to be involved in a side path of fatty acid oxidation in plants. Most attention was paid to the lipoxygenases by food scientists, since they were thought to have a role in lipid peroxidation of vegetable oils [for older work see reviews by Tappel (2,3) and Bergström and Holman (4)]. The occurence of lipoxygenases in animal tissues was questioned at that time, since lipid peroxidations were believed to be exclusively due to ubiquitous heme compounds (2,5). The unfolding prostaglandin research in the mid-1970s led to the discovery of lipoxygenase activity in platelets (6–8) and leukocytes (9). At the same time the lipoxygenase of rabbit reticulocytes was discovered in our laboratory (10). This protein was already observed in 1955 on account of its ability to inhibit the respiratory chain (11,12).

During the last decade enormous progress has been achieved in the knowledge of lipoxygenase metabolites in animal cells and of their biological functions. For a review the reader is referred to Hansson et al. (13). Highlights of this development were the discovery, identification, and chemical synthesis of leukotrienes as a novel class of bioregulators, which may play an important role in asthma bronchiale, inflammation, and other processes. In contrast, only a few studies have been done on the chemical structure, catalytic mechanism, enzyme kinetics, biological dynamics, and molecular biology of animal lipoxygenases. Some of these aspects are unexplored also with respect to the plant lipoxygenases. An obvious obstacle in the approach to these problems is the difficulty in the purification of lipoxygenases, in particular those from animal tissues. The first animal lipoxygenase so far purified to homogeneity and characterized is the enzyme from rabbit reticulocytes (14,15). A pure 5-lipoxygenase was recently obtained from RBL cells (18). Partial enrichments were achieved for the lipoxygenases from platelets (7,8,16,17), leukocytes (19–21), testicles (22,23), skin (24), and lung tissue (25,26). The aim of this article is to give a survey of the enzymology of animal lipoxygenases with special reference to reticulocyte lipoxygenase. Where necessary the animal enzymes will be compared with the plant lipoxygenases. For a detailed survey on plant lipoxygenases the reader is referred to the articles of Vliegenthart and co-workers (27–29). We include also a discussion on the catalytic activities of heme compounds related to lipid peroxidation with special emphasis on their quasi-lipoxygenase activity, which turned out to be an interesting model for lipoxygenases. Large progress was achieved during the last decade with respect to the positional and stereochemical aspects of the substrate specificity, since basic conclusions are available from analyses of reaction products and other approaches not requiring purified enzymes. The present knowledge of the stereochemical and positional specificity of the lipoxygenase reaction and the consequences for a comprehensive nomenclature of lipoxygenases and related enzymes are given in an accompanying article (30).

II. Lipoxygenase-Catalyzed Reactions and Lipoxygenase Assays

Lipoxygenases (EC 1.13.11.12) catalyze the dioxygenation of polyenoic fatty acids or their derivatives containing at least one 1,4-*cis,cis*-pentadiene system. In this reaction the corresponding product with a conjugated 1-hydroperoxy-2,4-*trans,cis*-pentadiene system is formed. A detailed description of the basic mechanism of the lipoxygenase reaction is given in Section IV.D.2. Three methods are most frequently recommended for the assay of lipoxygenases:

1. Measurement of the oxygen uptake by means of a Clark electrode.

2. Spectrophotometric recording of the formation of conjugated dienes absorbing at 234 nm (31).

3. Quantification of the labeled reaction products from 1-[14]C- or ^{3}H-labeled substrates after their separation by means of thin-layer chromatography or high-pressure liquid chromatography.

The latter method is more cumbersome than the two others but affords the advantages of higher sensitivity and permits lipoxygenase measurements in complex systems such as intact cells. This method is often the only approach to studying animal lipoxygenases. It does not allow, however, precise kinetic studies. For special purposes other methods of measuring lipoxygenase activities, especially in turbid samples, have been described (32–36).

A particular obstacle in all lipoxygenase assays is the poor watersolubility of the substrate fatty acids at neutral pH. For this reason, the manner of preparing the substrate solution may influence the enzyme kinetics. In the case of reticulocyte lipoxygenase an addition of 0.2% sodium cholate (final concentration) proved to be advantageous. The possibility must be considered, however, that the cholate may alter the enzyme kinetics in several respects. The activity of reticulocyte lipoxygenase with linoleic or arachidonic acid as substrate is strongly stimulated by cholate, which is not the case with soybean lipoxygenase under identical conditions (37). For the lipoxygenase of pea seeds we observed both a shift in pH optimum from 5.9 to 6.8 and an enhancement of self-inactivation (38). Another procedure to achieve better accessibility of fatty acids to lipoxygenases is their conversion to the potassium or ammonium salts before mixing with the assay buffer. A detailed review on aerobic lipoxygenase assays was given by Grossman and Zakut (39). Polyether detergents, which include the Triton, Tween, and Brij series, should not be used in studies on lipoxygenases, since these compounds tend to be peroxidized spontaneously (40), so that severe disturbances would occur owing to peroxides or their breakdown products not arising from a lipoxygenase reaction. This warning is also based on our own experiences. Difficulties in assaying lipoxygenases arise not only from the poor watersolubility of their sub-strates but also from the fact that fatty acids form multimolecular aggregates including acid soap dimers and micelles (41). The for-mation of such aggregates may depend on both the manner of pre-paring the substrate solution and on the presence of additives.

In addition to the limited watersolubility of the substrates, a va-riety of other factors may disturb the kinetics of lipoxygenase re-actions. One of them is the suicidal character of most lipoxygenases, that is, their tendency to self-inactivation (see Section IV.C.1). Since the self-inactivation is strongly temperature dependent, an assay temperature of 2°C is preferred. Other complications may arise from the occurrence of free radicals during the lipoxygenase reaction. Enzyme-bound radicals are obligatory intermediates in the lipoxy-genase catalysis (see Section IV.D.2). Under some conditions these radicals may dissociate from the enzyme and elicit side reactions such as cooxidations. In our experience 2,6-di-*t*-butyl-4-hydroxy-toluene (BHT) is a suitable antioxidant that prevents side reactions of free radicals without affecting the primary lipoxygenase reaction. In contrast, other well-known antioxidants such as 3-*t*-butyl-4-hy-droxyanisol (BHA) and nordihydroguaiaretic acid (NDGA) are at the same time strong inhibitors of lipoxygenases including reticu-locyte lipoxygenase (42).

For an exact measurement of lipoxygenase activity the parallel determination of oxygen uptake and formation of conjugated dienes that normally show a 1:1 stoichiometry to each other is recom-mended. Deviations from this stoichiometry may have the following reasons:

1. The occurrence of aerobic lipohydroperoxidase activity as shown for pea lipoxygenase (38).

2. The presence of other hydroperoxy fatty acid converting ac-tivities, in particular if crude lipoxygenase preparations are used.

3. The occurrence of secondary and other oxygen-consuming processes. For the purified reticulocyte lipoxygenase a nearly exact stoichiometry was observed with linoleic acid as substrate at tem-peratures up to 30°C provided that the concentrations of both linoleic acid and oxygen were not limiting; at low concentrations of the reac-tants, however, there occurred significant deviations from stoichi-ometry (38a).

In addition to the dioxygenation of polyenoic fatty acids lipoxygenases also catalyze other reactions. Under anaerobic conditions hydroperoxylinoleic acids are converted in the presence of linoleic acid to a variety of products including oxodienoic acids, epoxyhydroxy acids, dienoic fatty acid dimers, and pentane. These enzymatic conversions are initiated by a homolytic cleavage of the O–O bond of the hydroperoxide and are called lipohydroperoxidase activity (see Section IV.B.3). The formation of oxodienoic acids can be conveniently measured by their absorbance at 285 nm; pentane formation is determined by means of gas chromatography. Both anaerobic formation of oxodienoic acids and pentane formation have been demonstrated with purified reticulocyte lipoxygenase (43). To estimate the total lipohydroperoxidase activity it is necesary to determine the residual hydroperoxide content, for example, by iodometry.

A unique feature of reticulocyte lipoxygenase is its ability to attack biomembranes and to produce inhibition of the mitochondrial respiratory chain. The respiration-inhibitory activity was employed in our laboratory for the assay of this enzyme before its identification as lipoxygenase (14,15). This activity is not shown by the pure lipoxygenases from soybeans, pea seeds, or wheat. Respiration-inhibitory activity occurs, however, during the combined actions of soybean lipoxygenase and hemoglobin; hemoglobin also potentiates the respiration-inhibitory activity of reticulocyte lipoxygenase (44) (see Section V).

III.   Isolation and Molecular Properties of Reticulocyte Lipoxygenase

A.   PURIFICATION AND ISOLATION

Rabbit reticulocytes under normal conditions contain only little lipoxygenase. In bleeding anemia, however, large concentrations of lipoxygenase appear reaching up to 4 mg/mL cells as determined by immunoprecipitation (see Section VII.B). Therefore, reticulocytes of rabbits produced by bleeding are an excellent source for a lipoxygenase preparation. Their lipoxygenase content is comparable with that of soybeans, where 1 g of purified lipoxygenase can be obtained from 1 kg of dry soybeans (29). Owing to the high content of lipoxygenase in reticulocytes in bleeding anemia, a 100–200-fold enrichment is sufficient to achieve purity. The purification procedure, which is described in detail elsewhere (15), includes the following steps: After the cells have been washed three times with isotonic saline they are hemolyzed in twice the volume of distilled water for 10 min in the cold. The hemolysate is adjusted to pH 6.0 and the stroma (mainly consisting of plasma membrane ghosts and mitochondria) is centrifuged off. The stroma-free supernatant fluid is subjected to an ammonium sulfate precipitation at 0.55 saturation, which removes the bulk of hemoglobin that remains in the supernatant. The precipitate is dissolved in twice the volume of distilled water and dialyzed against heavy-metal-free 0.03 M Tris HCl, pH 7.2. Further fractionation proceeds on a Sephadex A-50 column under strict exclusion of oxygen and in the presence of 20 mM 2-mercaptoethanol and 1 mM $MgCl_2$; otherwise severe losses of activity occur. The last step is isoelectric focusing in ampholine pH 5–7 in a sucrose gradient. The active fractions are stored in liquid nitrogen. The ampholine–sucrose medium stabilizes the purified enzyme. For many purposes it is necessary to remove it by rapid desalting over a Sephadex G-25 column under a nitrogen atmosphere. The desalted purified enzyme loses its activity within a few hours, especially in the presence of oxygen (see last paragraph of this section).

Alternative methods for purifying the reticulocyte lipoxygenase, including affinity chromatography, were applied (14), but afforded no advantage as compared with the standard procedure. For any chromatographic purification step an absolute exclusion of traces of both oxygen and heavy metals proved to be necessary, which makes the procedure cumbersome; parallel preparations under normal aerobic conditions gave not only very low yields (10–20% of the initial activity), but also gave an enzyme preparation with altered enzymatic properties as judged from the analysis of the reaction products from arachidonic acid. The exclusion of oxygen is a prerequisite to achieve high yields and specific activities of purified enzyme not only for the reticulocyte lipoxygenase. We had the same experience in our laboratory in the isolation of the lipoxygenases from pea seeds, wheat kernels, and potato tubers (unpublished results).

The purified reticulocyte lipoxygenase exhibits a turnover number of about 25 s^{-1} in our standard assay (0.1 M phosphate buffer,

TABLE I

Molecular Masses and Isoelectric Points of Selected Lipoxygenases

Source of lipoxygenase	M_r	Reference	pI	Reference
Rabbit reticulocytes	78,500	14	5.5	14
Human platelets	100,000	16		
Rat basophilic leukemia cells (5-lipoxygenase)	73,000	18		
Rabbit leukocytes (15-lipoxygenase)	61,000	19		
Soybeans	102,000	27	5.6	51
Pea seeds	95,000	52	6.25	38
Potato tubers	95,000	53	4.5	53
Wheat kernels	86,000	54	5.9	55

pH 7.4 containing 0.2% cholate, 0.53 mM potassium linoleate; 2°C). This value is one order of magnitude lower than that reported for soybean lipoxygenase-1 at pH 9.0 (29), but it is similar to those for other plant lipoxygenases having a neutral pH optimum. Poly-acrylamide gel disc electrophoreses both under nondenaturing conditions and in sodium dodecylsulfate–mercaptoethanol reveal a single polypeptide, which coincides in native electrophoresis with the band stainable for lipoxygenase activity (45). The reticulocyte lipoxygenase proved to be homogeneous in immunological tests (46). The enzyme preparations that lost part of their activity, particularly in the presence of air and after removal of ampholine, exhibit additional oligomeric molecular species both in density gradient centrifugation (47) and in native disc electrophoresis (48). These oligomers are immunologically reactive but enzymatically inactive.

The inactivation of reticulocyte lipoxygenase during aerobic storage is accompanied by the formation of 1-mole of methionine sulfoxide per mole of enzyme (49) and by progressive changes in the CD spectrum (50).

B. MOLECULAR PROPERTIES

Table I shows a comparison of the M_r and pI values for a variety of animal and plant lipoxygenases (14,51–55). The molecular mass of reticulocyte lipoxygenase is 78,500 ± 3000 daltons according to three independent methods (poly-acrylamide gel electrophoresis in

TABLE II
Amino Acid Compositions of Reticulocyte Lipoxygenase
(57) and Soybean Lipoxygenase-1 (29)

| Amino acid | Number of residues | |
	Reticulocyte enzyme	Soybean enzyme
Asx	41	87
Thr	22	44
Ser	29	52
Glx	65	111
Pro	33	45
Gly	42	66
Ala	37	66
Cys	11	5
Val	40	52
Met	14[a]	17
Ile	24	50
Leu	74	82
Tyr	12	43
Phe	25	36
Trp	19	11
Lys	28	52
His	14	28
Arg	30	36
Total	560	883

[a] Determined by cyanogen bromide cleavage (49).

sodium dodecylsulfate/2-mercaptoethanol, sucrose density gradient centrifugation, and analytical ultracentrifuge (14). However, in recent studies on the biosynthesis and cloning of the reticulocyte lipoxygenase Thiele et al. observed a molecular mass of about 68,000 daltons for the translational product both in the cellular synthesis and in the cell-free system (56). This discrepancy must still be clarified. The exact value will be available only when the amino acid sequence is known. It is hoped that this can be obtained from sequencing of the cloned cDNA (56). The isoelectric point of reticulocyte lipoxygenase is 5.5.

The N-terminal amino acid is glycine, determined by the dansylation method (14). Analysis of the C-terminus revealed heterogeneity, since His, Ile, and Asn were found (57). This heterogeneity despite electrophoretical and immunological homogeneity may arise from endogenous carboxypeptidase activity in reticulocytes that generates this kind of isoprotein. The amino acid composition differs from that of soybean lipoxygenase (Table II); a higher content of Cys and Trp and a lower content of the sum of charged amino acid residues are noticeable. Reticulocyte lipoxygenase contains no cystine residue (57). The same holds for soybean lipoxygenase (29). Like soybean lipoxygenase-1 the reticulocyte enzyme contains one atom of Fe (57); the earlier report of two iron atoms (14) proved to be in error. Reticulocyte lipoxygenase is a glycoprotein; its carbohydrate content was determined to be 5% (14). The presence of a carbohydrate moiety indicates glycosylation in the absence of an endoplasmic reticulum and the Golgi apparatus, since the synthesis of the lipoxygenase occurs only at the reticulocyte stage (see Section VII.B), in which these organelles are absent.

IV.   Enzymology of Lipoxygenases

A.   SUBSTRATES AND PRODUCTS OF LIPOXYGENASES

Lipoxygenases are generally defined in terms of their ability to dioxygenate polyunsaturated fatty acids possessing a 1,4-*cis,cis*-pentadiene system. Among the naturally occurring fatty acids that meet this structural requirement the following are most frequently used as lipoxygenase substrates: linoleic acid (9,11-di-*cis*-octadecadienoic acid), α-linolenic acid (9,12,15-all-*cis*-octadecatrienoic acid), γ-linolenic acid (6,9,12-all-*cis*-octadecatrienoic acid), arachidonic acid (5,8,11,14-all-*cis*-eicosatetraenoic acid), bis(homo)-γ-linolenic acid (8,11,14-all-*cis*-eicosatrienoic acid), and 5,8,11-all-*cis*-eicosatrienoic acid. In addition, a variety of artificial fatty acids can serve as lipoxygenase substrates. They are valuable tools for the elucidation of the positional specificities of lipoxygenases. As far as animal lipoxygenases are concerned, arachidonic acid finds the greatest interest owing to its role as a main biological precursor of leukotrienes and other lipoxygenase products. In plants α-linolenic acid may play an analogous role, since this compound has been recently shown to be the precursor of jasmonic acid, a regulatory substance that inhibits growth and promotes senescence, via a lipoxygenase pathway (58,59).

The various lipoxygenases differ in three substantial features:

1. The site of primary hydrogen abstraction.

2. The direction of the double bond shift in the primary radical leading to the 1-hydroperoxy-2,4-*trans,cis*-pentadiene system.

3. The stereospecificity of both hydrogen abstraction and dioxygen insertion [for details see ref. (30)].

Owing to these features there are large differences in affinity between various polyenoic fatty acids and different lipoxygenases; different products may also arise. Linoleic acid is the simplest natural lipoxygenase substrate, since it possesses only one possible site of primary hydrogen abstraction (except for the prochiral character of this reaction step), so that only four regular primary lipoxygenase products are possible (9$_{DS}$-hydroperoxy-10,12-*trans,cis*-octadecadienoic acid, 9$_{LR}$-hydroperoxy-10,12-*trans,cis*-octadecadienoic acid, 13$_{DR}$-hydroperoxy-9,11-*cis,trans*-octadecadienoic acid, and 13$_{LS}$-hydroperoxy-9,11-*cis,trans*-octadecadienoic acid). From arachidonic acid 12 regular primary lipoxygenase products may be formed in principle [see also the companion paper, ref. (30)], even though not all of them were identified so far as real lipoxygenase products. In addition, even multiple dioxygenation products and hydrolysis products of leukotrienes A$_4$ can arise during the action of lipoxygenases on arachidonic acid, which are not possible in the case of linoleic acid. However, linoleic acid is not a good substrate for some lipoxygenases, such as those of platelets (7) and the arachidonate 5-lipoxygenase of leukocytes (20).

Many lipoxygenases attack not only unesterified polyenoic fatty acids but also their methyl esters. We found the rates of the dioxygenation of methyllinoleate to be consistently about one fourth those of free linoleic acid in the case of the lipoxygenases of reticulocytes, pea seeds, and wheat (38,45). This difference may be caused by the solubility of the methyl ester rather than by particular properties of these enzymes. Soybean lipoxygenase-1 attacks methyllinoleate not only with low reactivity but also with little regio- and stereospecificity as compared with the reaction with free linoleic acid (60). Phospholipids are also attacked by other 15-lipoxygenases such as that of polymorphonuclear leukocytes (217).

The actions of the lipoxygenases of platelets and leukocytes on cellular phospholipids must be preceded by a liberation of arachidonic acid, for example, by the action of a phospholipase A_2 or of other phospholipid-cleaving enzymes. The lipoxygenase from reticulocytes shows the unique property of attacking membranes to a sizeable extent even in the absence of a phospholipid-splitting enzyme (10). Nevertheless even reticulocyte lipoxygenase appears to prefer free polyenoic acids as a substrate over phospholipids, since it was observed that addition of snake venom phospholipase A_2 to a sample containing dilinoleyl phosphatidyl choline and reticulocyte lipoxygenase gives rise to a fourfold stimulation of oxygen uptake (61).

The products of the single dioxygenation reaction with polyenoic fatty acids as substrate are in turn substrates for other lipoxygenase-catalyzed reactions. They may be classified as three types: (a) multiple (double or triple) dioxygenations, (b) lipohydroperoxidase reactions, and (c) leukotriene A_4 synthase reactions [see Section IV.B as well as ref. (30)]. The common feature of all lipoxygenase-catalyzed reactions is that they are initiated by a radical-generating step. The dioxygenations involve a homolytic cleavage of a C–H bond, whereas lipohydroperoxidations are initiated by the homolytic cleavage of the O–O bond in hydroperoxy fatty acids (see Section IV.B.3).

Secondary lipoxygenase reactions seem to be more or less suppressed in animal cells, since the hydroperoxy eicosatetraenoic acids (HPETEs) are effectively converted to the corresponding hydroxy compounds (HETEs) via lipoxygenase-independent mechanisms such as glutathione peroxidases. The HETEs and DiHETEs including leukotriene B_4 belong to the most abundant lipoxygenase products found in animal tissues (13).

B. LIPOXYGENASE-CATALYZED REACTIONS

1. Positional Specificity and Stereospecificity of Single Dioxygenations Catalyzed by Reticulocyte Lipoxygenase

Single dioxygenation products with arachidonic acid from the reaction of purified reticulocyte lipoxygenase were identified by Bryant et al. (62). 15$_{LS}$-hydroperoxy-5,8,11,13-*cis,cis,cis,trans*-eicosatetraenoic acid (15$_{LS}$-HPETE) and 12$_{LS}$-hydroperoxy-5,8,10,14-*cis,cis,trans,cis*-eicosatetraenoic acid (12$_{LS}$-HPETE) were found in a ratio of about 15:1, whereas other monoHPETEs were absent. In intact reticulocytes the corresponding HETEs were observed in the same ratio. Thus the enzyme behaved identically in pure form and in the cell, with a subsequent conversion of the HPETEs in reticulocytes to the HETEs.

The formation of both 15_{L_S}-HPETE and 12_{L_S}-HPETE by reticulocyte lipoxygenase implies dual positional specificity, with a shift of the radical in the same direction and identical stereospecificity. the formation of 15-HPETE requires hydrogen removal at C-13 (n-8), that of 12-HPETE at C-10 (n-11), followed by a shift of the resulting radical in direction to the methyl terminus of the substrate fatty acid as well as a chiral dioxygen insertion in L_S position in both cases. The positional specificity of reticulocyte lipoxygenase was confirmed with various other fatty acids (Fig. 1). The results show that the positions of the primary H abstraction and of the O_2 insertion are recognized by the distance from the methyl terminus of the fatty acid chain. The same holds true for soybean lipoxygenase-1 (63). These two lipoxygenases show identical characteristics with respect to their main positional specificity [dioxygenation at position (n-6)].

The dual positional specificity of reticulocyte lipoxygenase is at variance with former assumptions concerning an absolute specificity of lipoxygenases. In the case of reticulocyte lipoxygenase it could be unequivocally shown that the (n-6) and (n-9) positional specificities reside in one enzyme species (64). A very low activity with (n-9) positional specificity was also observed with soybean lipoxygenase-1. A dual positional specificity also has been reported for the lipoxygenase from potato tubers that forms 5_{D_S}-HPETE from arachidonic acid and mainly 8-hydroperoxyeicosatrienoic acid from bis-homo-γ-linolenic acid (53). From the dual positional specificity of the enzymes it follows that the occurrence of several single dioxygenation products in a cell does not necessarily imply the involvement of different lipoxygenases. Separate lipoxygenases are, however, established for leukocytes owing to the partial purification of three enzymes possessing different characteristics (see Section VIII.A).

Reticulocyte lipoxygenase converts linoleic acid to 13-hydroperoxy-9,11-*cis,trans*-octadecadienoic acid (13-HPOD) and 9-hydroperoxy-10,12-*trans,cis*-octadecadienoic acid (9-HPOD) in a ratio of about 10:1 (65) despite the fact that hydrogen abstraction occurs at

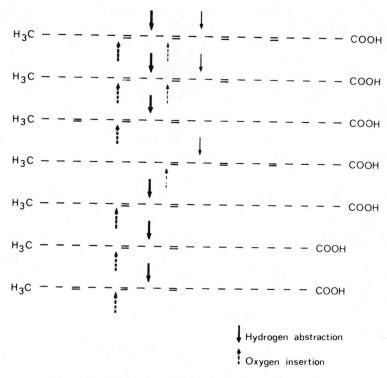

Figure 1.   Positional specificity of reticulocyte lipoxygenase.

a single site. A formation of 9-HPOD to a minor extent is also known for soybean lipoxygenase-1 (29). Analyses of the enantiomeric composition of the products of soybean lipoxygenase-1 (66) as well as those from reticulocyte enzyme (unpulished results), from linoleic acid revealed that 13-HPOD shows high stereospecificity (preponderance of 13_{L_S}-HPOD), whereas the 9-HPOD fraction consists of a racemic mixture. It is reasonable to assume that in the case of linoleic acid the intermediate radical may dissociate from the enzyme to a certain extent and combine with oxygen in a stochastic manner as is the case in nonenzymatic lipid peroxidation, so that all four possible HPODs are formed. In this manner, a reaction chain arises as a side path of the lipoxygenase catalysis that includes an enzymatic stereospecific hydrogen abstraction followed by a non-enzymatic oxygen insertion (see also Section IV.D.2). The share of racemic linoleic acid products varies, depending on the kind of lipoxygenase, pH, temperature, and other conditions [for a review see Vliegenthart and Veldink, ref. (29)].

The positional specificity and stereospecificity are partly lost after damage to the enzyme, as shown for soybean lipoxygenase-1 modified by organic mercurials (67). With arachidonic acid as substrate, stereospecific lipoxygenase products are normally found exclusively with native enzymes. Here again damage to the enzyme may lead to a loss of positional specificity and perhaps of stereospecificity. Thus, reticulocyte lipoxygenase damaged by aerobic storage forms 15-HPETE and 11-HPETE from arachidonic acid in a ratio of nearly 1:1. Therefore the absence of formation of 11-HPETE may be a criterion for native preparations of reticulocyte lipoxygenase.

2. Formation of Multiple Dioxygenation Products

Whereas reticulocyte lipoxygenase forms only single dioxygenation products from arachidonic acid at 2°C, at 37°C a variety of other products also occur, including 5,15-dihydroperoxy-6,8,11,13-eicosatetraenoic acid, (5,15-DiHPETE), and 8,15-dihydroperoxy-5,9,11,13-eicosatetraenoic acid, (8,15-DiHPETE) (68). The same compounds were observed with 15_{L_S}-HPETE as substrate. These metabolites were clearly identified as double dioxygenation products, since they were shown to have incorporated two molecules of $^{18}O_2$ and their formation was blocked by anaerobiosis (68). The same double dioxygenation products are also formed by soybean lipoxygenase-1 (69). The formation of 5,15-DiHPETE and 8,15-DiHPETE from 15-hydroperoxyeicosatetraenoic acid (15-HPETE) appears to contradict the positional specificity of the two lipoxygenases, since neither 5-HPETE nor 8-HPETE can be detected as single dioxygenation products from arachidonic acid. However, it must be taken into account that the oxygenated and the nonoxygenated polyenoic fatty acids may behave in a different manner as substrates. An inverse orientation of the substrate at the active center of the enzyme makes the formation of these double dioxygenation products plausible [see also ref. (30)].

5,15-DiHETE that arises from reduction of one of the double dioxygenation products of either reticulocyte or soybean lipoxygenase was shown to be in turn the substrate for reticulocyte lipoxygenase (70). This compound possesses still a double-allylic methylene group at C_{10} so that it meets the basic structural requirement of a lipoxygenase substrate. As discussed before, a hydrogen abstraction at this position occurs during both single and double dioxygenations catalyzed by reticulocyte lipoxygenase. As shown recently, 5,15-DiHETE is dioxygenated by reticulocyte lipoxygenase to a novel lipoxygenase metabolite possessing a conjugated tetraene system as judged from its characteristic UV spectrum. The structure of this triple dioxygenation product of arachidonic acid was tentatively defined as either 5,6,15-trihydroxy-7,9,11,13-trans,trans,cis,trans-eicosatetraenoic acid and/or 5,14,15-trihydroxy-6,8,10,12-eicosatetraenoic acid (70). The same compounds, called lipoxin A and lipoxin B, were also isolated and structurally identified from leukocytes after incubation with 15L$_S$-hydroperoxyeicosatetraenoic acid, Ca^{2+}, and ionophore A 23187 (71). Detailed investigations on the formation of lipoxins from 5,15-DiHETE catalyzed by reticulocyte lipoxygenase revealed two compounds after reduction by borohydride; one of them, amounting to about 90%, is in all likelihood identical with lipoxin A as judged from UV spectrum, retention times in high-pressure liquid chromatography, and gas chromatography/mass spectrometry (71a).

The reactivity of reticulocyte lipoxygenase with respect to single and multiple dioxygenations of arachidonic acid differs strongly in the following order: single dioxygenation $>>$ double dioxygenation $>>$ triple dioxygenation. Therefore, multiple dioxygenations occur only after the single dioxygenations have ceased owing to exhaustion of arachidonic acid and require fairly high concentrations of lipoxygenase. These considerations only hold true for the unesterified compounds. Surprisingly, a formation rate of lipoxin A two orders of magnitude higher was observed with 15-HETE methyl ester as substrate (71a).

3. Lipohydroperoxidase Activity

The anaerobic reaction of lipoxygenases was first described in detail in 1971 by Garssen et al. with soybean lipoxygenase-1 (72). A kinetic analysis has been published by Verhagen et al. (73). In this reaction the conversion of 1-mol of linoleic acid is strictly coupled to the conversion of 1-mol of 13L$_S$-hydroperoxylinoleic acid

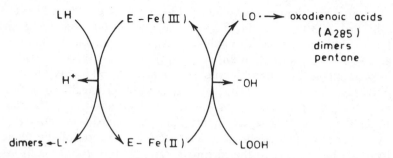

Figure 2.   Catalytic cycle of the anaerobic lipohydroperoxidase reaction of lipoxygenases. LH; linoleic acid; LOOH; hydroperoxylinoleic acid [from ref. (73)].

(13-HPOD). One half of 13-HPOD is converted to a 1:1 mixture of 13-oxooctadeca-9,11,-dienoic acid and 13-oxotrideca-9,11-dienoic acid. The oxodienoic acids show an absorbance maximum at 285 nm, which provides a convenient assay for that part of hydroperoxidase activity. In addition a variety of other compounds are formed during the anaerobic lipohydroperoxidase reaction, such as epoxyhydroxy octadecenoic acids, various conjugated fatty acid dimers and pentane. An alkoxy radical originating from a homolytic cleavage of the O–O bond of the hydroperoxy group has been proposed to be the common precursor of these products (74) (Fig. 2).

The formation of both pentane and of nonvolatile lipohydroperoxidase products by the lipoxygenases of soybeans and reticulocytes were compared and found to show similar characteristics and product patterns except for some quantitative differences (43,75). The reactions require the presence of linoleic acid and are completely inhibited by oxygen. The formation of pentane catalyzed by reticulocyte lipoxygenase was also demonstrated in intact reticulocytes, if they were stimulated by Ca^{2+} plus ionophore A 23187, or if arachidonic acid was added (43). This observation deserves particular attention in view of the wide use of pentane formation as a measure of nonenzymatic lipid peroxidation (76). The possibility should be taken into account that a part of the pentane formation of the body may arise from enzymatic reactions, for example, from lipoxygenase-containing cells in regions with partial anaerobiosis or under conditions of relative oversupply of hydroperoxy fatty acids.

From 15$_{LS}$-hydroperoxy-5,8,11,13-eicosatetraenoic acid (15-HPETE), 13-hydroxy-14,15-epoxy-5,8,11-eicosatrienoic acid is formed by the lipohydroperoxidase activity of reticulocyte lipoxygenase (68). Its formation differs from the conversion of 13-HPOD in two aspects: (a) it does not require the presence of nonoxygenated polyenoic fatty acid and (b) it proceeds also aerobically. The first point may be rationalized as follows: As discussed later (Section IV.D.2) the catalytic cycle of the lipohydroperoxidase activity of lipoxygenases includes as a crucial step a reduction of the ferric form of the enzyme to the ferrous form. The reductant is a lipoxygenase substrate that is able to undergo hydrogen abstraction. In the case of 13-HPOD, linoleic acid fulfills this function, whereas 15-HPETE is a lipoxygenase substrate itself owing to the presence of the double-allylic methylene group at C_{10}.

Soybean lipoxygenase-1 exhibits a linoleic acid-supported anaerobic lipohydroperoxidase activity only with 13$_{LS}$-HPOD, not with 9$_{DS}$-HPOD (77). The same is true for the reticulocyte lipoxygenase in contrast to an earlier report (78), whereas reticulocyte lipoxygenase shows this activity with both hydroperoxyoctadecadienoic acids (78).

In addition to the linoleic acid-supported conversion of HPODs that proceeds at a relatively high rate (one third to one fifth of the aerobic dioxygenase activity) there is a slow decomposition of HPODs in the absence of linoleic acid that requires at least 20-fold higher concentrations of lipoxygenase (77). 9-HPOD is decomposed faster than 13-HPOD by both soybean lipoxygenase-1 and reticulocyte lipoxygenase (77,78). This type of hydroperoxidase reaction apparently does not include the ferrous form of the enzyme and resembles in some characteristics the decomposition of hydroperoxides by heme compounds (44,79,80). It is assumed to be involved in the self-inactivation of reticulocyte lipoxygenase (78).

4. Leukotriene A₄ Synthase Activity

Bryant et al. demonstrated that reticulocyte lipoxygenase converts [1-^{14}C]arachidonic acid not only to products arising from single or double dioxygenations and lipohydroperoxidase activity, but also to a diastereomeric mixture of 8epi,15$_{LS}$-dihydroxy-5,9,11,13-cis,trans,trans,trans-eicosatetraenoic acid (68). These products were shown to arise from the spontaneous hydrolysis of 14,15-leukotriene A_4. The epoxyleukotriene could be trapped by acidic methanol with the resultant formation of 8-methoxy-15-hydroxy-5,9,11,13-eicosatetraenoic acid. In incubation experiments with $^{18}O_2$ gas the ^{18}O was introduced only at the C_{15} position (68). Thus, the reticulocyte lipoxygenase constitutes the first purified animal leukotriene synthase. A mechanism of this activity is proposed, as shown in Fig. 8 of the companion paper (30). One prerequisite is the ability of the enzyme to abstract hydrogen at C_{10} by virtue of its (n-9) dioxygenation activity (see Section IV.B.1). This reaction is the first step of the conversion of 15_{LS}-HPETE to either the double dioxygenation product 8,15-DiHPETE or 14,15-LTA$_4$. The second step may be the scission of the O–O bond of the 15-hydroperoxy group via a lipohydroperoxidase-type reaction giving rise to a putative bis-radical that is spontaneously stabilized to the epoxytriene. The iron atom of the active center of the lipoxygenase may be subjected to a valence change as is the case with the dioxygenase and hydroperoxidase activity (see Section IV.D.2). The LTA$_4$ synthase activity may involve partial steps of both dioxygenase and hydroperoxidase activity. The conversion of arachidonic acid to 14,15-LTA$_4$ becomes possible by virtue of a dual positional specificity of the dioxygenase activity. Quite similar results were obtained by Shimizu et al. using lipoxygenase from potato tubers (53). This plant lipoxygenase was shown to possess dual positional specificity and to convert 5-HPETE to 5,6-LTA$_4$, which was detected in the same way as in the case of reticulocyte lipoxygenase (53).

It may be concluded that the syntheses of LTA$_4$ isomers are generally lipoxygenase-catalyzed reactions. This should hold true also for the LTA$_4$ synthesis in neutrophils and other cells of the leukocyte family where the involvement of a stereospecific elimination of a hydrogen atom at C_{10} was established (81). Further evidence for the lipoxygenase nature of the conversion of 5-HPETE to LTA$_4$ comes from its high sensitivity towards 5,8,11,14-eicosatetraynoic acid (ETYA) (82), which is a powerful suicidal inactivator of lipoxygenases (see Section IV.C.2).

5. Arachidonic Acid Conversions by Reticulocyte Lipoxygenase

As shown, purified reticulocyte lipoxygenase converts arachidonic acid aerobically at 37°C and sufficiently high enzyme concentration to a very complex mixture of products consisting of single, double, and triple dioxygenation products, lipohydroperoxidase

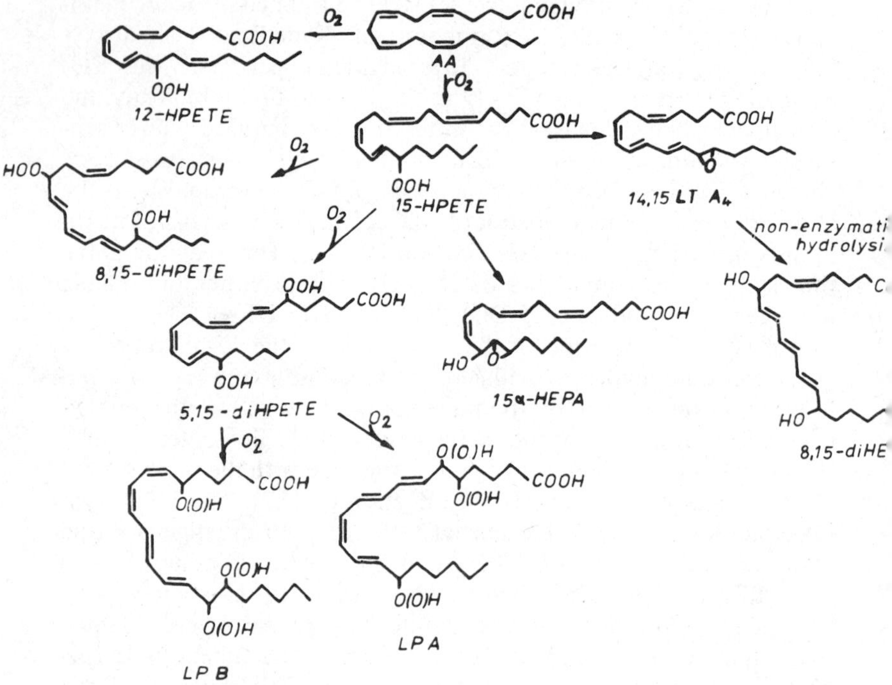

Figure 3. Conversions of arachidonic acid by reticulocyte lipoxygenase. LP, lipoxin; other abbreviations see text.

products, and compounds arising from hydrolysis of 14,15-leuko-triene A_4. Figure 3 shows a survey of these pathways.

6. Cooxidations

During the reaction of lipoxygenases with products there also occur conversions of compounds that are not lipoxygenase products by themselves. Lipoxygenase-catalyzed cooxidation of carotene has been known for a long time (2). Other substrates to be cooxidized are crocin (67,83) and 2,6-dichlorophenolindophenol (DCPIP) (84). Since these dyes are bleached during lipoxygenase-mediated coox-idations, they can be conveniently measured. The various lipoxy-genases differ considerably with respect to their aerobic cooxidative activity. Soybean lipoxygenase-1 exhibits such an activity only during anaerobic lipohydroperoxidase activity (84,85), whereas the lipoxygenase of pea seeds shows a particuarly high aerobic activity (38,52). Using this lipoxygenase two types of cooxidations could be differentiated; whereas the cooxidation of Cu-chlorophyllin was completely inhibited by the antioxidant 2,6-di-t-butyl-4-hydroxytoluene, this was not the case with the bleaching of DCPIP (38). One may conclude that the first type of cooxidation is mediated by free radicals dissociating from the enzyme, whereas the second type proceeds via enzyme-bound radicals of the catalytic cycle.

The cooxidizing properties of reticulocyte lipoxygenase have not been studied as yet in detail. It was observed, however, that glyceraldehyde phosphate dehydrogenase is inactivated in an aerobic system containing reticulocyte lipoxygenase and linoleic acid (14). Cooxidations may also be responsible for the inactivation of the respiratory chain and the destruction of the Fe–S proteins of the mitochondrial outer membrane after the action of reticulocyte lipoxygenase on beef heart submitochondrial particles (see Section VI).

C. INACTIVATION AND INHIBITION OF LIPOXYGENASES

1. Self-Inactivation of Lipoxygenases

Lipoxygenases show the phenomenon of self-inactivation. The instabilities of the lipoxygenases of pea seeds and wheat during catalysis were observed in earlier work (2) and were confirmed in recent studies from our laboratory (38,55). In their kinetic study of the reaction of soybean lipoxygenase with arachidonic acid as substrate Cook and Lands observed a self-catalyzed destruction of the enzyme that was only to a minor part due to a direct interaction of the hydroperoxy fatty acid with the enzyme (86). Similar observations were made by Regdel et al. with the dioxygenation of linoleic acid by pea lipoxygenase (38).

Self-inactivation of soybean lipoxygenase apparently does not occur with linoleic acid as substrate, at least under conditions where other lipoxygenases suffer self-inactivation. Reticulocyte lipoxygenase invariably shows self-inactivation at temperatures higher than 20°C independent of the kind of substrate (free polyenoic fatty acids, phospholipids, submitochondrial particles) (14). A measure of the self-inactivation is the suicide rate. It may be defined as the

reciprocal value of the number of turnovers of the dioxygenase reaction that had occurred when 50% of the enzyme activity is lost.
For reticulocyte lipoxygenase a suicide rate of 1/600 turnovers was
estimated with linoleic acid as substrate at 29°C.* This value depends
on a variety of experimental conditions.

The suicidal nature of the reaction with linoleic acid at 37°C was
shown to be due to a direct irreversible inactivation by the product
13$_{LS}$-hydroperoxylinoleic acid (75). Since a self-inactivation was
also observed during the anaerobic lipohydroperoxidase reaction at
this temperature it was first believed that this activity might be involved. However, in a later study (78) it was shown that the anaerobic inactivation of reticulocyte lipoxygenase by 13$_{LS}$-HPOD is
not stimulated by linoleic acid as reported before, whereas the presence of linoleic acid is necessary for the lipohydroperoxidase reaction (see Section IV.B.3). Since the rate of inactivation proved to
be strongly dependent on the concentration of 13$_{LS}$-HPOD, the apparent stimulation by linoleic acid observed in ref. (75) could have
been caused by incomplete anaerobiosis, so that the 13$_{LS}$-HPOD
concentration might have been enhanced by the dioxygenase reaction. It appears more likely that the inactivation proceeds at the
level of the ferric enzyme–HPOD complex, the occurrence of several species of which have been established for soybean lipoxygen-
ase-1 on the basis of EPR spectral data (87).

In a recent study on the mechanism of the anaerobic inactivation
of reticulocyte lipoxygenase by 13$_{LS}$-HPOD it was found that this
process is paralleled by the formation of 1-mol of methionine sulfoxide per mole of enzyme (49). The same was observed after long-
term aerobic inactivation of reticulocyte lipoxygenase in the absence
of HPOD. Soybean lipoxygenase-1 did not show either inactivation
or formation of methionine sulfoxide under identical conditions.

To establish that a distinct methionine of reticulocyte lipoxygenase is monooxygenated during inactivation, the following strategy
was adopted: Reticulocytes that are able to synthesize the lipoxygenase (see Section VII.B) were preincubated with [^{35}S]methionine.
The labeled lipoxygenase was isolated and subjected to cyanogen
bromide cleavage before and after inactivation. The cleavage takes
place specifically at the methionines in such a manner that the ^{35}S

---

* The value of 600,000 given in ref. (14) is a misprint.

is volatilized as methyl thiocyanate, so that the radioactivity will be lost in the peptide fragments. The cleavage cannot take place, however, with methionine sulfoxide, so that this part of radioactivity is retained. The peptide map of the native enzyme revealed 15 cyanogen bromide fragments corresponding to the presence of 14 methionines in reticulocyte lipoxygenase, which is nearly consistent with amino acid analyses where 13 methionines were found (57). After cyanogen bromide cleavage of the inactivated lipoxygenase, exactly one fourteenth (7.2%) of the radioactivity was retained. In the peptide map the disappearance of two fragments and the appearance of a new one was clearly seen. The new fragment contained nearly the whole amount of the retained activity. These results show that a single methionine sulfoxide is formed during the inactivation of the lipoxygenase by HPOD.

Since HPOD is in all likelihood bound to the Fe in the active center of lipoxygenases, as shown for the purple complex of soybean lipoxygenase-1, the susceptible methionine may be located at the active center. This holds not only for reticulocyte lipoxygenase but also for soybean lipoxygenase, since the inactivation of this enzyme by 5,8,11,14-eicosatetraynoic acid is also accompanied by the formation of 1 mol of methionine sulfoxide per mol of enzyme (see Section IV.C.2) and the alkylation of one methionine of soybean lipoxygenase leads to inactivation as well (88). Preliminary experiments showed that the self-inactivation of pea lipoxygenase is also paralleled by the formation of methionine sulfoxide; however the characteristics of the suicidal reaction of this enzyme differs from those of reticulocyte lipoxygenase. Pea lipoxygenase is preferably inactivated syncatalytically in the dioxygenase reaction rather than by the interaction of HPOD with the enzyme (38). The high sensitivity of reticulocyte lipoxygenase to HPOD at 37°C, which would fully explain the self-inactivation during the dioxygenase reaction, does not exclude the possibility of an additional syncatalytical inactivation for reticulocyte lipoxygenase. Figure 4 shows the proposed mechanism of self-inactivation of reticulocyte lipoxygenase. There is a probable relation to the anaerobic linoleic acid-independent decomposition of 9- and 13-HPOD (see Section IV.B.3). It is assumed than an oxyferryl radical is formed from the liganded ferric enzyme–HPOD complex that may be an intermediate in both formation of hydroperoxidase products (e.g., epoxyhydroxyoctade-

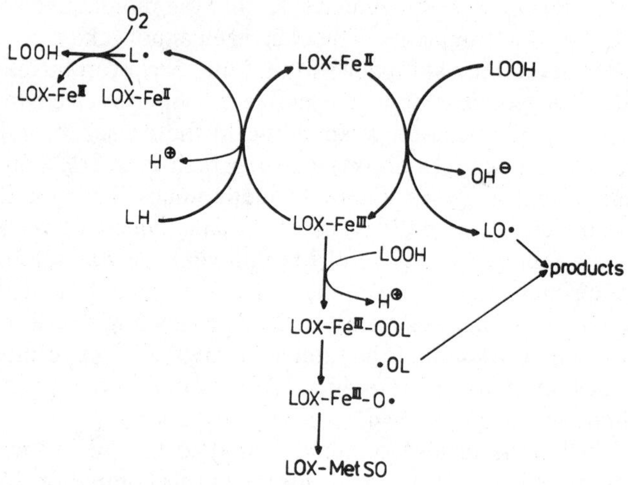

Figure 4. Lipohydroperoxidase activities and self-inactivation of reticulocyte lipoxygenase.

cenoic acids) and monooxygenation of the susceptible methionine leading to self-inactivation. In line with such a mechanism is the observation that the purple soybean ferrilipoxygenase–13L_S-HPOD complex decomposes easily forming, among others, an epoxyhydroxy compound via a true isomerization reaction, since both oxygen atoms in this product were shown to stem from the peroxy group as judged from experiments with ^{18}O-labeled hydroperoxides (89). The decay of the purple complex of soybean lipoxygenase is, however, not accompanied by a loss of enzyme activity. It is conceivable that a special assembly of the active center of soybean lipoxygenase renders this enzyme resistant to the attack by HPODs in contrast to other lipoxygenases. The suicidal inactivation of soybean lipoxygenase by acetylenic fatty acids and arachidonic acid that form hydroperoxides possessing a spatial structure other than HPODs show that even soybean lipoxygenase-1 is in principle susceptible to inactivation.

The self-inactivation of reticulocyte lipoxygenase is not accompanied by global structural changes. The immunological reactivity was retained and no change of the CD spectrum was observed except for secondary long-term alteration after the inactivation by aerobic storage (50).

Self-inactivation was also observed for both lipoxygenase and cyclooxygenase in platelets (90) indicating the suicidal nature of the reactions of the dioxygenases of the arachidonic acid metabolism in animal cells. The self-inactivation of the cyclooxygenase from sheep vesicular glands is however not accompanied by formation of methionine sulfoxide (91). Nevertheless, a related mechanism may be expected, since a variety of cooxidative monooxygenations are known, some of which occur on the level of the hydroperoxidase step of the prostaglandin H synthesis (92,93).

2. Suicidal Substrates

Downing et al. were the first to observe the inactivation of both soybean lipoxygenase and prostaglandin synthase by 5,8,11,14-eicosatetraynoic acid (ETYA), the acetylene analog of arachidonic acid (94). These authors postulated that the acetylenic fatty acid serves as substrate for these enzymes forming a highly reactive hydroperoxyallene that might react covalently with the amino acid residues of the active center. This assumption seemed to be consistent with the deductions of Bloch et al. who studied the inactivation of D-3-hydroxy-decanoyl-(acyl carrier protein)dehydratase by 3-decenoyl-N-cysteamine that proceeds via an allenic intermediate (95). Meanwhile other examples for acetylenic compounds as suicidal substrates were reported (96). Moreover, other acetylenic fatty acids have been described that inhibit lipoxygenases more selectively than ETYA (see the following section). These compounds are even now widely used as powerful inhibitors of arachidonic acid metabolism in animal cells (13,97). The mechanism of action of ETYA was elucidated in a recent study (98). The allene mechanism previously predicted (99) was ruled out, since a covalent binding of labeled ETYA was not observed under the conditions of inactivation. Experimental evidence was adduced, however, which established that ETYA serves as substrate for the lipoxygenase: (a) the conversion of radioactively labeled methyl ester of ETYA to more polar products by lipoxygenase, (b) the oxygen requirement of the inactivation, (c) the competetive protective effect of linoleic acid, (d) the similarity of the activation energies for the dioxygenation of linoleic acid and the inactivation by ETYA, and (e) the formation of 1 mol of methionine sulfoxide/mol enzyme during the reaction with ETYA.

From the evidence adduced and the kinetics of the reaction it was concluded that both the inactivation by acetylenic fatty acids and the self-inactivation of lipoxygenases with physiological substrates proceed according to basically the same mechanisms. An obvious difference exists in the much higher relative rate of inactivation compared with the enzymatic activity in the case of acetylenic fatty acids that exceeds that with polyenoic fatty acids by about three orders of magnitude. The high relative suicidal rate of the reaction with acetylenic fatty acids approaches that expected if each enzymatic turnover gives rise to inactivation. This circumstance and the very low rate of the enzymatic conversion permitted the study of the kinetics of the reaction of lipoxygenases with acetylenic fatty acids. With the rate of the binding of ETYA much greater than that of the enzymatic conversion, the following reaction scheme could be applied:

$$E + S \underset{k_{-1}}{\overset{k_1}{\rightleftharpoons}} ES \overset{k_2}{\longrightarrow} EP \underset{k_{-3}}{\overset{k_3}{\rightleftharpoons}} E + P$$
$$k_4 \downarrow$$
$$E_iP$$
(inactivated)

with k_1, $k_{-1} >> k_2$ and $k_4 >> k_3$.

The rate equations resulting from these premises gave an excellent fit with the experimental data by means of nonlinear regression (98). On the basis of this approach a K_m value for ETYA of 1.3 μM, which is about one tenth of that for arachidonic acid and a reaction constant k_2 of 0.006 s^{-1}, four orders of magnitude lower than in the reaction with arachidonic acid, were estimated (98). Reticulocyte lipoxygenase was shown to obey the same kinetics; its inactivation rate is however much higher than that of soybean lipoxygenase. For this reason, reticulocyte lipoxygenase is more sensitive by two orders of magnitude to ETYA than soybean lipoxygenase if the concentrations for half-inactivation are compared under identical conditions (98).

The different sensitivities of various lipoxygenases to ETYA are of interest in light of the frequently found differential effects of this inhibitor on the synthesis of arachidonic acid metabolites. For polymorphonuclear leukocytes a selective inhibition of the formation of leukotrienes from 5_{DS}-hydroperoxy-6,8,11,14-eicosatetraenoic acid was reported, whereas the formation of 5_{DS}-HETE from arachidonic acid was not inhibited under these conditions (82). Even though some controversal results were reported for neutrophils and eosinophils of some species [for a review see ref. (13)] it may be concluded that the synthesis of biologically active leukotrienes via the 5-lipoxygenase pathway includes two sequential lipoxygenase steps, that is, 5-lipoxygenase and leukotriene A_4 synthase. In contrast to the synthesis of leukotriene, the formation of 5_{DS}-HETE requires only the 5-lipoxygenase, which seems to be relatively resistant to ETYA in some cells of the leukocyte family.

In addition to ETYA a variety of other acetylenic fatty acids are known as lipoxygenase inhibitors (100). 5,8,11-Eicosatriynoic acid, which has been reported to inhibit selectively the 12-lipoxygenase pathway in platelets (101), also inhibits the reticulocyte lipoxygenase with an efficacy by one order of magnitude lower than ETYA, whereas soybean lipoxygenase is resistant (64). This behavior corresponds to the positional specificity of lipoxygenases for polyenoic fatty acids as substrate (see Section IV.B.1). Obviously 5,8,11-eicosatriynoic acid inactivates only those lipoxygenases that also dioxygenate the corresponding trienoic acid or arachidonic acid at C_{12} to a sizable extent. Both conditions are met with reticulocyte and platelet lipoxygenase. 5,8,11-Eicosatriynoic acid also inhibits the formation of 12-HETE in cultured aortic endothelial cells (102).

3. Lipoxygenase Inhibitors

Lipoxygenase inhibitors are valuable tools to study the role of lipoxygenases and their products in biological systems. A selection of inhibitors hitherto known (101–124) is complied in Table III. The mechanism of action of acetylenic fatty acids as suicidal substrates has been discussed in the preceding section. Catechols and hydroxamic acids act as ferric chelators. Most information is available for the interaction of soybean lipoxygenase-1 with 4-nitrocatechol (103). It has been shown that 4-nitrocatechol forms a green 1:1 complex with the ferric form of lipoxygenase and that this inhibitor and 13_{LS}-hydroperoxylinoleic acid compete for the same binding site. For this

TABLE III
Classes of Lipoxygenase Inhibitors

Class	Examples	References
Acetylenic fatty acids	5,8,11,14-Eicosatetraynoic acid	94
	5,8,11-Eicosatriynoic acid	101
	4,7,10,13-Eicosatetraynoic acid	100
Catechols	4-Nitrocatechol	103
	Nordihydroguajaretic acid (NDGA)	104, 105
	Propylgallate	3, 111
	Quercetin and related flavonoids	112–115
	Esculetin and related coumarins	116, 117
	Caffeic acid	118
Hydroxamic acids	Salicylhydroxamic acid	107
	2-Hydroxy-1-naphthylhydroxamic acid	107
	2-Naphthoxyacethydroxamic acid	107
Phenols	3-t-Butyl-4-hydroxyanisol (BHA)	119
	1-Naphthol	2, 107
Pyrazolines	3-Amino-1-(3-trifluoromethylphenyl)-pyrazoline (BW 755C)	120
Phenylhydrazones	Phenidone	121
	Acetone phenylhydrazone	122, 123
Disulfides	Diphenyl disulfide	124

reason the catechols may be regarded as active site probes. In line with this conclusion is the observation that nordihydroguaiaretic acid (NDGA) inhibits the dioxygenation reaction competetively (104,105). For 4-nitrocatechol the state of equilibrium of the enzyme–inhibitor complex is reached very slowly so that a pseudononcompetitive inhibition was observed in steady state kinetic experiments; moreover, the reversible green complex is converted to an irreversible brown one on prolonged incubation (103). Preliminary experiments suggest a similar mechanism for 4-nitrocatechol also with reticulocyte lipoxygenase (106). The actions of both catechols and hydroxamic acids are abolished by complexation with ferric salts prior to addition to reticulocyte lipoxygenase (107). In contrast, subsequent addition of Fe^{3+} did not reverse the inhibition of reticulocyte lipoxygenase by salicylhydroxamic acid; however, methodical difficulties may have arisen from the very poor solubility of ferric salts at neutral pH. A complex between reticulocyte lipoxygenase and 2-hydroxynaphthyl-1-hydroxamic acid was detected spectrophotometrically (107). In addition to catechols and phenolic hydroxamic acids a variety of other phenols inhibit the reticulocyte lipoxygenase, but the concentration of their half-inhibition are much higher. Examples are the monomethylethers of catechols as well as 1-naphthol, 2-naphthol, and the antioxidant 3-*t*-butyl-4-hydroxyanisol (BHA).

The ability of some antioxidants to inhibit lipoxygenases deserves particular discussion. In the earlier literature the inhibitory effects of NDGA, propylgallate, and related compounds were ascribed to their property of being radical scavengers, which was believed to be in line with the occurrence of intermediate radicals in the catalytic cycle of lipoxygenases (2). However, we found that 2,6-di-*t*-butyl-4-hydroxy-toluene (BHT) which is one of the most powerful antioxidants for nonenzymatic lipid peroxidation, does not inhibit reticulocyte lipoxygenase at all. Obviously the intermediate radicals arising during the catalytical cycle remain enzyme bound and, hence, inaccessible for a radical scavanger. With pea lipoxygenase there exist, however, special conditions in which partial inhibition by BHT occurs (38). Apparently in this case the intermediate radicals dissociate from the enzyme and cause secondary reactions. An example is the cooxidative destruction of chlorophyll in the aerobic system pea lipoxygenase–linoleic acid, which is completely inhibited by BHT (38). In this manner BHT may be a probe to discriminate secondary radical reactions concomitant to primary lipoxygenase-catalyzed reactions. However, soybean lipoxygenase-1 is completely inhibited by low concentrations of BHT (22). As far as NDGA and BHA are concerned, these compounds seem to combine in their molecules the property of an active site probe for lipoxygenases via their ferric chelating properties and that of radical scavangers. This dual nature may be a disadvantage in studies of the mechanism of the lipoxygenase reaction. On the other hand, these compounds are universal inhibitors of both enzymatic and nonenzymatic lipid peroxidation.

The mechanism of action of pyrazolines and related compounds is not yet established. From the different shapes of the titration curves of the inhibition it may be concluded that the mechanisms of action of ferric chelators and pyrazolines are different.

TABLE IV

Inhibitory Potencies of Some Compounds with Reticulocyte Lipoxygenase

Compound	Concentration of half-inhibition (μM)
5,8,11,14-Eicosatetraynoic acid	0.12
5,8,11-Eicosatriynoic acid	1.3
Nordihydroguaiaretic acid	0.5
Propylgallate	13
4-Nitrocatechol	4.6
2-Hydroxynaphthyl-1-hydroxamic acid	3.7
2-Naphthoxyacethydroxamic acid	2.5
Salicylhydroxamic acid	47
1-Naphthol	500
2-Naphthol	650
3-t-Butyl-4-hydroxyanisol	160
3-Amino-1-(3-trifluoromethylphenyl)-pyrazoline (BW 755C)	15

Most of the lipoxygenase inhibitors act universally, that is, they inhibit all lipoxygenases independent of their origin and reaction specificity. However, there are great quantitative differences in the inhibitory potencies towards various lipoxygenases. For many inhibitors reticulocyte lipoxygenase proved to be more sensitive by two orders of magnitude than soybean lipoxygenase-1. Among other plant lipoxygenases the pea lipoxygenase shows inhibitor sensitivity comparable to the reticulocyte enzyme. The IC_{50} values for reticulocyte lipoxygenase with a selection of inhibitors are listed in Table IV. For some inhibitors these values are in the range of those reported for the inhibition of the 5-lipoxygenase pathway of the arachidonic acid metabolism in leukocytes and related cells. Apparently the inhibitor sensitivity is not necessarily determined by the positional specificity (except for acetylenic fatty acids and possible other suicidal substrates). Subtle differences in the ligand-binding site may be responsible for the different inhibitor sensitivity of lipoxygenases. The only compound that is reported to be an inhibitor of the 5-lipoxygenase pathway (108), but does not inhibit reticulocyte lipoxygenase, is benoxaprofen. Since this compound was not yet tested on the purified 5-lipoxygenase, there is a lack of evidence that the site of attack is the lipoxygenase itself. This lipoxygenase requires activation by calcium, and it is conceivable that the activation process could be an alternative site of drug action.

Another aspect of inhibitors of the arachidonic acid metabolism is the selectivity for lipoxygenases and cyclooxygenase. Whereas a variety of selective cyclooxygenase inhibitors are known (e.g., acetylsalicylic acid, indomethacine, and other nonsteroidal antiinflammatory drugs), the most frequently used lipoxygenase inhibitors such as ETYA, NDGA, propylgallate, and BW 755C are dual inhibitors of both the lipoxygenase and cyclooxygenase pathway with the lipoxygenase being more sensitive. This is somewhat surprising, since the active center of the cyclooxygenase, which includes essential heme, differs from that of lipoxygenases. Unfortunately the mode of action of catechols and pyrazolines on the cyclooxygenase (prostaglandin synthase) is not yet established.

The inactivation of reticulocyte lipoxygenase by hydroperoxy fatty acids was already mentioned in relation to the self-inactivation (see Section IV.C.1). Hydrogen peroxide is also a strong inactivator of both soybean and reticulocyte lipoxygenase. From CD spectral investigations on soybean lipoxygenase-1 it was concluded that hydrogen peroxide acts near the iron chromophore (29). Moreover, 13_{L_S}-hydroperoxylinoleic acid protects the enzyme from inactivation by hydrogen peroxide. From these observations it may be concluded that hydrogen peroxide acts in an identical way as hydroperoxy fatty acids.

Cyanide is a poor inhibitor for both reticulocyte and soybean lipoxygenase. Reticulocyte lipoxygenase is inhibited by high concentrations of cyanide, if the enzyme is converted to the ferric form by addition of equimolar hydroperoxylinoleic acid (unpublished results). In contrast, a variety of other lipoxygenases such as those of platelets (109), and testicles (22) are reported to be sensitive to cyanide. In contrast to certain ferric chelators, ferrous ligands inhibit lipoxygenases only weakly if at all. Reticulocyte lipoxygenase is partly inhibited by a high partial pressure of carbon monoxide (unpublished results). A NO complex is reported for soybean lipoxygenase-1 (110); unfortunately, this complex is destroyed by either hydroperoxy fatty acids or oxygen. α,α'-Dipyridyl does not inhibit reticulocyte lipoxygenase at 1 mM concentrations. 8-Hydroxyquinoline inhibits at this concentration, but the inhibition is accompanied by a precipitation of the enzyme so that unspecific effects cannot be excluded.

1. Lipoxygenase Kinetics

Studies on the kinetics of lipoxygenases are complicated by a variety of facts. One is the limited solubility of the substrate. Most lipoxygenases have their optimum near to the neutral pH (except for soybean lipoxygenase-1). In this range the fatty acid monomers tend to form multimolecular aggregates, acid soap dimers, and micelles. The formation of acid soaps is favored by the presence of hydroperoxy fatty acids. Whereas soybean lipoxygenase is believed to act exclusively on the monomers it cannot be excluded that reticulocyte lipoxygenase is also able to attack the aggregates. The activity of reticulocyte lipoxygenase towards phospholipid liposomes and biological membranes would be in line with such an assumption.

A characteristic feature of all lipoxygenases so far studied kinetically is the obligatory activation of the enzyme by the product hydroperoxy fatty acid. The hydroperoxy group appears to be essential for the activating effect. The activation may be in part due to the oxidation of the ferrous form of the enzyme to a higher oxidized state (for a detailed discussion see Section IV.D.2). For activation, micromolar concentrations of hydroperoxy fatty acid are sufficient, which are normally present in commercial preparations of polyenoic fatty acids owing to autoxidation. If pure substrate is used for measurement of lipoxygenase activity, a lag phase occurs that is overcome either after a sufficient amount of product has been produced by the lipoxygenase itself or by addition of product.

Lipoxygenases are also inhibited by the substrate fatty acid dependent on the concentration of oxygen, the second substrate of these enzymes.

For soybean lipoxygenase-1 a variety of kinetic studies were conducted, for example, by Cook and Lands (86), Egmond et al. (125), Gibian and Galaway (126), and Lagocki et al. (127). In most kinetic models it was assumed that there are two binding sites for which substrate fatty acid and its product compete. One of them is the catalytic site, the other the activator site. With such models it was possible to describe the kinetics of soybean lipoxygenase-1 including product activation and substrate inhibition on the basis of experimental data [for a review see refs. (27) and (29)].

An analysis of progress curve kinetics of reticulocyte lipoxygenase was conducted by Ludwig et al. (128). In this study the dioxygenation of linoleic acid was measured spectrophotometrically at pH 7.4 in the presence of 0.2% cholate that allows the measurement of maximal rates. The experimental conditions were so chosen that a self-inactivation of the lipoxygenase (see Section IV.C.1) did not occur to a sizable extent. Several models were fitted to the experimental data that also include two binding sites for both substrate linoleic acid and product, $13L_S$-HPOD, at the catalytic center and a putative regulatory site. The occurrence of two redox forms of the enzyme during the catalytic cycle was considered that were designated as E (ferrous enzyme) and E* (ferric activated enzyme). It was presumed further that only the ferrous form possesses a binding site for oxygen and that only the ferric form is catalytically competent for the step of hydrogen abstraction. Sequential irreversible steps of hydrogen abstraction and peroxide formation were assumed (see Section IV.D.2). The simplest kinetic model to meet these requirements is shown in Fig. 5. The parameters obtained by this model are compiled in Table V. The following conclusions may be drawn: (a) The product activates the ferric enzyme independent of the preceding oxidation of the ferrous enzyme. (b) The hydrogen abstraction is the rate-limiting step as expected; it is stimulated 10-fold if the putative regulatory site is occupied by the product. (c) The affinity for oxygen is fairly low. The latter fact was somewhat surprising, since a very low apparent K_m value is observed under conditions of a normal lipoxygenase assay. This discrepancy is due to the much faster (two orders of magnitude) product formation as compared with the hydrogen abstraction, so that an oxygen concentration below $10\mu M$ is sufficient to maintain a maximal rate of the whole catalytical cycle despite the low oxygen affinity. The model also reflects the apparant inhibition by substrate at low oxygen concentrations (Fig. 6) as experimentally observed.

Although the model affords a satisfactory description of the experimental data, there is lack of evidence for the existence of two fatty acid binding sites. Our laboratory plans experiments to determine the number of fatty acids bound to lipoxygenase. There are,

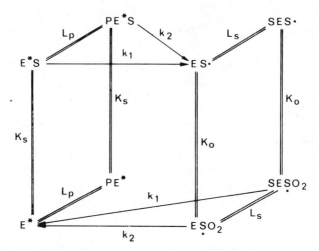

Figure 5. Simplified kinetic two-site model for reticulocyte lipoxygenase. E, ferrous lipoxygenase; E*, ferric lipoxygenase; S, linoleic acid; P, 13_{LS}-hydroperoxylinoleic acid; O, oxygen. Symbols on the left side of E or E* refer to ligands of the regulatory site and those on the right side of E or E* to ligands to the catalytic site.

TABLE V

Kinetic Parameters of the Dioxygenation of Linoleic Acid by Reticulocyte Lipoxygenase According to the Model in Fig. 5

K_m values		
K_S (linoleic acid, catalytic site)		6.36 μM
K_O (O_2, catalytic site)	2290	μM
L_S (linoleic acid, regulatory site)		50.6 μM
L_P (13_{LS}-HPOD, regulatory site)		4.84 μM
Rate constants		
k_1 (H abstraction, free regulatory site)		0.016
k_2 (H abstraction, 13_{LS}-HPOD at the regulatory site)		0.17
k_1' (product formation, free regulatory site)	21.5	
k_2' (product formation, linoleic acid at the regulatory site	4.1	

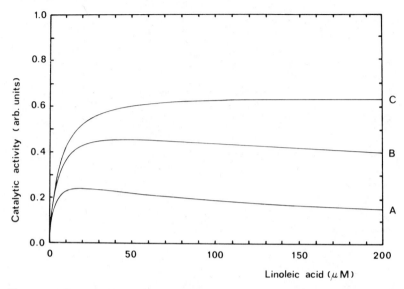

Figure 6. Computer-fitted curves for the catalytic activity of reticulocyte lipoxygenase as a function of the concentration of linoleic acid at a fixed concentrations of 13_{LS}-hydroperoxylinoleic acid (10 μM) and various concentrations of oxygen (A, 1 /μM; B, 5/μM; C, 50/μM).

however, some arguments in favor of only one fatty acid binding site. The first one comes from the anaerobic lipohydroperoxidase activity of soybean lipoxygenase-1. Verhagen et al. could describe satisfactorily the steady state kinetics by a simple scheme representing a substituted-enzyme mechanism (ping-pong) with double substrate inhibition (73). The model presumes the occurrence of only one site for the alternative binding of either linoleic acid or 13_{LS}-HPOD. The experimental conditions were so chosen that no formation of acid soaps and micelles occurred (pH 10, low concentrations of reactants). A strong argument in favor of such a mechanism is the fact that the partial steps of the reaction cycle can be measured separately so that a compulsory ternary complex of enzyme, linoleic acid, and 13_{LS}-HPOD can be excluded. One may conclude that at least under anaerobic conditions only one fatty acid binding site is operative. This fact does not exclude, however, the possibility that

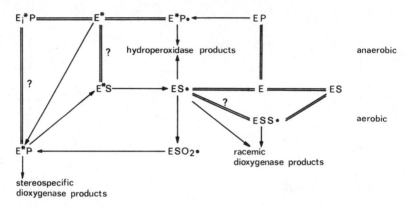

Figure 7.   Kinetic one-site model for both aerobic and anaerobic reactions of lipoxygenases. Symbols as in Fig. 5.

a binding of oxygen to ferrous lipoxygenase (see Section IV.D.2) opens an additional binding site that has a regulatory role only aerobically.

A one-site model extending that for the anaerobic reaction of lipoxygenase (73), including the premise of the two-site model of Ludwig et al. (128), also fits the kinetics of the aerobic reaction of reticulocyte lipoxygenase in a satisfactory manner (128). This model (Fig. 7) must include a dissociation or displacement by substrate of the intermediate linoleic acid radical, giving rise to an inactive ferrous enzyme–substrate complex. This step is proposed to be the basis for the substrate inhibition of lipoxygenases. The radical should be converted nonenzymatically to racemic hydroperoxylinoleic acids. Indeed, a share of at least 10% of racemic products from linoleic acid was reported for a variety of lipoxygenases (29). The percentages vary with pH, temperature, substrate concentration, and other conditions. Unfortunately no data are available to indicate whether the formation of racemic products is enhanced under conditions of substrate inhibition such as at low oxygen concentrations as would be predicted by the one-site model in Fig. 7.

In the aerobic one-site model the activation by the product, which is also required if the reaction starts from ferric enzyme (see Section IV.D.2), is explained by the assumption that S binds faster to E* by a displacement mechanism if the fatty acid binding site is occupied by P. The binding of S to E*P may be preceded by a chemical activation by P of the ferric iron to a reactive species competent for hydrogen abstraction (see Section IV.D.2). Since P was also reported to inhibit the anaerobic hydroperoxidase reaction, two distinct E*P complexes have to be proposed, which were indeed detected recently for soybean lipoxygenase-1 by means of EPR spectroscopy (87).

2. Mechanism of Lipoxygenase Catalysis

As stated before, lipoxygenases catalyze two types of reactions:

1. The dioxygenase reaction that is initiated by a homolytic cleavage of a C–H bond.

2. The lipohydroperoxidase reaction, which involves a homolytic scission of an O–O bond.

The two reactions may be combined as consecutive reactions. Most information concerning the mechanism of catalysis has been obtained from experiments with soybean lipoxygenase-1 (27–29). It was shown that at least three forms of the enzyme occur, which can be interconverted: (a) the ferrous form, which is silent in EPR spectroscopy and does not show particular features in the absorption spectrum, (b) the yellow ferric form showing an EPR signal at g 6 and a broad shoulder in the near-UV region of the absorption spectrum, and (c) the purple complex between the ferric enzyme and 13_{LS}-hydroperoxylinoleic acid (13_{LS}-HPOD) showing an additional EPR signal at g 4.3 and an absorption maximum at 580 nm. A high-spin state of iron was established for all forms (129–131).

The ferrous lipoxygenase is converted to the ferric enzyme on addition of an equimolar amount of 13_{LS}-HPOD. Stopped-flow kinetic measurements of this transition have revealed that the rate constant of this process is only about one fifth of that of the dioxygenation of linoleic acid (132). In addition to the EPR and optical spectral changes the formation of ferric lipoxygenase is accompanied by a quenching of the fluorescence maximum at 328 nm (133). The spectral changes upon addition of 13_{LS}-HPOD were also observed with reticulocyte lipoxygenase (75).

Ferric soybean lipoxygenase-1 is reduced to the ferrous form by linoleic acid under anaerobic conditions. This process is accompanied by a hydrogen abstraction at C_{11} of linoleic acid; accordingly the rate constant of the corresponding reaction with $[11\text{-}^2H_2]$linoleic acid amounts to only one fourth of that with unlabeled linoleic acid (132). The rate of the reduction of ferric lipoxygenase by linoleic acid is higher than that of the oxidation of ferrous lipoxygenase by 13_{LS}-HPOD, but does not reach the rate of the overall dioxygenation of linoleic acid, which is 2.6-fold higher (132). The oxidation of ferrous lipoxygenase by 13_{LS}-HPOD and the reduction of ferric lipoxygenase by linoleic acid, if coupled sequentially, give rise to the catalytic cycle of the anaerobic lipohydroperoxidase activity first proposed by de Groot et al. (74) (see Section IV.B.3).

Addition of 13_{LS}-HPOD to ferric soybean lipoxygenase-1 gives rise to the formation of several species of an equimolar complex that differ in the ligand symmetry of their EPR signals of the ferric iron (87,129). The formation of this complex proceeds very quickly, exceeding the limits of the stopped-flow kinetic analysis (132).

There are some indications that ferrous lipoxygenase is able to bind oxygen reversibly in the absence of fatty acids. One argument is the enhancement of the fluorescence of ferrous lipoxygenase on addition of oxygen (134). The same effect was also observed with reticulocyte lipoxygenase (unpublished results). As mentioned, the fluorescence of lipoxygenases is due to tryptophan(s) situated in a nonpolar region and is also altered during the redox change of the enzyme iron (133,134). Therefore it may be concluded that the tryptophan(s) responsible for the fluorescence changes are located at or in the neighborhood of the iron of the active center and, consequently, that oxygen is bound to ferrous iron. A second indication for the binding of oxygen is the existence of a nitric oxide complex of ferrous soybean lipoxygenase-1 under anaerobic conditions as detected by EPR spectroscopy (110). The assumed binding of oxygen to ferrous lipoxygenase may be analogous to that of ferrohemoproteins. In contrast, complexes of molecular oxygen with ferric compounds are generally not known. Moreover, there are no indications for the existence of such a complex with ferric lipoxygenase.

From all the experimental data including those of kinetic studies (see Section IV.D.1) a general picture of the lipoxygenase catalysis may be deduced. The following premises must be considered: (a) Hydrogen abstraction from the substrate is caused by one of the ferric species of the lipoxygenase, which is reduced to a ferrous species. (b) The species competent for hydrogen abstraction must be a highly reactive one, since normal ferric compounds cannot catalyze the homolytic cleavage of an C–H bond. (c) The activation requires the presence of a hydroperoxy fatty acid. (d) Activation can start from either ferric lipoxygenase or ferrous lipoxygenase in the presence of oxygen. (e) Oxygen is bound with low affinity to the enzyme or is introduced into the fatty acid molecule only if the lipoxygenase is in a ferrous state. (f) Hydrogen abstraction and oxygen insertion do not proceed independently of each other owing to their antarafacial character.

Unfortunately, it is not possible to measure the partial reactions of the dioxygenase reaction as was done with the anaerobic lipohydroperoxidase reaction (132). Nevertheless it is reasonable to assume similar steps so that a catalytic cycle arises as shown in Fig. 8. In the initial step of the cycle the hydrogen is removed from the substrate by a ferric species of the enzyme in an energetically favorable way as H^+ rather than as $H\cdot$. The stereochemical aspects of the single steps of the cycle are discussed in the companion article of Kühn et al. (30). The hydrogen abstraction proceeds in such a manner that the fatty acid radical formed remains enzyme bound. In addition, the enzyme acquires the ferrous state enabling oxygen binding as a consecutive step of the catalytic cycle. It should be emphasized, however, that the enzyme-bound fatty acid radical may possess by itself a sufficiently high reactivity with oxygen so that a specific binding site for oxygen located on the enzyme itself and operative in the catalytic cycle appears to be not absolutely mandatory. The low association constant for oxygen despite a very high rate of oxygen introduction as calculated from the kinetic model for reticulocyte lipoxygenase (see Section IV.D.1) would be in line with such an assumption. Nevertheless, the active center of the enzyme must be involved in the step of oxygen introduction, since it proceeds stereospecifically in contrast to nonenzymatic lipid peroxidations.

The reaction of oxygen with the enzyme-bound fatty acid radical may give rise to the formation of a hydroperoxy radical that picks up an electron from ferrous iron yielding the anion of the hydroperoxy fatty acid, the final product of the dioxygenase reaction. By this inner electron transfer the ferric state of the enzyme is recovered. It should be stressed that the proposed catalytic cycle of

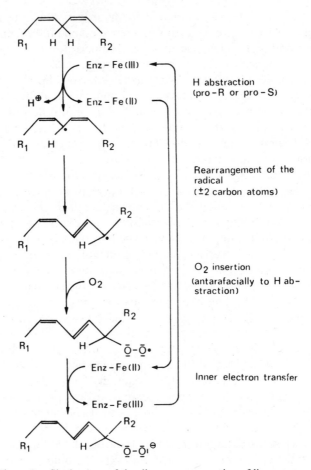

Figure 8.   Single steps of the dioxygenase reaction of lipoxygenases.

the dioxygenase reaction (Fig. 8) does not necessarily imply that these steps proceed sequentially. An alternative possibility is a simultaneous abstraction of hydrogen and insertion of oxygen via a multicenter mechanism. In any case the hydrogen abstraction is the prerequisite for the oxygen introduction. In this way the lipoxygenase catalysis is initiated by the activation of the fatty acid, whereas a variety of other oxygenation reactions such as those catalyzed by the cytochrome P_{450} system are initiated by the activation of the dioxygen molecule (135,136). The occurrence of a catalytic cycle as shown in Fig. 8 is strongly supported by the fact that under conditions of exhaustion of oxygen in the reaction mixture there is a formation of fatty acid dimers arising from the fatty acid radical via the anaerobic lipohydroperoxidase reaction (see Section IV.B.3).

The catalytic cycle in Fig. 8 does not consider the question of the ferric species of lipoxygenase competent for hydrogen abstraction. One of the ferric lipoxygenase–hydroperoxy fatty acid complexes, perhaps the purple enzyme, which was proposed to show charge transfer interactions of the enzyme with ligands (87), is a likely candidate for the active species or a precursor of it. Another possibility is the occurrence of a higher valency of iron during the catalytic cycle. Higher valencies of iron (e.g., a "ferryl" state) were proposed to be involved in the catalysis of heme peroxidases and catalase (137), cytochrome P_{450} (135,136), and in the iron–bleomycin catalyzed destruction of DNA (138). The fact that both ferrous and ferric lipoxygenase require hydroperoxy fatty acid for activation and do not differ in their kinetic characteristics on the one hand, and that the anaerobic oxidation of ferrous lipoxygenase by hydroperoxylinoleic acid is too slow to be involved in the dioxygenase reaction on the other, may infer that there exist two different pathways of activation of ferrous and ferric lipoxygenase each with a very high reaction rate. The activation of ferrous lipoxygenase may start from the putative complex with oxygen (see the preceding paragraphs of this section) and bypass the normal (nonactivated) ferric state.

Unfortunately the chemistry of iron binding in lipoxygenases is not yet established. The iron seems to be bound very tightly, since it cannot be removed without denaturation of the enzyme. There are some indications for the presence of methionine (49,88,98) and tryptophan at or in the neighborhood of the catalytically active iron.

In Fig. 9 the catalytic cycles of both dioxygenase [Reactions (1)–(3)] and lipohydroperoxidase activity [Reactions (5) and (6)] are shown. For simplicity the phases of activation are omitted. The scheme may also explain the interrelationship between the two activities under aerobic conditions. The lipohydroperoxidase activity of both reticulocyte and soybean lipoxygenase-1 with $13L_S$-HPOD and linoleic acid as substrates is completely inhibited aerobically, whereas pea lipoxygenase forms oxodienoic acids aerobically

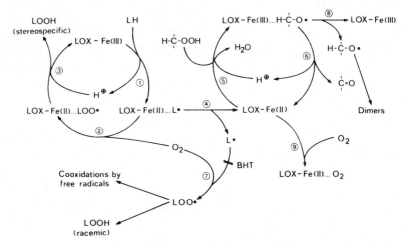

Figure 9.   Proposed reaction scheme for both aerobic and anaerobic reactions of lipoxygenases with linoleic acid as primary substrate. LH, linoleic acid; LOOH, hydroperoxylinoleic acid(s); LOX, lipoxygenase; BHT, 2,6-di-*t*-butyl-4-hydroxytoluene.

(38,52). As shown in Fig. 9 Reaction (5) requires the presence of free ferrous enzyme, which may originate from the dissociation of the LOX–Fe^{II}···L· complex [Reaction (4)]. For most lipoxygenases such as those from soybeans and reticulocytes the rate of Reaction (2) is much greater than that of Reaction (4), so that the latter is suppressed under aerobic conditions, whereas Reaction (2) cannot take place under anaerobic conditions so that Reactions (4)–(6) become possible. The pecularity of pea lipoxygenase may consist in a high tendency of the LOX–Fe^{II}···L· complex to dissociate so that a sufficient steady state concentration of the free ferrous enzyme arises even under aerobic conditions. At the same time free linoleic acid radicals are formed that yield hydroperoxy radicals [Reaction (7)], which may be responsible for both free radical-mediated cooxidations (see Section IV.B.6) and the formation of racemic products (29). The requirement of the anaerobic lipohydroperoxidase reaction for linoleic acid is based on Reaction (1), which yields free ferrous enzyme via Reaction (4) to initiate the catalytic cycle involving reactions (5) and (6). Under anaerobic conditions the free L· radical arising from Reaction (4) may react with either LOX–Fe$^{II}\cdots$L· forming L–L dimer or LOX-Fe$^{III}\cdots$LO· to LOL dimer. Indeed both dimers were found with soybean lipoxygenase-1 (72) and reticulocyte lipoxygenase (43). With pea lipoxygenase no dimers were observed aerobically despite formation of oxodienoic acids, which is in line with the scheme in Fig. 9.

V. Quasi-Lipoxygenase Activity of Hemoproteins

The ability of hemoglobin and other heme compounds to catalyze lipid peroxidation has been known since the pioneering work of Tappel and others [(3) and literature cited therein, (139,140)]. This classic hematin-catalyzed lipid peroxidation proceeds via radical chain reactions and differs in principle from lipoxygenase reactions. Some years ago, however, we could detect a new catalytic effect of hemoglobin, the quasi-lipoxygenase activity (141). This activity was unequivocally discriminated from other catalytic effects of hemoglobin (142). It resembles in many respects true lipoxygenase activities. For this reason, the quasi-lipoxygenase activity of hemoglobin is a valuable model for the general mechanism of the lipoxygenase catalysis in view of the fact that the structure of the prosthetic group of hemoproteins is exactly known in contrast to that of lipoxygenases. An additional interest in such systems arises from the fact that cyclooxygenase, the key enzyme of prostaglandin synthesis (143–145), and the lipoxygenase from the fungus *Fusarium oxysporum* (146,147) contain heme or require it for their activity.

A remarkable feature of the quasi-lipoxygenase activity is the high substrate specificity for di-*cis*-dienoic fatty acids. Among a variety of polyenoic fatty acids only linoleic acid and 11,14-*cis,cis*-eicosadienoic acids are attacked, whereas trienoic and tetraenoic acids and other fatty acids are competitive inhibitors of this reaction (141,142). A free carboxylic group is not obligatory, since glycerol-1-monolinoleate and linoleyl alcohol are also attacked. The specificity for dienoic acids may be explained on steric grounds. Apparantly only dienoic fatty acids are substrates that attach to the heme in such a manner that the hydrogen atom to be abstracted is situated in the neighborhood of the central iron atom, whereas in the complexes with trienoic or tetraenoic fatty acids this is not the case.

Preliminary spatial models of these complexes and quantum-chemical calculations support this assumption (148).

The quasi-lipoxygenase activity is exerted by hemoglobin, myoglobin, and cytochrome P_{450}LM2 but not by cytochrome c and catalase; free heme shows one tenth of the activity of hemoglobin. Accordingly, heat denaturation of hemoglobin causes 90% loss of activity. One may assume that the ability of the ferro form of the hemoprotein to bind oxygen is a prerequisite for the quasi-lipoxygenase activity. Ferrohemoglobin and ferrihemoglobin do not differ in their activity. This behavior is understandable as both redox states of iron are involved in the catalytic cycle as is the case with true lipoxygenases (see Section IV.D.2). The valency change of iron during the quasi-lipoxygenase reaction was established by trapping ferrohemoglobin as a carbon monoxide complex after the reaction with linoleic acid had been started from ferrihemoglobin (141). Accordingly the quasi-lipoxygenase activity is inhibited by both cyanide and carbon monoxide while other catalytic activities of hemoglobin are inhibited only by cyanide (3).

The kinetics of the quasi-lipoxygenase activity shares many features with true lipoxygenases. Here again an activation by product is required. For this reason a long lag period is observed that is abolished by about $8\mu M$ 13_{LS}-hydroperoxylinoleic acid, whereas 13_{LS}-hydroxylinoleic acid prolongs the lag period. The quasi-lipoxygenase activity leads to a self-catalyzed heme destruction as indicated by the disappearance of the Soret peak in the absorption spectrum. The dependence on substrate concentration of the quasi-lipoxygenase activity of myoglobin obeys exactly Michaelis–Menten kinetics, while hemoglobin shows cooperativity with Hill coefficients of about 2.0–2.5 and apparant $S_{0.5}$ values of 0.3–0.5 mM dependent on the experimental conditions. In contrast, other catalytic activities of hemoglobin such as hemin-catalyzed lipid peroxidation show a linear dependence on the square root of the catalyst concentration that is typical for radical chain reactions (3). The pH optimum of the quasi-lipoxygenase activity is 8.5. A convenient measurement is only possible at a catalyst concentration of 0.01–0.1 μM. Under these conditions linoleic acid is converted predominately to a mixture of 9- and 13-hydroperoxylinoleic acid in equal amounts; the composition of the enantiomers has not yet been analyzed. Less than 10% of hydroxylinoleic acids and oxodienoic acids were found

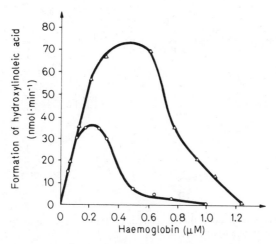

Figure 10. Quasi-lipoxygenase activity as a function of hemoglobin concentration. △, oxygen uptake; ○, formation of conjugated dienes.

indicating that aerobic lipohydroperoxidase activity is negligible at this concentration of hemoglobin. At higher concentrations of hemoprotein the share of lipohydroperoxidase activity increases considerably as judged from the nonstoichiometry of oxygen uptake and formation of conjugated dienes (Fig. 10). The lipohydroperoxidase activity leads to a decomposition of hydroperoxylinoleic acids that are necessary for the product activation of the quasi-lipoxygenase activity. For this reason the quasi-lipoxygenase activity is completely inhibited at concentrations of hemoglobin higher than 1 μM.

The quasi-lipoxygenase activity is inhibited by a variety of lipoxygenase inhibitors, in particular by dual inhibitors of both lipoxygenase and cyclooxygenase such as propylgallate and salicylhydroxamic acid. The inhibitory action of salicylhydroxamic acid is, however, not abolished by complexation with ferric salts in contrast to the inhibition of true lipoxygenases (see Section IV.C.3). Moreover, the quasi-lipoxygenase activity is inhibited by 2,6-di-t-butyl-4-hydroxytoluene (BHT). The BHT sensitivity may imply that the intermediate radicals formed during the quasi-lipoxygenase reaction are somehow accessible.

A scheme for the catalytic cycle of the quasi-lipoxygenase activity

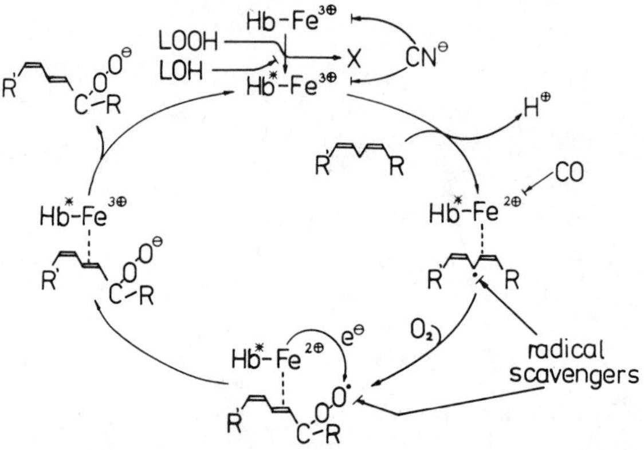

Figure 11. Proposed scheme for the catalytic cycle of the quasi-lipoxygenase activity of oxygen-binding hemoproteins.

has been proposed that is consistent with experimental data (Fig. 11). It is closely similar to that of the dioxygenase activity of lipoxygenases. An activated species of ferrihemoglobin, (Hb*Fe^{3+}) is presumed that may be formed by interaction with hydroperoxylinoleic acid and recovered in the catalytic cycle. Such an activated species must possess a reactivity high enough to cause hydrogen abstraction. The concomitant heme destruction is in line with the occurrence of a reactive species. An oxyheme radical as proposed by Tappel (3) may be a possible candidate for it. The cooperative character of the quasi-lipoxygenase reaction suggests that the in-

TABLE   VI
Common Characteristics of Genuine Lipoxygenases and
Quasi-lipoxygenase Activity

Loss of activity on denaturation
Autoactivation by product
Self-catalyzed destruction
Participation of both ferric and ferrous states of iron in the
    catalytic cycle
Identical activation energy
Kinetics typical for enzymatic reactions

TABLE VII

Comparison of Catalytic Properties of Hemoproteins Related to Lipid Peroxidation

	Quasi-lipoxygenase activity	Hemin-catalzyed lipid peroxidation	Lipohydroperoxidase activity
Catalyst	Hemoglobin, myoglobin, cytochrome P_{450}, hemin (not cytochrome c; catalase)	All heme compounds	All heme compounds
Substrates	Dienoic acids	All unsaturated lipids	Organic hydroperoxides
Requirement for hydroper-oxides	+	+	+ (as substrate)
Valency states involved	Fe^{II}, Fe^{III}	Fe^{III}	Fe^{III}
Inhibitors	CN^-, CO, phenolic antioxidants, tri- and tetraenoic fatty acids	CN^-, phenolic antioxidants	CN^-, phenolic antioxidants
Self-catalyzed heme destruction	+	+	+
Activation energy (kJ/mol)	19	21	55
Dependence of rate (v) on catalyst concentration (c)	$v \propto c$ (below 0.2 μM)	$v \propto c^{1/2}$	$v \propto c^{1/2}$

teractions among the subunits of hemoglobin are involved and that the substrate is bound to the sixth ligand position of the heme iron. This implies further the existence of one binding site for both linoleic acid and hydroperoxylinoleic acid rather than an additional regulatory site (see Section IV.D.2).

The common characteristics of true lipoxygenases and of the

quasi-lipoxygenase activity are listed in Table VI. The agreement of many of their properties prompted us to use the term quasi-lipoxygenase. In contrast, the other catalytic properties of heme compounds differ in many respects; therefore, the names pseudolipoxygenase and pseudolipohydroperoxidase, respectively, should be preferred for these activities to designate their nonenzymatic nature. A comparison of the three catalytic effects of hemoglobin is compiled in Table VII.

The biological importance of the quasi-lipoxygenase activity of hemoglobin can be excluded, since it is completely inhibited at physiological concentrations. For cytochrome P_{450} it is conceivable that hydroperoxylinoleic acids formed by this activity may serve as an alternative oxygen donor for cytochrome P_{450}-dependent monooxygenations or as an activator for the cyclooxygenase or true lipoxygenases.

According to present knowledge there are three different ways by which heme compounds may catalyze the formation of hydroperoxy lipids: (a) the quasi-lipoxygenase activity, (b) radical chain oxidation initiated by the interaction of heme compounds with hydroperoxylipids (presumably via an alkoxy radical) (3), and (c) by formation of an activated oxygen species via a reaction chain including oxyhemoglobin → superoxide radical → hydroxyl radical and singlet oxygen (149–151).

Another catalytic property of hemoglobin that is of interest for lipoxygenase research is its ability to convert hydroperoxypolyenoic fatty acids to epoxyleukotrienes (63,152). This type of reaction is also catalyzed by true lipoxygenases including reticulocyte lipoxygenase (see Section IV.B.4). Whether the hemoglobin-catalyzed reaction is related to the quasi-lipoxygenase activity remains to be clarified.

VI. Interaction of Lipoxygenases with Membranes

Reticulocyte lipoxygenase has the unique property of attacking biological membranes even in the absence of a lipid-hydrolyzing-enzyme. At least the lipoxygenases from soybeans, pea seeds, and wheat kernels do not share this property, as shown in comparative studies in our laboratory. Only indirect indications for a possible direct atack of lipoxygenases on complex lipids in plants have been reported for potato (153) and cauliflower (154), so that the reticulocyte enzyme is the only clear-cut example of a membrane-attacking lipoxygenase until now. This fact does not exclude the possibility of the secondary involvement of lipoxygenases in membrane breakdown after liberation of free polyenoic fatty acids by a lipid-hydrolyzing enzyme.

With rat liver mitochondria or beef heart submitochondrial particles at 37°C a strong oxygen uptake is produced by reticulocyte lipoxygenase, which is prevented by the inhibitors salicylhydroxamic acid or nordihydroguaiaretic acid (Fig. 12). The rate of oxygen uptake drops rapidly, since here again self-inactivation of the lipoxygenase occurs. It is noteworthy that the extent of oxygen consumption exceeds considerably the calculated amount of lipoxygenase-susceptible fatty acids in the membranes. This discrepancy may suggest consecutive oxygen-dependent degradation of the hydroperoxylipids primarily formed. An involvement of the respiratory chain can be excluded, since antimycin A does not inhibit the lipoxygenase-mediated oxygen uptake.

The action of reticulocyte lipoxygenase on rat liver mitochondria causes drastic structural damages as seen from electron micrographs (Fig. 13). Moreover, a release of malate dehydrogenase, which is located in the mitochondrial matrix and formation of malonyl dialdehyde, a secondary product arising from hydroperoxylipids, are observed (10,155). These changes are apparently not accompanied by a hydrolysis of the peroxidized lipids, since hydroperoxyphospholipids were detected in rat liver mitochondria treated with reticulocyte lipoxygenase (unpublished results). The susceptibility of biological membranes to attack by reticulocyte lipoxygenase depends on a variety of conditions, such as type of membrane, temperature, and functional state of the membrane. Various membranes differ in their susceptibility according to the following order: mitochondrial membranes > membranes of the endoplasmic reticulum > erythrocyte ghosts (156). The relative resistance of plasma membranes may have two reasons: first, the protective effect of cholesterol that also was demonstrated with phospholipid liposomes (157), second, a different type of protein–lipid interaction, since heat denaturation of the membranes partly abolished the difference (157). Both effects may shield the phospholipids in plasma membranes from lipoxygenase action. The relative resistance of red cell plasma

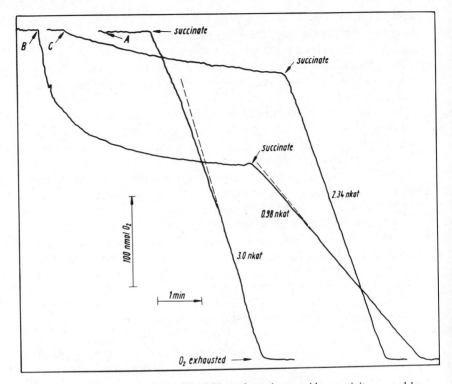

Figure 12. Oxygen uptake and inhibition of succinate oxidase activity caused by the action of reticulocyte lipoxygenase on beef heart submitochondrial particles (SMP). A, without treatment; B, SMP and lipoxygenase mixed in the cold and added to the reaction vessel at 37°C; C, the same as B, but in the presence of 0.7 mM salicylhydroxamic acid.

membranes causes a selective action of the lipoxygenase on mitochondria. This selectivity is a prerequisite for the role of this enzyme in the maturation of reticulocytes (see Section VII.D).

With rat liver mitochondria a strong protective effect of ATP or ADP plus succinate was demonstrated (10). One may assume that the energization of the mitochondrial inner membrane, which alters the state of protein–lipid interactions, surface charge, and other properties, renders the mitochondria resistant to lipoxygenase action. The protection by energization may be the cause that the mi-

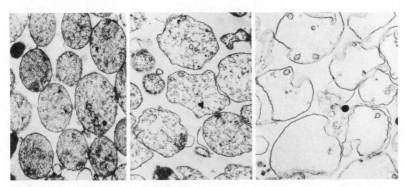

Figure 13. Action of reticulocyte lipoxygenase on isolated rat liver mitochondria. From the left to the right: mitochondria without treatment; after 5 min, and after 15 min incubation with lipoxygenase in 0.3 M buffered sucrose at 37°C. Positive staining, magnification 50,000:1.

tochondria in reticulocytes are susceptible to lipoxygenase only after an intrinsic functional damage has occurred in a fairly advanced maturational stage of the cell (see Section VII.C).

The action of reticulocyte lipoxygenase on membranes shows a much stronger temperature dependence than that on polyenoic fatty acids or phospholipids. At 20°C the enzyme exhibits only very low reactivity on rat liver mitochondria or beef heart submitochondrial particles. The reaction is strongly enhanced by the addition of 0.2% sodium cholate, reaching an oxygen uptake comparable to that at 37°C in the absence of cholate. At 37°C cholate does inhibit rather than activate the lipoxygenase-mediated oxygen uptake. These observations suggest that the temperature dependence of the lipoxygenase action on membranes is determined by the susceptibility of the membrane lipids rather than by the activity of the lipoxygenase itself. The susceptibility of mitochondria appears also to vary with their source, since mitochondria from certain fungi exhibited a high oxygen uptake with reticulocyte lipoxygenase even at 20°C that was not influenced by the presence of cholate (158). Generally any kind of mitochondria seems to be susceptible, since an effect of reticulocyte lipoxygenase has been reported for the mitochondria of cauliflower (159) and of sea urchin eggs (160).

The primary action of reticulocyte lipoxygenase on mitochondrial

TABLE VIII

Secondary Actions of Reticulocyte Lipoxygenase on
Mitochondria

Action	References
Lysis of mitochondria	168
Formation of malonyl dialdehyde	155
Inhibition of cytochrome oxidase	161
Inactivation of the respiratory chain at the level of complexes I and II	162
Uncoupling of oxidative phosphorylation	167
Cooxidative destruction of the Fe–S centers of the outer membrane	162
Transformation of monoaminoxidase	unpublished
Triggering of ATP-dependent proteolysis	169

lipids has a variety of consequences for the mitochondrial functions that are listed in Table VIII. A crucial event is the inactivation of the respiratory chain. This effect led to the discovery of the lipoxygenase in reticulocytes long before its identification (11,12). The lipoxygenase exerts two types of inhibitory actions on the respiratory chain. One effect is directed on cytochrome c oxidase and is restricted to the phospholipids, it is reversed, therefore, by subsequent substitution of native phospholipids for the peroxidized ones (161). The attack of lipoxygenase on phospholipids has two consequences: (a) a lowering of hydrophobicity owing to the introduction of hydroperoxy or hydroxy groups into the unpolar fatty acid chains and (b) a decrease of membrane fluidity. Both effects interfere with the electron transfer through the cytochrome c oxidase and possibly other subsystems of the mitochondrial electron transfer system.

The second action of lipoxygenase on the respiratory chain proved to be irreversible and is located in the respiratory chain between the proximate Fe–S centers and ubiquinone (162). This effect leads to an inactivation of both NADH oxidase and succinate oxidase activities. The irreversibility indicates secondary damage to proteins. From a study of various partial systems of the respiratory chain as well as from EPR spectroscopic investigations it has been proposed that ubiquinone-binding proteins may be the sites of the

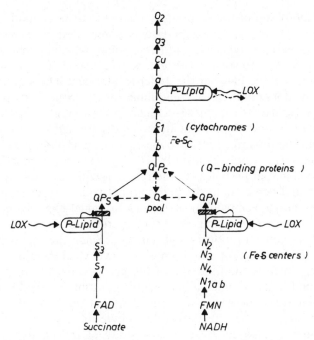

Figure 14. Sites of inhibitory actions of reticulocyte lipoxygenase on the mitochondrial respiratory chain.

damage (162). The inhibitory effects on NADH oxidase and succinate oxidase activities of submitochondrial particles are prevented by EDTA, which is by itself not a lipoxygenase inhibitor. EDTA does not affect the lipoxygenase-mediated oxygen uptake with the particles. The effect of EDTA is indicative of the involvement of metal ions in the lipoxygenase-mediated respiratory inhibition. Possible candidates are Ca^{2+}, which is known to accumulate in mitochondria, or Fe, which is released owing to the destruction of the Fe–S proteins of the mitochondrial outer membrane (162). The actions of reticulocyte lipoxygenase on the mitochondrial electron transfer system are shown in Fig. 14. The electron transfer system of both animal and plant mitochondria is affected by reticulocyte lipoxygenase, whereas that of *Escherichia coli* membranes is resistant (163).

The inhibition of the respiratory chain of submitochondrial particles by reticulocyte lipoxygenase is potentiated by hemoglobin (44). Soybean lipoxygenase and hemoglobin, which themselves are without effect on the respiratory chain, exert in combination a strong inhibition. On the basis of the experimental data it was concluded that hemoglobin is the proximate inhibitor by virtue of its pseudo-lipohydroperoxidase activity (see Section V), which becomes effective after hydroperoxy fatty acids have been formed by the soybean enzyme. The substrate for the soybean lipoxygenase are presumably the nonesterified polyenoic fatty acids present in preparations of submitochondrial particles. In the reaction with reticulocyte lipoxygenase and hemoglobin the lipohydroperoxidase reaction is catalyzed by either hemoglobin, as shown in model experiments with 13L_S-hydroperoxylinoleic acid (44) or by reticulocyte lipoxygenase itself if its concentration is high enough. Since the lipohydroperoxidase of reticulocyte lipoxygenase is lower than the dioxygenase activity, which is the first step of the process, the full respiratory inhibition is only obtained if hemoglobin contributes additional lipohydroperoxidase activity. The crucial role of the lipohydroperoxidase reaction in the effects of reticulocyte lipoxygenase on mitochondria was also established in a study of the passive electric properties of beef heart submitochondrial particles (164). Lipoxygenase-treatment caused a fourfold increase in the membrane capacity and a twofold increase in the membrane conductivity. Since the experimental setup required samples containing very high concentrations of both particles and lipoxygenase, oxygen was exhausted in the measuring cell within the first minute of the reaction; nevertheless, both changes of passive electric properties and appearance of the respiratory inhibition continued under anaerobic conditions indicating the involvement of lipohydroperoxidase activity that is possible or even favored anaerobically. The occurrence of an anaerobic lipohydroperoxidase reaction in the reaction of reticulocyte lipoxygenase with beef heart submitochondrial particles was also established by pentane formation during the anaerobic period following an aerobic preincubation (43).

Lipoxygenases also cause damage to the photosynthetic electron transport system in chloroplasts. The action of soybean lipoxygenase on wheat chloroplasts was demonstrated by both a slow but significant oxygen uptake and a selective loss of linolenic acid (165).

From studies on various subsystems of the photosynthetic electron transport it was deduced that soybean lipoxygenase specifically inhibits a site in Photosystem II other than the water-splitting component, whereas reticulocyte lipoxygenase treatment interrupts the electron transfer additionally at least at two other sites (Photosystem I and the water-splitting component of Photosystem II) (166). In contrast to the effect on submitochondrial particles, hemoglobin does not potentiate the action of lipoxygenases on chloroplasts; it is conceivable that the chlorophyll of chloroplasts fulfills the same function, since a lipohydroperoxidase activity of chlorophyll was demonstrated in a model system with $13L_S$-hydroperoxylinoleic acid.

The action of reticulocyte lipoxygenase on membranes produces hydrophilic clusters within the lipid bilayer that are in all likelihood responsible for the drastic changes of the passive electric properties mentioned before. For the same reason an increase in the proton permeability of both inner mitochondrial membranes and thylakoid membranes occurs leading to uncoupling of oxidative phosphorylation (167) and photophosphorylation (166), respectively.

In addition to the actions on the mitochondrial inner membrane reticulocyte lipoxygenase also produces drastic changes in the mitochondrial outer membrane. The outer membranes are no longer visible in electron micrographs of lipoxygenase-treated rat liver mitochondria (168). The Fe–S clusters of the outer membrane are completely destroyed in contrast to those of the inner membrane (162). Preliminary data showed that the monoaminooxidase of brain mitochondria is converted from type A to type B showing diaminooxidase activity. An identical change has been reported after nonenzymatic lipid peroxidation. In contrast, no effect was observed on the rotenone-insensitive NADH–cytochrome c oxidoreductase that is also located in the outer membrane. The triggering of ATP-dependent proteolysis (169) is discussed in Section VII.C.

Various membranes such as submitochondrial, endoplasmic, and plasma membranes as well as liposomes exert a curious stimulatory effect on the activity of reticulocyte lipoxygenase, but not of soybean lipoxygenase with fatty acid substrates (170). Various fractions of lipoproteins of the blood plasma exert the same effect (171). The stimulation is not related to an action on the membranes or lipoproteins as substrates, since liposomes made from dioleyl lecithin exhibited the same effect. A preliminary suggestion that membranes and lipoproteins might act as allosteric effectors of reticulocyte lipoxygenase (170) could not be established in further experiments. It would rather appear that the substrate fatty acids are imbedded in the membranes in a physical state better accessible to the lipoxygenase than in water. The effect of membranes resembles that of sodium cholate below its critical micellar concentration (172).

VII.   Physiology of Reticulocyte Lipoxygenase

A.   CELL BIOLOGY OF THE RETICULOCYTE; THE ROLE OF LIPOXYGENASE

The reticulocyte is the last intermediate stage of the differentiation and maturation of erythroid cells preceding the normocyte. It still possesses mitochondria and ribosomes and, concomitantly, active cell respiration and protein synthesis. These organelles disappear during the transition to the mature erythrocyte (normocyte). In contrast to the preceding stages, the reticulocyte is also characterized by the absence of the nucleus or its inactivity in lower vertebrates. For this reason, the protein synthesis in reticulocytes occurs in the absence of any nuclear RNA synthesis. There are also only vestigial remnants of endoplasmic reticulum and few secondary lysosomes left. The maturation of the reticulocyte involves the degradation of mitochondria and ribosomes, and consequently the loss of the respiratory capacity and of the protein synthesis as well as the disappearance of most of the receptors and transport proteins of the cell membrane [for details see ref. (173)].

The differentiation and maturation processes from the hematopoetic stem cell to the mature erythrocyte require about 6 days; half of this time is spent in the reticulocyte stage that begins in the bone marrow and continues in the peripheral blood. Therefore one can distinguish more or less mature reticulocytes. Their maturity can be gauged by their hemoglobin content that allows separation of the reticulocytes according to age by means of density gradient centrifugation techniques. In the separated cell populations the concentration of RNA, mainly determined by the content of ribosomes, and the capacity of respiratory enzymes decrease with increasing buoyant density. In addition, a variety of other cell physiological and biochemical parameters change with the maturational stage of the reticulocyte (174).

The lipoxygenase specific for erythroid cells occurs in the more mature stages of the reticulocyte (see the following section). There is ample evidence for the involvement of the lipoxygenase in the maturational degradation of the mitochondria. A strong argument is the inhibition of the decline of respiratory enzymes during *in vitro* maturation of reticulocytes in the presence of the lipoxygenase inhibitors salicylhydroxamic acid or 5,8,11,14-eicosatetraynoic acid (175). Addition of purified lipoxygenase to isolated rat liver mitochondria or beef heart submitochondrial particles leads to a variety of deleterious effects that were discussed in Section VI. The action of the lipoxygenase on mitochondria concerns not only the lipids, it also triggers the action of a ubiquitin ATP-dependent proteolytic system in reticulocytes (169,176,177), the physiological substrate of which are the mitochondrial proteins (176). In this manner, the synthesis and actions of the cell-specific lipoxygenase are key events in the maturation of reticulocytes. This biological role differs basically from that of most other animal lipoxygenases, the action of which appears to be determined by their products from arachidonic acid (see Section VIII). Some of the products obtained with purified reticulocyte lipoxygenase such as $8_{DS},15_{LS}$-dihydro(per)oxy-5,9,11,13-*cis,trans,cis,trans*-eicosatetraenoic acid and lipoxins could possibly exert regulatory actions as shown with leukocytes (71,178); however, there are no indications for a role of lipoxygenase products in the maturation process of erythroid cells.

B. BIOLOGICAL DYNAMICS OF THE LIPOXYGENASE; REGULATION OF LIPOXYGENASE SYNTHESIS IN RETICULOCYTES

The reticulocyte lipoxygenase is the only enzyme of this class for which some detailed information on the regulation of its synthesis is available. The lipoxygenase in red cells shows marked biological dynamics of appearance and disappearance during an experimental anemia (Fig. 15). Earlier studies have revealed that the respiration-inhibitory activity of the lipoxygenase is absent both in bone marrow and in mature erythrocytes of nonanemic rabbits. The activity starts to increase on the third day to reach peak levels on the sixth day of an experimental anemia provoked by strong daily bleeding (14,179). During this period the share of reticulocytes increases from

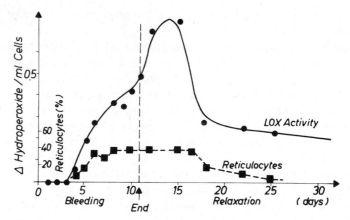

Figure 15.   Percentage of reticulocytes and lipoxygenase activity in the course of an experimental bleeding anemia of a rabbit.

1–3% to approximately 40% of the total red cells. With cell populations obtained by buoyant density centrifugation it has been shown that very young reticulocytes of low density and high RNA content exhibit only little or no lipoxygenase activity (179). A sharp maximum of lipoxygenase is found with more mature reticulocytes followed by a steep decline during the transition to the normocyte. The maximum coincides with a period of decrease of mitochondrial activities such as cytochrome c oxidase. The biological dynamics of increase and decrease of lipoxygenase activity were further assessed by experiments in which rabbit reticulocytes of the 6th day of bleeding anemia were allowed to mature *in vitro* (180). Parallel determinations of the concentration of lipoxygenase protein by means of antiserum against reticulocyte lipoxygenase showed the same profile indicating that the maturational changes of lipoxygenase result from synthesis and degradation of the enzyme (181).

The synthesis of lipoxygenase was measured by means of the incorporation of labeled amino acids both in intact cells and in a cell-free system using messenger ribonucleoproteins, (mRNP), isolated from the reticulocytes and subjected to proteinase K digestion [Fig. 16 (182,183)]. With intact cells the synthesis of lipoxygenase occurs only in the more mature populations of reticulocytes, whereas in the cell-free system with isolated free mRNA synthesis

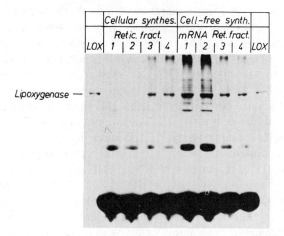

Figure 16. Cellular (lines 2–5) and cell-free (lines 6–9) synthesis of lipoxygenase in reticulocytes of different ages. The numbers refer to cell fractions of increasing density, that is, of increasing age. For the cell-free synthesis the mRNA isolated from the same cell fraction was used [from ref. (183)].

of lipoxygenase was observed in all reticulocyte fractions. It is evident from these experiments that all maturational stages of the reticulocyte and in all likelihood also its precursors contain mRNA coding for lipoxygenase; in erythroblast and immature reticulocytes this mRNA is, however, masked presumably via interaction with proteins in the form of free cytoplasmic mRNPs.

Under conditions of bleeding anemia the lipoxygenase is one of the most abundant nonhemoglobin proteins synthesized in reticulocytes, accounting for about one third of the newly made nonhemoglobin proteins and for about 3–4% of the total amount of newly synthesized cytosolic proteins (182). The concentration of lipoxygenase reaches levels of approximately 4 mg/mL of cells in the mixed reticulocyte suspension and surpasses even that of catalase (14). The rate of lipoxygenase synthesis was estimated to be about 1.4 nmol/hxmL of reticulocytes or 2.5-mg of lipoxygenase/day xmL reticulocytes, so that the peak level is reached within 2 days, which corresponds to the life span of reticulocytes in the peripheral blood up to their conversion to normocytes.

In reticulocytes from nonanemic rabbits both synthesis and cellular concentration of lipoxygenase are much lower than in those from anemic animals. The lipoxygenase mRNA could be detected also in these cells by hybridization of total cellular RNA with a probe obtained from a clone of cDNA derived from the purified lipoxygenase mRNA (56). By use of the same approach the specificity of this lipoxygenase for erythroid cells and their occurrence in mice and rats could be demonstrated. The amount of lipoxygenase mRNA was found to be approximately the same in the bone marrow of anemic and nonanemic rabbits. This observation appears to exclude the possibility of a depression of the transcription of lipoxygenase mRNA in bone marrow cells as the reason for the strongly diminished synthesis of lipoxygenase in reticulocytes appearing under nonanemic conditions in rabbits and rats and even in mice with anemia. Two possible explanations may be assumed: (a) degradation of the translationally inactive mRNPs before unmasking has occurred, (b) accelerated breakdown of translationally active mRNA as compared with anemic conditions. Unfortunately there is a lack of information as to the breakdown of mRNA in reticulocytes. Reticulocytes contain several maturation-dependent ribonucleases. There also occurs an inhibitory protein for ribonucleases in reticulocytes (184) that may modulate the degradation of mRNA.

The mechanism of the unmasking of the mRNPs is under investigation. The involvement of a limited proteolytic attack is conceivable. This example of a posttranscriptional control of the protein synthesis is of general biological interest, since masked mRNA species have been described in plant seeds before germination, oocytes of amphibia, embryonal tissues, and other systems, but the mechanism of this type of posttranscriptional control is not yet established. The reticulocyte is an appropriate model to study the switching on of a cell-specific protein synthesis because of the absence of transcription and the possibility of separating preparatively a cell population in which lipoxygenase mRNA is completely masked.

The synthesis of lipoxygenase in reticulocytes is apparently not limited by the supply of iron that is a constituent of this enzyme. Addition of an iron–transferrin complex to the cells during *in vitro* maturation does not enhance the increase in concentration and activity of the lipoxygenase (181). It may be assumed that an inner iron economy exists for the synthesis of lipoxygenase. The mito-

chondria that are degraded during the maturation process of reticulocytes are rich in iron-containing proteins such as cytochromes
and Fe–S proteins and may supply the iron for lipoxygenase synthesis during the late maturational stage of the reticulocyte. This
supply does not suffice for the optimal synthesis of hemoglobin,
which requires extracellular iron.

The degradation of the lipoxygenase may be governed by at least
two processes: (a) self-inactivation during the actions on mitochondria (see Sections IV.C.1 and VI) and (b) by thiol-containing pro-
teinase(s) as indicated by the suppression of the maturational decline
of lipoxygenase by the inhibitor of thiol proteinases leupeptin (14).
Addition of iron transferrin to the cells gives rise to an accelerated
disappearance of lipoxygenase activity (unpublished results) possibly via an enhancement of the susceptibility of the mitochondria
to the lipoxygenase attack (see Section VII.C) and hence to increased self-inactivation.

Whereas no lipoxygenase activity can be measured either in the
osmotic hemolysate of mature erythrocytes or in the intact cells by
measuring the formation of 15-hydroxy-5,8,11,13-eicosatetraenoic
acid from [1-^{14}C]arachidonic acid, sizable activity was detected after
incubation in the presence of Ca^{2+} and ionophore A 23187 or after
freezing and thawing. Under these conditions lipoxygenase activity
was also detected in an ammonium sulfate precipitate from the membrane-free hemolysate. The nature of this phenomenon is under
investigation.

C. THE ACTIVITY OF RETICULOCYTE LIPOXYGENASE *IN VIVO*

The share of lipoxygenase in the nonrespiratory oxygen uptake
was determined in nonseparated reticulocyte suspensions in the
presence of the inhibitor of the mitochondrial electron transfer antimycin A (185); it amounted to about 20–30% corresponding to 5–
7% of the total oxygen uptake in the absence of antimycin A. Consistent results were obtained with five different lipoxygenase inhibitors. The lipoxygenase-mediated oxygen uptake amounted under
these conditions to 1.2 μmol O_2 per mL of packed cells per hour.
From a rough estimate of the amount of mitochondrial phospholipids
in reticulocytes and considering the fatty acid composition of the
phospholipids of their mitochondria it may be calculated that the
mitochondria as lipoxygenase substrate may maintain this oxygen

uptake for about 10 h. This time period is in the same order of magnitude as that of the maturation process of reticulocytes, which is accompanied by a complete degradation of mitochondria. In experiments on the maturation of reticulocytes *in vitro* a half-life of the mitochondrial respiration of 12 h was observed for nonseparated cell suspensions obtained on the sixth day of bleeding anaemia (186). It is likely that the share of lipoxygenase in the oxygen uptake is lower in the absence of antimycin A than in its presence by virtue of the fact that with the inhibition of the respiration both the β oxidation of the fatty acids and their incorporation in phospholipids are inhibited, that is, reactions that compete with the lipoxygenase for free polyenoic acids.

The endogenous action of the lipoxygenase may proceed in two ways: (a) a direct attack on mitochondrial phospholipids (10), (b) the preceding liberation of free fatty acids that are dioxygenated by reticulocyte lipoxygenase at a higher rate (156). For nonstimulated rabbit reticulocytes there are some arguments in favor of the absence of a preceding phospholipase A_2 involvement: (a) No free hydroxypolyenoic fatty acids are detectable in reticulocytes both in the presence and in the absence of antimycin A (185). (b) The intracellular Ca^{2+} level in reticulocytes is expected to be too low to stimulate a Ca^{2+}-dependent phospholipase A_2.

On the other hand, free polyenoic fatty acids appear to be the preferred substrates of the lipoxygenase in reticulocytes stimulated by Ca^{2+} plus ionophore A 23187 (185). Under these conditions a doubling of the lipoxygenase-mediated oxygen uptake and the occurrence of both nonoxygenated and oxygenated free fatty acids (185) as well as pentane formation (43) were observed.

In the maturation of reticulocytes there is a certain stage in which lipoxygenase and functionally intact mitochondria occur simultaneously. Such a situation appears to be necessary, since synthesis of lipoxygenase and that of hemoglobin require an abundant supply of ATP, and therefore presuppose mitochondrial energy conservation. One may expect the mitochondria of immature cells not to be intrinsically susceptible to lipoxygenase attack. A protection appears to be afforded by conditions maintaining an energized state as shown for isolated rat liver mitochondria (see Section VI). The presence of both susceptible and nonsusceptible mitochondria in the same reticulocyte is also evident from electron micrographs of re-

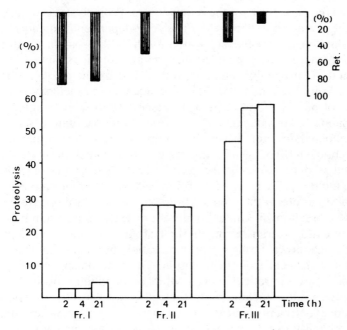

Figure 17.   Maturation dependence of proteolysis and decline of reticulocyte count. Fractions I, II, and III, refer to the most immature, the intermediate, and the most mature reticulocytes, separated by density fractionation. The times refer to periods of *in vitro* maturation.

ticulocyte mitochondria in situ (187,188). The different susceptibility of mitochondria to lipoxygenase attack enables the reticulocyte to degrade a part of its mitochondria and to utilize these amino acids for the synthesis of hemoglobin under conditions in which mito-chondrial ATP supply is still maintained. These relations raise the question as to the events that render the mitochondria susceptible to the attack of lipoxygenase and subsequently to ubiquitin–ATP-dependent proteolysis.

In Fig. 17 the extent of ATP-dependent proteolysis in separated reticulocyte populations and its behavior during *in vitro* maturation in the absence of external iron supply is shown. As may be seen, proteolysis is very low in young cells in which it does not increase during *in vitro* maturation. The proteolysis increases strongly with the degree of maturity of the reticulocytes. Proteolysis is completely absent in mature erythrocytes. The ATP-dependent proteolysis is a time-limited process that stops after 2–3 h (189). From reconstitution experiments it was concluded that the proteolysis is limited by the amount of susceptible substrate rather than by the enzyme system (190). From these data one may conclude that in reticulocytes three types of mitochondria occur with respect to their susceptibility to degrading factors dependent on the maturational state of the cell: (a) mitochondria not susceptible to lipoxygenase, (b) mitochondria susceptible to lipoxygenase that triggers the proteolytic system, and (c) mitochondria already damaged by lipoxygenase and susceptible to ubiquitin–ATP-dependent proteolysis. Addition of lipoxygenase inhibitors protects the second group but not the third one from proteolytic breakdown. Under conditions of external iron supply, preferably as ferric transferrin, the first group is rendered susceptible to the breakdown during *in vitro* maturation; here again lipoxygenase inhibitors protect from degradation. The mechanism by which an uptake of iron in reticulocytes confers susceptibility towards lipoxygenase to all mitochondria of the cell is not yet fully understood. Recent experiments indicate the synthesis of a protein which confers susceptibility to lipoxygenase on mitochondria of immature reticulocytes (unpublished). Iron is thus one of the signals that initiate a sequence of events leading to maturational degradation of mitochondria. The other signal is the unmasking of the lipoxygenase mRNA leading to the onset of the synthesis of lipoxygenase.

The molecular basis of the triggering of the ubiquitin–ATP-dependent proteolysis by the action of lipoxygenase on mitochondria is not yet established. This proteolytic system was formerly known to attack only denatured proteins (191).

The mitochondrial proteins in reticulocytes were identified as physiological substrates (176). Surprisingly heat denaturation of the mitochondria greatly reduces the susceptibility to the proteolysis. The oxygenation of the membrane phospholipids by lipoxygenase may give rise to alterations in the protein–lipid interactions in such a manner that α-amino groups which are involved in the molecular mechanism of action of the ubiquitin–ATP-proteinase (192), become accessible. Recent experiments have shown that heat-denatured stroma exposes fewer binding sites for ubiquitin (193). The endogenous action of the ubiquitin–ATP-proteinase in reticulocytes can

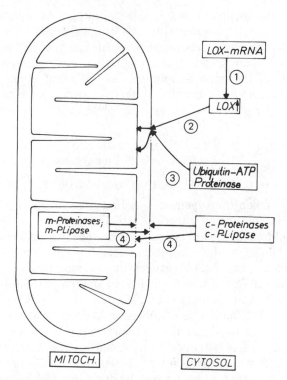

Figure 18.   Scheme for the degradation of mitochondria during the maturation process of reticulocytes.

also be triggered independently by the mitochondrial damage of lipoxygenase (175).

D.   DEGRADATION OF MITOCHONDRIA IN RETICULOCYTES

In Fig. 18 a scheme of the events instrumental in the degradation of mitochondria during the maturation process of reticulocytes is given. The first step is assumed to be the unmasking of the mRNA for lipoxygenase by processes hitherto unidentified (see Section VII.B). Unmasking is succeeded by synthesis of lipoxygenase. There must follow a change of conformation of the mitochondria rendering them susceptible to lipoxygenase. Now the stage is set for the various deleterious effects of lipoxygenase (see Section VI).

They include disruption of the physical integrity of mitochondrial outer and inner membranes with loss of matrix enzymes as well as changes in the physicochemical properties of the membranes, inactivation of the respiratory chain as well as destruction of the Fe–S proteins of the outer membrane. At this point ubiquitin–ATP-dependent proteolysis becomes effective. Other proteinases, peptidases, and phospholipases complete the breakdown of mitochondria.

VIII.   Other Animal Lipoxygenases

A.   ARACHIDONATE 12-LIPOXYGENASE OF PLATELETS

A mammalian lipoxygenase was first recognized in platelets in 1974 (6). Despite many efforts (7,8,16,17,109) this enzyme has not yet been obtained in a homogeneous form. The main difficulties arise from the lability of the enzyme as is also the case with most other lipoxygenases. The soluble platelet lipoxygenase tends to aggregate and to lose activity, if repeatedly frozen and thawed; during aerobic storage 50% of the activity is lost after 20 min at 37°C and after 17 h at 0°C (109). Conflicting results have been published concerning the intracellular localization of the enzyme; it was found either in the cytosol (7,16,194) or in the microsomal fraction (8,16). In a recent study with human platelets it was shown that two thirds of the total activity is localized in the cytosol, whereas one fifth is found in the membrane fraction (195). The membrane-bound part of the activity could not be removed by extensive washing or sonication; separation of the membranes by means of free-flow electrophoresis led to an enrichment of enzyme activity in the intracellular membrane fraction. The membrane-bound enzyme did not differ from the cytosolic one with respect to the positional specificity (195). The cytosolic part of activity could be separated by either gel chromatography or anion exchange chromatography in two fractions: one of them with an apparent molecular mass of 100,000 daltons and preferential lipoxygenase activity, the other with an apparent M_r of 160,000 daltons exhibiting both lipoxygenase and peroxidase activity as judged from the prominent formation of 12-hydroxy-5,8,10,13-eicosatetraenoic (12-HETE) acid from arachidonic acid instead of the primary hydroperoxide (16). It was not assessed whether the hydroperoxidase activity in the absence of an extra hydrogen donor of the

160,000 daltons fraction is an intrinsic property of the lipoxygenase or due to a distinct enzyme present in this fraction. The relations between the various fractions of platelet lipoxygenase have not yet been clarified. Platelet lipoxygenase shows a marked substrate specificity (7,196,197). Polyenoic fatty acids possessing a bis-allylic methylene group in the ($n - 11$) position (C-10 in eicosapolyenoic acids) are preferred substrates; the dioxygenation proceeds stereospecifically in the ($n - 9$) position. The methyl terminus of the fatty acid chain appears to serve as a signal for hydrogen abstraction and oxygen insertion. 12_{LS}-hydroperoxy-5,8,10,14-*cis,cis,trans,cis*-eicosatetraenoic acid (12-HPETE) is the exclusive dioxygenation product from arachidonic acid. The K_m values for arachidonic acid reported in the literature range from 3.4 (109) to 80 μM (16). Like other lipoxygenases the platelet lipoxygenase requires activation by a suitable hydroperoxy fatty acid, to overcome a lag period (197,198). 12-HPETE, 11-HPETE, and apparently also PGG_2 but not 8- or 9-HPETE, activate at micromolar concentrations (197,199). Most compounds known as universal lipoxygenase inhibitors (see Section IV.C.3), such 5,8,11,14- and 4,7,10,13-eicosatetraynoic acid, 5,8,11-eicosatriynoic acid, nordihydroguaiaretic acid, 3-*t*-butyl-4-hydroxyanisol, BW 755C, some phenylhydrazones, and high concentrations of cyanide inhibit the platelet enzyme. Platelet lipoxygenase is also inhibited by 15-HETE, the product of the arachidonate 15-lipoxygenase action and the succeeding peroxidase reaction (200). In accordance with other lipoxygenases, the platelet enzyme suffers self-inactivation in intact platelets after their stimulation (90). With soluble enzyme preparations the formation of 12-HPETE ceases after 10 min at 37°C (195). In line with the suicidal property is the strong inhibition by 12-HPETE or 15-HPETE (200).

Although soluble enzyme preparations appear to contain peroxidase activity that converts the 12-HPETE in part to 12-HETE in the absence of extra hydrogen donors (7,16), this reaction in intact platelets requires a functional selenium-containing glutathione peroxidase as judged from the need for the glucose supply via the hexose–monophosphate shunt, the inhibition by a diamide-induced depletion of reduced glutathione, or by selenium deficiency (201,202). Under conditions of inhibition or nonfunctioning of the Se–glutathione peroxidase there is both an abnormal release of 12-HPETE from the platelets and a formation of 10-hydroxy-11,12-epoxy-

5,8,14-all-*cis*-eicosatrienoic acid and two isomeric trihydroxyeicos-atrienoic acids that may arise from hydrolysis of another epoxy-hydroxyeicosatrienoic acid (202,203). These products may be formed by lipohydroperoxidase activity, which may reside either in the lipoxygenase itself (see Section IV.B.3) or in hemoproteins (204). The inhibition of 12-HPETE peroxidase activity in platelets by some nonsteroidal antiinflammatory drugs such as the cyclooxygenase in-hibitors acetylsalicylic acid, indomethacin, phenylbutazone, ibu-profen, naproxen, and sulindac as well as sodium salicylate, which does not inhibit either cyclooxygenase or lipoxygenase activity, may be of pharmacological interest (16).

The biological significance of the lipoxygenase pathway in plate-lets is still debated. A possible regulatory role in the activation and aggregation of platelets is proposed by many researchers (109), but there are conflicting results as to the mode of action of the lipox-ygenase pathway and its products. Dutilh et al. concluded from their results that the lipoxygenase pathway may render the thromboxane-induced aggregation irreversible (205). In contrast, the same and other authors observed an inhibition of platelet aggregation by 12-HPETE, 12-HETE, and 5,8,11-all-*cis*-eicosatrienoic acid that serves as substrate for platelet lipoxygenase but not for the cyclooxygenase (109,199,206). Another possible biological function of 12-HETE de-rived from platelets may consist in the modulation of the interactions with other cells. In this connection the selective high potency of 12-HETE to trigger migration of aortic smooth muscle cells (207) de-serves some attention.

B. THE LIPOXYGENASES OF LEUKOCYTES

1. Arachidonate-5-Lipoxygenase

Leukocytes contain at least three lipoxygenases that have been partially purified. The arachidonate 5-lipoxygenase has attracted most interest, since this pathway is the biosynthetic route to the biologically active leukotrienes LTB_4, LTC_4, LTD_4, and LTE_4 [for review see ref. (13)]. This lipoxygenase is strongly dependent on Ca^{2+} as shown for basophils (208) and neutrophils (209). As far as the mechanism of activation by Ca^{2+} is concerned contradictory results were published; a former report of the formation of an active dimer was not confirmed (see ref. 18). The arachidonate 5-lipoxygenase dioxygenates also 5,8,11,14,17-all-*cis*-eicosapenaenoic acid at C_5 and to a lesser extent 8,11,14-all-*cis*-eicosatrienoic acid at C_8, but not 12-HETE, the product of arachidonate 12-lipoxygenases (20). From arachidonic acid 5_{DS}-hydroperoxy-6,8,11,14-*trans,cis,cis,cis*-eicosatetraenic acid, (5-HPETE), is formed. Like in platelets the succeeding formation of 5-HETE appears to require glutathione peroxidase activity (20). 5-HPETE is also the proximate precursor of leukotriene A_4 that is in turn the common precursor for the other leukotrienes of this pathway. The arachidonate 5-lipoxygenase is reported to be stimulated by several nucleotides, most prominently by ATP (20); the significance and the mechanism of this effect are not clear. ATP also partially protects the enzyme from inactivation during aerobic storage. The enzyme suffers self-inactivation with the activity ceasing 3–5 min after the initiation of the reaction at 30°C (20).

Purification of arachidonate 5-lipoxygenase has been reported for the enzyme from guinea pig peritoneal polymorphonuclear leukocytes (207) and for that from rat basophilic leukemia cells (18). In the first case chromatography on aminobutyl-Sepharose 4B was applied, which removed the bulk of other proteins, but the specific activity could not be elevated in comparison to the cytosol owing to strong inactivation of the enzyme. Goetze et al. applied a 4-step procedure including anion exchange HPLC on two different media and obtained from 10^9 rat basophilic leukemia cells 33 µg of electrophoretically homogeneous enzyme, that converts arachidonic acid exclusively to 5-HPETE (18). Their molecular activity amounted to about 15 s^{-1} at 37°C.

2. Arachidonate 12-Lipoxygenase

This enzyme was purified about 30-fold from the cytosol of porcine blood polymorphonuclear leukocytes (21). The purification included ammonium sulfate fractionation and DEAE-Sephadex chromatography. An origin from contaminating platelets could be excluded. The partially purified enzyme converts arachidonic acid to 12-HPETE with 15-HPETE as minor product. The enzyme also reacts with 5-HETE as substrate; a mixture of a $5_{DS},12_{LS}$-dihydroxy- and $5_{DS},15_{LS}$-dihydroxyeicosatetraenoic acid was isolated after reduction of the enzymatic products by sodium borohydride. It was not assessed whether the products oxygenated in the C_{15} position of arachidonic acid arose from a dual positional specificity as shown for reticulocyte lipoxygenase (see Section IV.B.1) or from contaminating arachidonate 15-lipoxygenase. The dihydroxyeicosatetraenoic acids observed are also produced by intact leukocytes. The biological role of the arachidonate 12-lipoxygenase pathway and of the double dioxygation products formed by it remains to be established. An involvement of this or another arachidonate 12-lipoxygenase in the synthesis of epoxyleukotrienes (5,6-LTA$_4$ and 14,15-LTA$_4$) is conceivable, since the leukotriene A$_4$ synthase activities require a hydrogen abstraction at C$_{10}$ that is also the primary step of an arachidonate 12-lipoxygenase activity (see Section IV.B.4). Another possibility is the involvement in the synthesis of lipoxins, the youngest family of newly discovered bioregulatory lipoxygenase products (71). In case that the lipoxins produced by leukocytes are also triple dioxygenation products of arachidonic acid, which is clearly established for purified reticulocyte lipoxygenase (see Section IV.B.4), their synthesis would involve all three types of lipoxygenases present in leukocytes.

The enzyme partially purified by Yoshimoto et al. (27) did not require Ca^{2+}. In contrast, the arachidonate 12-lipoxygenase activity of the cytosol of rat basophilic leukemia cells was reported to be strongly stimulated by Ca^{2+} (210).

3. Arachidonate 15-Lipoxygenase

The presence of such a lipoxygenase was to be expected, since leukocytes produce a variety of arachidonic acid products that are oxygenated at C$_{15}$ (13,178,211). An arachidonate 15-lipoxygenase was partially purified from rabbit peritoneal polymorphonuclear leukocytes (19). Even though a 250-fold purification as compared with the cytosol and a specific activity corresponding to an approximate turnover number of 45 s^{-1} at 30°C were achieved, the enzyme preparation was still highly impure. In the last purification step the activity peak coincided with a valley of the elution profile of the proteins. The enzyme preparation still contained heat-stable pseudolipohydroperoxidase activity, possibly arising from hemoproteins, which converted the enzymatically formed 15-HPETE to a multitude of nonenzymatic secondary products. Therefore, the characteristics described for this enzyme must be regarded as preliminary. Besides 15-HPETE, no further enzymatic dioxygenation product was observed. The apparent K_m value for arachidonic acid was estimated to be about 28 μM, the pH optimum is around 6.5. An apparent molecular mass of 61,000 daltons was estimated by means of a Sephadex G-150 column. Like the other animal lipoxygenases the enzyme loses its activity during aerobic storage even in the cold and shows self-inactivation. It is sensitive to the common lipoxygenase inhibitors 5,8,11,14-eicosatetraynoic acid and BW 755C. The intracellular localization of the enzyme is not clear. The arachidonate 15-lipoxygenase was only observed in the membrane-free supernatant fluid after vigorous sonication, but not after freezing and thawing.

The biological importance of the arachidonate 15-lipoxygenase in leukocytes may consist in the synthesis of lipoxins (see the preceding section). Another role could arise from the actions of 15-HETE that is formed by the glutathione peroxidase reaction following the action of 15-lipoxygenase on arachidonic acid. This metabolite is reported to be an inhibitor of both arachidonate 5- and 12-lipoxygenases (200,212). The possible biological significance of this inhibition remains, however, to be assessed.

C.   Lipoxygenases from Various Sources

Lipoxygenase-derived dioxygenation products were detected in many mammalian cells and tissues. The most widely distributed appear to be the arachidonate 12-lipoxygenases [for a review see Table 2 of ref. (13)]. In some tissues a release of lipoxygenase products can be only observed under special conditions. Using cultured endothelial cells from calf aorta it was observed that damage to the cells by freezing and thawing gives rise to a strong enhancement of the share of lipoxygenase products in the total arachidonic acid metabolism; at the same time the spectrum of products changes from exclusively 12-HETE to a mixture of 12- and 15-HETE, the latter prevailing (102). It would appear that lipoxygenases occur in all mammalian cells, but in some cases they may be masked.

Apart from the lipoxygenases of blood cells described in the foregoing sections, only a few efforts have been undertaken to purify the animal lipoxygenases of other origin. An arachidonate 12-lipoxygenase has been purified about 30-fold from the rat lung cytosol (213). This enzyme has been shown to be part of an enzyme system that converts arachidonic acid to 8,11,12-trihydroxyeicosatrienoic acid (214).

The partly purified lipoxygenase was free of glutathione peroxidase activity and converted arachidonic acid to 12-HPETE and 5,8,11,14,17-eicosapentaenoic acid to the corresponding 12-hydroperoxy derivative. The activity was found both in the cytosol and in the microsomes and other particulate fractions. The enzyme resembles in many respect that of platelets.

A lipoxygenase was also partly purified from the epidermal layer of guinea pig skin (24). This enzyme forms a HPETE from arachidonic acid. Although the positional isomerism of the product was not assessed, the enzyme was assumed to be an arachidonate 12-lipoxygenase, since the skin of other species is known to produce 12-HETE (215).

A lipoxygenase was purified almost to homogeneity by affinity chromatography from the microsomal fraction of rat testicles (22,23). The enzyme converts linoleic acid to a mixture of 9- and 13-hydroperoxylinoleic acids in a ratio of 2:1 as well as to two secondary carbonyl group-containing products, one of them identified as 13-hydroxy-12-oxo-octadec-*cis*-9-enoic acid. The latter compound is unusual for a lipoxygenase-catalyzed reaction; it is also formed by linoleate hydroperoxide isomerases present in a variety of plants, therefore a heterogeneity of the enzyme preparation with respect to the enzymatic activities was taken into consideration. The action of the lipoxygenase on arachidonic acid was not studied.

The only animal lipoxygenase from an invertebrate source was demonstrated in the gorgonian coral *Pseudoplexaura porosa* (216). This enzyme converts arachidonic acid to 15-HPETE. Surprisingly linoleic acid and α-linolenic acid are poor substrates for this enzyme in contrast to other arachidonate 15-lipoxygenases such as those of reticulocytes and soybeans.

IX. Concluding Remarks

Lipoxygenases are a family of related nonheme iron-containing enzymes that are widely distributed in man, animals, and plants. They exhibit some diversity, for instance, with respect to their substrate specificity, but they exhibit a uniform behavior with respect to such properties as sensitivity to inhibitors, self-inactivation during their reaction, obligatory activation by micromolar concentrations of hydroperoxy fatty acid, anaerobic lipohydroperoxidase activities, and in regard to the basic mechanism of catalysis and kinetics. Lipoxygenases are involved in the synthesis of a broad spectrum of products ranging from simple monohydroperoxy and monohydroxypolyenoic acids, the various classes of dihydroxypolyenoic fatty acids and leukotrienes, epoxy compounds to the lipoxins. Some of the lipoxygenase products exert fundamental biological actions; for many others the biological role remains to be clarified. A second basic function of lipoxygenases is the triggering of the degradation of biological membranes. This function has been established only for reticulocyte lipoxygenase up till now. The enzyme from rabbit reticulocytes is the only animal lipoxygenase purified to homogeneity that has been extensively characterized with respect to molecular characteristics, enzymology, including kinetics and self-inactivation, and molecular biology. This knowledge may help us to understand the enzymology and molecular biology of other lipoxygenases.

Despite the great progress in lipoxygenase research during the last decade there are still many blank spots on the map of our knowledge. The mechanism of catalysis remains to be unequivocally ascertained. The structure of the active center of lipoxygenases, in particular the ligands of the catalytically active iron are unknown, as is the amino acid sequence. The recent success in the cloning of the reticulocyte lipoxygenase cDNA may open new possibilities in the elucidation of the structure and the molecular biology of lipoxygenases.

References

1. Theorell, H., Holman, R.T., and Åkeson, Å., *Arch. Biochem. Biophys.*, **14**, 250–252 (1974).
2. Tappel, A.L., in *The Enzymes*, P.D. Boyer, H. Lardy, and K. Myrback, Eds., 2nd ed., Vol. 8, Academic, New York, 1963 pp. 275–283.
3. Tappel, A.L., in *Autoxidation and Antioxidants*, W.O. Lindberg, Ed., Vol. 1, Wiley-Interscience, New York, 1961, pp. 325–366.
4. Bergström, S. and Holman, R.T., in *Advances in Enzymology*, Vol. 8, F.F. Nord, Ed., Interscience, New York, 1948, pp. 425–457.
5. Boyd, D.H. and Adams, G.A., *Can. J. Physiol.*, **23**, 191–198 (1955).

6. Hamberg, M. and Samuelsson, B., *Proc. Natl. Acad. Sci. USA*, **71**, 3400–3405 (1974).

7. Nugteren, D.H., *Biochim. Biophys. Acta*, **380**, 299–307 (1975).

8. Ho, P.P.K., Walters, C.P., and Sullivan, H.R., *Biochem. Biophys. Res. Commun.*, **76**, 398–405 (1977).

9. Borgeat, P., Hamberg, M., and Samuelsson, B., *J. Biol. Chem.*, **251**, 7816–7820 (1976).

10. Schewe, T., Halangk, W., Hiebsch, C., and Rapoport, S.M., *FEBS Lett.*, **60**, 149–152 (1975).

11. Rapoport, S.M. and Gerischer-Mothes, W., *Hoppe-Seyler's Z. Physiol. Chem.*, **302**, 167–178 (1955).

12. Rapoport, S.M. and Nieradt-Hiebsch, C., *Hoppe-Seyler's Z. Physiol. Chem.*, **302**, 179–185 (1955).

13. Hansson, G., Malmsten, C., and Rådmark, O., in *Prostaglandins and Related Substances*, C.R. Pace-Asciak and E. Granström, Eds., Elsevier, Amsterdam, 1983, pp. 127–169.

14. Rapoport, S.M., Schewe, T., Wiesner, R., Halangk, W., Ludwig, P., Janicke-Höhne, M, Tannert, C., Hiebsch, C., and Klatt, D., *Eur. J. Biochem.*, **96**, 545–561 (1979).

15. Schewe, T., Wiesner, R., and Rapoport, S.M., *Methods Enzymol.*, **71**, 430–441 (1981).

16. Siegel, M.I., McConnel, R.T., Porter, N.A., and Cuatrecasas, P., *Proc. Natl. Acad. Sci. USA.*, **77**, 308–312 (1980).

17. Wallach, D.P. and Brown, V.R., *Biochem. Biophys. Acta*, **663**, 361–372 (1981).

18. Goetze, A.M., Fayer, L., Bouska, J., Bornemeier, D., and Carter, G.W., *Prostaglandins* **29**, 689–701 (1985).

19. Narumiya, S., Salmon, J.A., Cottee, F.H., Weatherley, B.C., and Flower, R.J., *J. Biol. Chem.*, **256**, 9583–9592 (1981).

20. Ochi, K., Yoshimoto, T., Yamamoto, S., Taniguchi, K., and Miyamoto, T., *J. Biol. Chem.*, **258**, 5754–5758 (1983).

21. Yoshimoto, T., Miyamoto, Y., Ochi, K., and Yamamoto, S., *Biochim. Biophys. Acta*, **713**, 638–646 (1982).

22. Shahin, I., Grossman, S., and Sredni, B., *Biochim. Biophys. Acta*, **529**, 300–308 (1978).

23. Grossman, S., Shahin, I., and Sredni, B., *Biochim. Biophys. Acta*, **572**, 293–297 (1979).

24. Ruzicka, T., Vitto, A., and Printz, M.P., *Biochim. Biophys. Acta*, **751**, 369–374 (1983).

25. Pace-Asciak, C.R., Mizuno, K., and Yamamoto, S., in *Leukotrienes and Other Lipoxygenase Products*, B. Samuelsson and R. Paoletti, Eds., Raven, New York, 1982, pp. 71–76.

26. Yokoyama, C., Mizuno, K., Mitachi, H., Yoshimoto, T., Yamamoto, S., and Pace-Asciak, C.R., *Biochim. Biophys. Acta*, **750**, 237–243 (1983).

27. Veldink, G.A., Vliegenthart, J.F.G., and Boldingh, J., *Prog. Chem. Fats Other Lipids*, **15**, 131–166 (1977).

28. Vliegenthart, J.F.G., Veldink, G.A., and Boldingh, J., *J. Agric. Food Chem.*, **27**, 623–626 (1979).

29. Vliegenthart, J.F.G. and Veldink, G.A., in *Free Radicals in Biology*, Vol. 5, W.A. Pryor, Ed., Academic, New York, 1982, pp. 29–64.

30. Kühn, H., Schewe, T., and Rapoport, S.M., this volume.

31. Holman, R.T., *Arch. Biochem. Biophys.* **10**, 519–529 (1946).

32. Snipes, W., Kenny, K., and Keit, A., *Anal. Biochem.*, **81**, 425–431 (1977).

33. Boveris, A., Cadenas, E., and Chance, B., *Photobiochem. Photobiol.*, **1**, 175–182 (1980).

34. Lilius, E.M. and Laakso, S., *Anal. Biochem.*, **119**, 135–141 (1982).

35. Nwanze, E.A.C., *J. Chromatogr.*, **202**, 313–316 (1980).

36. Mitsuda, H., Yasumoto, K., Yamamoto, A., and Kusamo, T., *Agric. Biol. Chem.*, **31**, 115–121 (1976).

37. Lankin, V.Z., Kühn, H., Hiebsch, C., Schewe, T., Rapoport, S.M., Tikhaze, A.K., and Gordeeva, N.T., *Biomed. Biochim. Acta*, **44**, 655–664 (1985).

38. Regdel, D., Schewe, T., and Rapoport, S.M., *Biomed. Biochim. Acta*, **44**, 1411–1428 (1985).

38a. Kühn, H., Salzmann-Reinhardt, U., Pönicke, K., Schewe, T., and Rapoport, S.M., *Biochim. Biophys. Acta*, in press.

39. Grossman, S. and Zakut, R., *Methods Biochem. Anal.*, **25**, 303–329 (1979).

40. Lever, M., *Anal. Biochem.* **83**, 274–284 (1977).

41. Verhagen, J., Vliegenthart, J.F.G., and Boldingh, J., *Chem. Phys. Lipids*, **22**, 255–259 (1978).

42. Slapke, J., Schewe, T., Hummel, S., Winkler, J., and Kopf, M., *Biomed. Biochim. Acta*, **42**, 1309–1318 (1983).

43. Salzmann, U., Kühn, H., Schewe, T., and Rapoport, S.M., *Biochim. Biophys. Acta*, **795**, 535–542 (1984).

44. Schewe, T., Hiebsch, C., Ludwig, P., and Rapoport, S.M., *Biomed. Biochim. Acta*, **42**, 789–903 (1983).

45. Heydeck, D. and Schewe, T., *Biomed. Biochim. Acta*, **44**, 1261–1263 (1985).

46. Wiesner, R., Tannert, C., Hausdorf, G., Schewe, T., and Rapoport, S.M., *Acta Biol. Med. Ger.*, **36**, 393–403 (1977).

47. Höhne, H. and Andree, H., *Acta Biol. Med. Ger.*, **39**, 1241–1242 (1980).

48. Heydeck, D. and Ludwig, P., unpublished results.

49. Rapoport, S.M., Härtel, B., and Hausdorf, G., *Eur. J. Biochem.*, **139**, 573–576 (1984).

50. Ludwig, P., Tordi, G., and Colosimo, A., *Biochim. Biophys. Acta*, **830**, 136–139 (1985).

51. Diel, E. and Stan, H.J., *Planta*, **142**, 321–328 (1978).

52. Yoon, S. and Klein, B., *Agric. Food Chem.*, **27**, 955–962 (1979).

53. Shimizu, T., Rådmark, O., and Samuelsson, B., *Proc. Natl. Acad. Sci. USA*, **81**, 689–693 (1984).

54. Wallace, J.M. and Wheeler, E.L., *Phytochemistry*, **81**, 389–393 (1979).

55. Heydeck, D., Thesis, Humboldt University, Berlin, GDR, 1984.

56. Thiele, B., Black, E., and Harrison P.R., *Nucl. Acid Res.*, in press.

57. Wiesner, R., Hausdorf, G., Anton, M., and Rapoport, S.M., *Biomed. Biochim. Acta*, **42**, 431–436 (1983).

58. Vick, B.A. and Zimmerman, D.C., *Biochem. Biophys. Res. Commun.*, **111**, 470–477 (1983).

59. Vick, B.A. and Zimmerman, D.C., *Plant Physiol.*, **75**, 458–461 (1984).

60. Hatanaka, A., Kajiwara, T., Sekiya, J., and Asano, M., *Z. Naturforsch.*, **39c**, 171–173 (1984).

61. Kagan, V. and Schewe, T., unpublished results.

62. Bryant, R.W., Bailey, J.M., Schewe, T., and Rapoport, S.M., *J. Biol. Chem.*, **257**, 6050–6055 (1982).

63. Hamberg, M., *Prostagland Leuk. Med.*, **13**, 27–33 (1984).

64. Kühn, H., Wiesner, R., Schewe, T., and Rapoport, S.M., *FEBS Lett.*, **153**, 353–356 (1983).

65. Kühn, H. and Wiesner, R., unpublished results.

66. Hamberg, M., *Anal. Biochem.*, **43**, 515–526 (1971).

67. Spaapen, L.J., Verhagen, L., Veldink, G.A., and Vliegenthart, J.F.G., *Biochim. Biophys. Acta*, **618**, 153–162 (1980).

68. Bryant, R.W., Schewe, T., Bailey, J.M., and Rapoport, S.M., *J. Biol. Chem.*, **260**, 3548–3555 (1985).

69. van Os, C.P.A., Rijke-Schilder, G.P.M., van Halbeek, H., Verhagen, J., and Vliegenthart, J.F.G., *Biochim. Biophys. Acta*, **663**, 177–193 (1981).

70. Kühn, H., Wiesner, R., and Stender, H., *FEBS Lett.* **177**, 255–259 (1984).

71. Serhan, C.N., Hamberg, M., and Samuelsson, B., *Proc. Natl. Acad. Sci. USA*, **81**, 5335–5339 (1984).

71a. Kühn, H., Wiesner, R., Nekrasov, A.S., and Lankin, V.Z., submitted.

72. Garssen, G.J., Vliegenthart, J.F.G., and Boldingh, J., *Biochem. J.*, **130**, 435–442 (1972).

73. Verhagen, J., Veldink, G.A., Egmond, M.R., Vliegenthart, J.F.G., Boldingh, J., and van der Star, *J. Biochim. Biophys. Acta*, **529**, 369–379 (1978).

74. de Groot, J.J.M.C., Veldink, G.A., Vliegenthart, J.F.G., Boldingh, J., Wever, R., and van Gelder, B.F., *Biochim. Biophys. Acta*, **377**, 71–79 (1975).

75. Härtel, B., Ludwig, P., Schewe, T., and Rapoport, S.M., *Eur. J. Biochem.*, **126**, 353–357 (1982).

76. Tappel, A.L., in *Free Radicals in Biology*, Vol. 4, W.A. Pryor, Ed., Academic, New York, 1980, pp. 1–47.

77. Verhagen, J., Bouman, A., Vliegenthart, J.F.G., and Boldingh, J., *Biochim. Biophys. Acta*, **486**, 114–120 (1977).

78. Härtel, B., Kühn, H., and Rapoport, S.M., in *Proceedings of the 16th FEBS Congress*, Pt. A, VNU Science, Utrecht, 1985, pp. 299–304.

79. Maier, V.P. and Tappel, A.L., *J. Am. Oil Chem. Soc.*, **36**, 8–12 (1959).

80. Hamberg, M., *Lipids*, **10**, 87–92 (1975).

81. Maas, R.L., Ingram, C.D., Taber, D.F., Oates, J.A., and Brash, A.R., *J. Biol. Chem.*, **257**, 13515–13519 (1982).

82. Bokoch, G.M. and Reed, P.W., *J. Biol. Chem.*, **256**, 4156–4159 (1981).

83. Weber, F. and Grosch, W., *Z. Lebensm. Unters. Forsch.*, **161**, 223–230 (1976).

84. Ikediobe, C.O., *Agric. Biol. Chem.*, **41**, 2369–2375 (1977).

85. Klein, B.P., Grossman, S., King, D., Cohen, B.S., and Pinski, A., *Biochem. Biophys. Acta*, **793**, 72–79 (1984).

86. Cook, H.W. and Lands, *Can. J. Biochem.*, **53**, 1220–1231 (1975).

87. Slappendel, S., Veldink, G.A., Vliegenthart, J.F.G., Aasa, R., and Malmström, B.G., *Biochim. Biophys. Acta*, **747**, 32–36 (1983).

88. Zakut, R., Grossman, S., Pinsky, A., and Wilchek, M., *FEBS Lett.*, **71**, 107–110 (1976).

89. Garssen, G.J., Veldink, G.A., Vliegenthart, J.F.G., and Boldingh, J., *Eur. J. Biochem.*, **62**, 33–36 (1976).

90. Lapetina, E.G. and Cuatrecasas, P., *Proc. Natl. Acad. Sci. USA*, **76**, 121–125 (1979).

91. Ludwig, C. and Ludwig, P., unpublished results.

92. Marnett, L.J., Wlodawer, P., and Samuelsson, B., *J. Biol. Chem.*, **250**, 8510–8517 (1975).

93. Gross, H., Nastainczyk, W., and Ullrich, V., in *Oxygen Radicals in Chemistry and Biology*, W. Bors, M. Saran, and D. Tait, Eds., de Guyter, Berlin, 1984, pp. 435–439.

94. Downing, D.T., Ahern, D.G., and Bachta, M., *Biochem. Biophys. Res. Commun.*, **40**, 218–223 (1970).

95. Endo, K., Helmcamp, G.M., and Bloch, K., *J. Biol. Chem.*, **245**, 4293–4296 (1970).

96. Seiler, N., Jung, M.I., and Koch-Weser Ed., *Enzyme-Activated Irreversible Inhibitors*, Elsevier/North-Holland Biomedical Press, Amsterdam, 1978.

97. Tobias, L.D. and Hamilton, J.G., *Lipids*, **14**, 181–193 (1978).

98. Kühn, H., Holzhütter, H.G., Schewe, T., Hiebsch, C., and Rapoport, S.M., *Eur. J. Biochem.*, **139**, 577–583 (1984).

99. Gibson, K.H., *Chem. Soc. Rev..*, **6**, 489–510 (1977).

100. Sun, F.F., McGuire, J.C., Morton, D.R., Pike, J.E., Sprecher, H., and Kunau, W.H., *Prostaglandins*, **21**, 333–343 (1981).

101. Hammarström, S., *Biochim. Biophys. Acta*, **487**, 517–519 (1977).

102. Kühn, H., Pönicke, K., Halle, W., Wiesner, R., Schewe, T., and Förster, W., *Prostagland Leuk. Med.*, **17**, 291–303 (1985).

103. Spaapen, L.J.M., Verhagen, J., Veldink, G.A., and Vliegenthart, J.F.G., *Biochim. Biophys. Acta*, **617**, 132–140 (1980).

104. Siddiqi, A.M., and Tappel, A.L., *J. Am. Oil Chem. Soc.*, **34**, 529–543 (1975).

105. Yasumoto, K., Yamamoto, A., and Mitsuda, H., *Agric. Biol. Chem.*, **34**, 1162–1168 (1970).

106. Kühn, H., unpublished results.

107. Schewe, T., Kühn, H., and Rapoport, S.M., to be submitted.

108. Harvey, J., Parish, H., Ho, P.P.K., Boot, J.R., and Dawson, W., *J. Pharm. Pharmacol.*, **35**, 44–45 (1983).

109. Aharony, D., Smith, J.B., and Silver, J.M., in *The Leukotrienes*, L.W. Cahkwin and D.M. Bailey, Eds., Academic, New York, 1984, pp. 103–123.

110. Galpin, J.R., Veldink, G.A., Vliegenthart, J.F.G., and Boldingh, J., *Biochim. Biophys. Acta*, **536**, 356–362 (1978).

111. van Wauwe, J. and Goossens, J., *Prostaglandins*, **26**, 725–730 (1983).

112. Baumann, J., Bruchhausen, F., and Wurm, G., *Prostaglandins*, **20**, 627–639 (1980).

113. Sekiya, K. and Okuda, H., *Biochem. Biophys. Res. Commun.*, **105**, 1090–1095 (1982).

114. Yoshimoto, T., Furukawa, M., Yamamoto, S., Horie, T., and Watanabe-Kohno, S., *Biochem. Biophys. Res. Commun.*, **116**, 612–618 (1983).

115. Nakadate, T., Yamamoto, S., Aizu, E., and Kato, R., *Gann*, **75**, 214–222 (1984).

116. Sekiya, K., Okuda, H., and Arichi, S., *Biochim. Biophys. Acta*, **713**, 68–72 (1982).

117. Neichi, T., Koshihara, Y., and Murota, S., *Biochim. Biophys. Acta*, **753**, 130–132 (1983).

118. Koshihara, Y., Neichi, T., Murota, S., Lao, A., Fujimoto, Y., and Tatsuno, T., *Biochim. Biophys. Acta*, **792**, 92–97 (1984).

119. Agradi, E., Petroni, A., Socini, A., and Galli, C., *Prostaglandins*, **22**, 255–266 (1981).

120. Higgs, G.A., Flower, R.J., and Vane, J.R., *Biochem. Pharmacol.*, **28**, 1959–1961 (1979).

121. Blackwell, G.J. and Flower, R.J., *Prostaglandins*, **16**, 417–425 (1978).

122. Sun, F.F., McGuire, J.C., Wallach, D.P., and Brown, V.R., in *Advances in Prostaglandin and Thromboxane Research*, Vol. 6, B. Samuelsson, P.W. Ramwell, and R. Paoletti, Eds., Raven, New York, 1980, pp. 111–113.

123. Baumann, J. and Wurm, G., *Agents Actions*, **12**, 360–364 (1982).

124. Egan, R.W., Tischler, A.N., Baptista, E.M., Ham, E.A., Soderman, D.D., and Gale, P.H., in *Advances in Prostaglandin and Thromboxane Research*, Vol. 11, B. Samuelsson, P.W. Ramwell, and R. Paoletti, Eds., Raven, New York, 1983, pp. 151–157.

125. Egmond, M.R., Brunori, M., and Fasella, P.M., *Eur. J. Biochem.*, **61**, 93–100 (1976).

126. Gibian, M.J. and Galaway, R.A., *Biochemistry*, **15**, 4209–4214 (1976).

127. Lagocki, J.W., Emken, E.A., Law, J.H., and Kezdy, F.J., *J. Biol. Chem.*, **251**, 6001–6006 (1976).

128. Ludwig, P., Holzhütter, H.G., and Colosimo, A., submitted for publication.

129. Slappendel, S., Aasa, R., Malmström, B.G., Verhagen, J., Veldink, G.A., and Vliegenthart, J.F.G., *Biochim. Biophys. Acta*, **708**, 259–265 (1982).

130. Slappendel, S., Malmström, B.G., Petersson, L., Ehrenberg, A., Veldink, G.A., and Vliegenthart, J.F.G., *Biochem. Biophys. Res. Commun.*, **108**, 673–677 (1982).

131. Cheesbrough, T.M. and Axelrod, B., *Biochemistry*, **22**, 3837–3840 (1983).

132. Egmond, M.R., Fasella, P.M., Veldink, G.A., Vliegenthart, J.F.G., and Boldingh, J., *Eur. J. Biochem.*, **76**, 469–479 (1977).

133. Egmond, M.R., Finazzi-Agrò, A., Fasella, P.M., Veldink, G.A., and Vliegenthart, J.F.G., *Biochim. Biophys. Acta*, **375**, 43–49 (1975).

134. Finazzi-Agrò, A., Avigliano, L., Veldink, G.A., Vliegenthart, J.F.G., and Boldingh, J., *Biochim. Biophys. Acta*, **326**, 462–470 (1973).

135. Coon, M.J., *Nutr. Rev.*, **36**, 319–328 (1978).

136. Ullrich, V., *Top. Curr. Chem.*, **83**, 68–100 (1979).

137. Chance, B., Sies, H., and Boveris, A., *Physiol. Rev.*, **59**, 527–605 (1979).

138. Peisach, J., Burger, R., and Horwitz, S.B., in *Oxidative Damage and Related Enzymes—EMBO Workshop 1983*, G. Rotilio and J.V. Bannister, Eds., Life Chemistry Reports Suppl. Ser., Suppl. 2, Harwood Academic, New York, 1984, pp. 73–80.

139. Eriksson, C.E., Olsson, P.A., and Svensson, S.G., *J. Am. Oil Chem. Soc.*, **48**, 442–447 (1971).

140. Kaschnitz, R.M. and Hatefi, Y., *Arch. Biochem. Biophys.*, **171**, 292–304 (1975).

141. Kühn, H., Götze, R., Schewe, T., and Rapoport, S.M., *Eur. J. Biochem.*, **120**, 161–168 (1981).

142. Kühn, H., Götze, R., Schewe, T., and Rapoport, S.M., *Biomed. Biochim. Acta*, **43**, S35–S36 (1984).

143. Miyamoto, T., Ogino, N., Yamamoto, S., and Hayaishi, O., *J. Biol. Chem.*, **251**, 2629–2636 (1976).

144. van der Ouderaa, F.J., Buytenhek, M., Nugteren, D.H., and van Dorp, D.A., *Biochim. Biophys. Acta*, **487**, 315–331 (1977).

145. Kulmacz, R.J. and Lands, W.E.M., *J. Biol. Chem.*, **259**, 6358–6363 (1984).

146. Matsuda, Y., Beppu, T., and Arima, K., *Biochim. Biophys. Acta*, **530**, 439–450 (1978).

147. Matsuda, Y., Beppu, T., and Arima, K., *Agric. Biol. Chem.* **43**, 189–190 (1979).

148. Kühn, H., Hache, A., Sklenar, H., Schewe, T., and Rapoport, S.M., *Biomed. Biochim. Acta*, **42**, S175–S176 (1983).

149. Weiss, J.J., *Nature*, **202**, 83–84 (1964).

150. Carrell, R.W., Winterbourn, C.C., and French, J.K., *Hemoglobin*, **1**, 815–827 (1977).

151. Ribarov, S.R., Benov, L.C., Marcova, V.I., and Benchev, I.C., *Chem. Biol. Interact.* **45**, 105–112 (1983).

152. Sok, D.E., Chung, T., and Sih, C.J., *Biochem. Biophys. Res. Commun.*, **110**, 273–279 (1983).

153. Berkeley, H.D. and Galliard, T., *Phytochemistry*, **15**, 1481–1484 (1976).

154. Dupont, J., *Physiol. Plant.*, **52**, 225–232 (1981).

155. Halangk, W., Schewe, T., Hiebsch, C., and Rapoport, S.M., *Acta Biol. Med. Ger.*, **36**, 405–410 (1977).

156. Lankin, V.Z., Tikhaze, A.K., Osis, Yu.G., Vikhert, A.M., Schewe, T., and Rapoport, S.M., *Dok. Akad. Nauk SSSR* **281**, 204–207 (1985).

157. Fritsch, B., Maretzki, D., Hiebsch, C., Schewe, T., and Rapoport, S.M., *Acta Biol. Med. Ger.*, **38**, 1315–1321 (1979).

158. Lyr, H., Grunwald, D., and Schewe, T., unpublished results.

159. Schewe, T., Hiebsch, C., Garcia Parra, M., and Rapoport, S.M., *Acta Biol. Med. Ger.*, **32**, 419–426 (1974).

160. Rapoport, S.M., Hofmann, E.C.G., and Ghiretti-Magaldi, A., *Experientia*, **14**, 169–172 (1958).

161. Wiesner, R., Ludwig, P., Schewe, T., and Rapoport, S.M., *FEBS Lett.*, **123**, 123–126 (1981).

162. Schewe, T., Albracht, S.P.J., and Ludwig, P., *Biochim. Biophys. Acta*, **636**, 210–217 (1981).

163. Schewe, T. and Hiebsch, C., *Acta Biol. Med. Ger.*, **36**, 961–966 (1977).

164. Kühn, H., Pliquett, F., Wunderlich, S., Schewe, T., and Krause, W., *Biochim. Biophys. Acta*, **735**, 283–290 (1983).

165. Köckritz, A., Schewe, T., Hieke, B., and Hass, W., *Phytochemistry*, **24**, 381–303 (1985).

166. Köckritz, A., Hieke, B., Hoffmann, P., and Schewe, T., *Photosynthetica*, in press.

167. Schewe, T. and Rapoport, S.M., *Acta Biol. Med. Ger.*, **40**, 591–596 (1980).

168. Krause, W., Schewe, T., and Behrisch, D., *Acta Biol. Med. Ger.*, **34**, 1609–1620 (1975).

169. Dubiel, W., Müller, M., and Rapoport, S.M., *Biochem. Internat.* **3**, 165–171 (1981).

170. Lankin, V.Z., Tikhaze, A.K., Gordeeva, N.T., Schewe, T., and Rapoport, S.M., *Biokhimiya*, **48**, 2009–2015 (1983).

171. Lankin, V.Z., Gordeeva, N.T., Osis, Yu. G., Vikhert, A.M., Schewe, T., and Rapoport, S.M., *Biokhimiya*, **48**, 914–921 (1983).

172. Lankin, V.Z., Kühn, H., Hiebsch, C., Schewe, T., Rapoport, S.M., Tikhaze, A.K., and Gordeeva, N.T., *Biomed. Biochim. Acta*, **44**, 655–664 (1985).

173. Rapoport, S.M., *The Reticulocyte*, CRC Press Inc., in press.

174. Rapoport, S.M., Rosenthal, S., Schewe, T., Schulze, M., and Müller, M., in *Cellular and Molecular Biology of Erythrocytes*, H. Yoshikawa and S.M. Rapoport, Eds., University of Tokyo Press, Tokyo, 1974, pp. 93–141.

175. Schmidt, J., Prehn, S., and Rapoport, S.M., *Biomed. Biochim. Acta*, **44**, 1429–1434 (1985).

176. Rapoport, S.M., Dubiel, W., and Müller, M., *Acta Biol. Med. Ger.*, **40**, 1277–1283 (1981).

177. Rapoport, S.M., Dubiel, W., and Müller, M., *FEBS Lett.*, **180**, 249–252 (1985).

178. Shak, S., Perez, D.H., and Goldstein, I.M., *J. Biol. Chem.*, **258**, 14948–14953 (1982).

179. Wiesner, R., Rosenthal, S., and Hiebsch, C., *Acta Biol. Med. Ger.*, **30**, 631–646 (1973).

180. Höhne, M., Bayer, D., Prehn, S., Schewe, T., and Rapoport, S.M., *Biomed. Biochim. Acta*, **42**, 1129–1134 (1983).

181. Krause, K., Münn, A., Thesis, Humboldt University, Faculty of Medicine, Berlin (1983).

182. Thiele, B.J., Belkner, J., Andree, H., Rapoport, T.A., and Rapoport, S.M., *Eur. J. Biochem.*, **96**, 563–569 (1979).

183. Thiele, B.J., Andree, H., Höhne, M., and Rapoport, S.M., *Eur. J. Biochem.* **129**, 133–141 (1982).

184. Priess, H. and Zillig, W., *Hoppe-Seyler's Z. Physiol. Chem.* **348**, 817–822 (1967).

185. Salzmann, U., Ludwig, P., Schewe, T., and Rapoport, S.M., *Biomed. Biochim. Acta*, **44**, 211–219 (1985).

186. Thilo, C., Schewe, T., Belkner, J., and Rapoport, S.M., *Acta Biol. Med. Ger.*, **39**, 1431–1440 (1979).

187. Krause, W., David, H., Uerlings, I., and Rosenthal, S., *Acta Biol. Med. Ger.*, **28**, 779–786 (1972).

188. Gasko, O. and Danon, D., *Exp. Cell Res.* **75**, 159–163 (1972).

189. Müller, M., Dubiel, W., Rathmann, J., and Rapoport, S.M., *Eur. J. Biochem.*, **109**, 405–410 (1980).

190. Rapoport, S.M., Schmidt, J., and Prehn, S., *FEBS Lett.*, **183**, 370–374 (1985).

191. Hershko, A., Ciechanover, A., and Rose, I., *Proc. Natl. Acad. Sci. USA*, **76**, 3107–3110 (1979).

192. Hershko, A. and Ciechanover, A., *Annu. Rev. Biochem.*, **51**, 335–364 (1982).

193. Dubiel, W. and Rapoport, S.M., unpublished results.

194. Chang, W.C., Nakao, J., Orimo, H., and Murota, S., *Biochem J.* **202**, 771–776 (1982).

195. Lagarde, M., Croset, M., Authi, K.S., and Crawford, N., *Biochem. J.*, **222**, 495–500 (1984).

196. Hamberg, M., *Biochim. Biophys. Acta*, **431**, 651–654 (1976).

197. Hamberg, M. and Hamberg, G., *Biochem. Biophys. Res. Commun.*, **95**, 1090–1097 (1981).

198. Funk, M.O., Kim, S.H., and Alteneder, A.W., *Biochem. Biophys. Res. Commun.*, **98**, 923–929 (1981).

199. Siegel, M.I., McConnell, R.T., Abrams, S.L., Porter, N.A., and Cuatrecasas, P., *Biochem. Biophys. Res. Commun.*, **89**, 1273–1280 (1979).

200. Vanderhoek, J.Y., Bryant, R.W., and Bailey, J.M., *J. Biol. Chem.*, **255**, 5996–5998 (1980).

201. Bryant, R.W. and Bailey, J.M., *Biochem. Biophys. Res. Commun.*, **92**, 268–276 (1980).

202. Bryant, R.W., Simon, T.C., and Bailey, J.M., in *Oxidative Damage and Related Enzymes—EMBO Workshop 1983*, G. Rotilio and J.V. Bannister, Eds., Life Chemistry Reports, Suppl. Ser., Suppl. 2, 1984, pp. 304–308.

203. Bryant, R.W. and Bailey, J.M., *Prostaglandins*, **17**, 9–18 (1979).

204. Bryant, R.W. and Bailey, J.M., *Prog. Lipid Res.*, **20**, 279–281 (1981).

205. Dutilh, C.E., Haddeman, E., and ten Hoor, F., *Adv. Prostaglandin Thromboxane Res.*, **6**, 101–115 (1980).

206. Dutilh, C.E., Haddeman, E., van Dorp, J.A., and ten Hoor, F., *Prostaglandins Med.*, **6**, 111–126 (1981).

207. Nakao, J., Ito, H., Ooyama, T., Chang, W.C., and Murota, S., *Atherosclerosis*, **46**, 309–316 (1983).

208. Jakschik, B.A., Sun, F.F., Lee, L., and Steinhoff, M.M., *Biochem. Biophys. Res. Commun.*, **95**, 103–110 (1980).

209. Siegel, M.I., McConnell, R.T., Bonser, R.W., and Cuatrecasas, P., *Prostaglandins*, **21**, 123–132 (1981).

210. Hamasaki, Y. and Tai, H.H., *Biochim. Biophys. Acta*, **793**, 393–398 (1984).

211. Maas, R.L. and Brash, A.R., *J. Biol. Chem.*, **80**, 2884–2888 (1983).

212. Vanderhoek, J.Y., Bryant, R.W., and Bailey, J.M., *J. Biol. Chem.*, **255**, 10064–10066 (1980).

213. Yokoyama, C., Mizuno, K., Mitachi, H., Yoshimoto, T., Yamamoto, S., and Pace-Asciak, C.R., *Biochim. Biophys. Acta*, **750**, 237–243 (1983).

214. Pace-Asciak, C.R., Mizuno, K., and Yamamoto, S., *Biochim. Biophys. Acta*, **712**, 142–145 (1982).

215. Hammarström, S., Lindgren, J.A., Marcello, C., Duell, E.A., Anderson, T.F., and Voorhees, J.J., *J. Invest. Dermatol.*, **73**, 180–183 (1979).

216. Doerge, D.R. and Corbett, M.D., *Experientia*, **38**, 901–902 (1982).

217. Jung, G., Yang, D.C., and Nakao, A., *Biochem. Biophys. Res. Commun.* **130**, 559–566 (1985).

THE STEREOCHEMISTRY OF THE REACTIONS OF LIPOXYGENASES AND THEIR METABOLITES. PROPOSED NOMENCLATURE OF LIPOXYGENASES AND RELATED ENZYMES

By HARTMUT KÜHN, TANKRED SCHEWE, AND SAMUEL M. RAPOPORT, *Institute of Biochemistry, Humboldt University, DDR-1040 Berlin, GDR*

CONTENTS

I. Introduction

Lipoxygenases are widely distributed in plants and animals (1–8). For general reviews of lipoxygenases the reader is referred to the articles of the Vliegenthart group (1) and to the accompanying article by Schewe et al. (2) in this volume.

In recent years enormous progress has been achieved in the identification and sterochemical characterization of a variety of lipoxygenase products such as hydroperoxyeicosanoids, leukotrienes, and lipoxins. The stereochemical analysis of the lipoxygenase products, the use of stereospecifically labeled substrates, and of selected substrate derivatives have yielded insights into the stereochemistry of the lipoxygenase reactions. This paper gives an overview on the stereochemistry of lipoxygenase products. Moreover some conclusions concerning the orientation of the lipoxygenase substrates at the active site of the enzyme combined with a proposal for a comprehensive nomenclature of lipoxygenases and related enzymes are presented.

II. Steric Structure of Lipoxygenase Substrates

Lipoxygenases dioxygenate polyunsaturated fatty acids containing all-*cis*-methylene-interrupted double bonds such as 9,12-all-*cis*-octadecadienoic acid (linoleic acid), 9,12,15-all-*cis*-octadecatrienoic acid (α-linolenic acid), 5,8,11,14-all-*cis*-eicosatetraenoic acid (arachidonic acid), or 5,8,11,14,17-all-*cis*-eicosapentaenoic acid. The precise steric structure of these fatty acids during the reaction is difficult to establish because of the flexibility of the molecules. For arachidonic acid it has been estimated that there are more than 10^7 low-energy staggered conformations owing to its 14 rotatable C–C single bonds, 5 of which have a threefold barrier (sp^3–sp^3) and 9 a

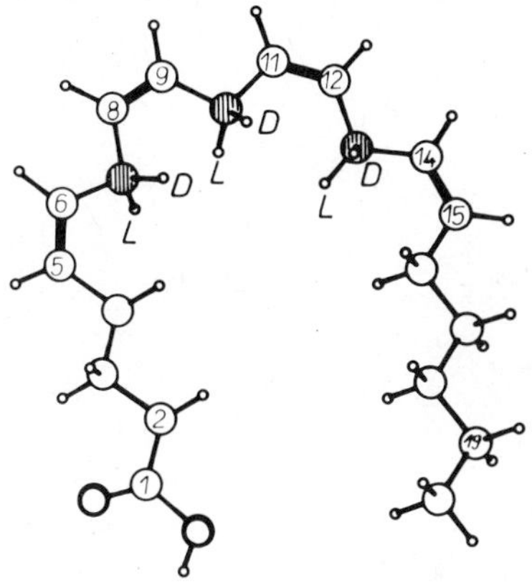

Figure 1.   Horseshoe-like structure of arachidonic acid. The prochiral centers (double allylic carbon atoms) are hatched.

sixfold barrier (sp^2–sp^3) (9). However, this estimate is excessive, since the assumption that any combination of two torsion minima gives rise to a low-energy conformer is not true. Nonbonded interactions reduce the number of possible low-energy conformers (10). Moreover, it is not clear whether the minimum energy conformer or a conformer less stable but more likely in terms of the substrate binding to the active site of the enzyme should be assumed.

The conformation of the fatty acids, in which all double bonds present in the fatty acid are situated in a plane and all torsions about the single bounds are trans, thus leading to a planar molecule, corresponds to such a low energy conformer and may serve as a suitable model for the fatty acid conformation during the lipoxygenase reaction. For arachidonic acid this structure resembles a horseshoe (Fig. 1). The hydrogen atoms at the double allylic methylene carbon atoms are localized above and below the plane determined by the double bonds.

Fatty acids containing a trans double bond, such as linolelaidic acid, are not oxygenated but some of them are competitive inhibitors of the linoleate oxygenation by soybean lipoxygenase (1). That means that both cis and trans fatty acids are bound to the enzyme but the lipoxygenase reaction, probably the initial hydrogen removal, is sterically hindered by the trans structure. Moreover, fatty acids containing a conjugated diene system are strong competitive inhibitors of linoleate oxygenation.

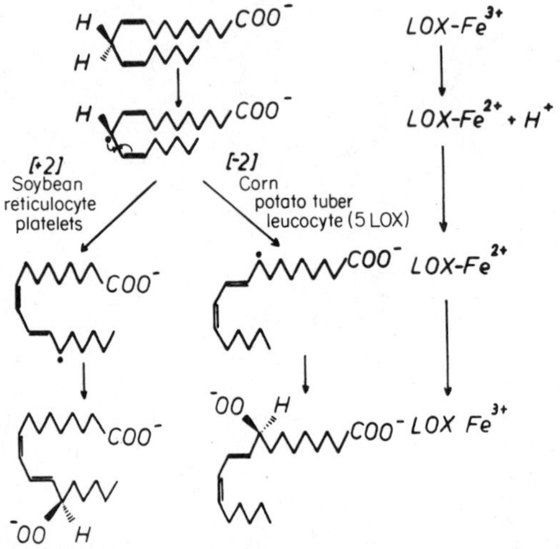

Figure 2. Mechanism of the dioxygenase reaction catalyzed by lipoxygenases. Linoleic acid as substrate. The little arrows indicate the [+ 2] rearrangement of the fatty acid radical. The valency change of the enzyme bound nonheme iron, proposed to be involved in the reaction is shown on the right.

Lipoxygenases in general prefer free cis polyunsaturated fatty acids. An exception is the lipoxygenase from reticulocytes, which is able to oxygenate esterified fatty acids, phospholipids, and even biological membranes (11,12).

III. The Mechanism of Lipoxygenase Reaction

The dioxygenation of polyunsaturated fatty acids by a lipoxygenase formally consists of three steps (Fig. 2).

1. Removal of one hydrogen atom from the double allylic methylene carbon atom forming a fatty acid radical.

2. Conjugation of the double bonds with a trans isomerization of the shifted double bond. The conjugation is accompanied by a rearrangement of the radical electron.

3. Insertion of dioxygen.

For the soybean lipoxygenase it has been proposed that a valency change of the nonheme iron present in the lipoxygenase is involved in the catalytic cycle (13).

To establish the complete structure of lipoxygenase products three parameters have to be examined: (a) positional isomerism (positional specificity of the lipoxygenase), (b) optical isomerism (stereospecificity of the lipoxygenase), and (c) cis–trans isomerism.

The lipoxygenase reaction was originally only understood as the formation of monohydroperoxypolyenoic fatty acids from the corresponding polyenoic fatty acids and molecular dioxygen. However, according to the present knowledge there is a variety of other reactions that are catalyzed by lipoxygenases and related enzymes, some of which have great biological importance:

1.   Dioxygenation of monohydroperoxy- or monohydroxypolyenoic fatty acids forming dihydroperoxy derivatives.

2.   Formation of triple oxygenation products (hydroperoxylipoxins) from 5,15-DiHETE (see Section VII.G.)

3.   Prostaglandin G synthase reaction (first step of the cyclooxygenase reaction).

4.   Breakdown of hydroperoxypolyenoic fatty acids via the lipohydroperoxidase reaction to a variety of secondary products including conjugated oxopolyenoic fatty acids, epoxyhydroxy- and trihydroxy derivatives.

5.   Formation of epoxyleukotrienes from $15L_S$-HPETE or $5D_S$-HPETE (leukotriene A synthase reaction).

The Reactions (1), (2) and (3) depend on the presence of dioxygen, whereas Reactions (4) and (5) do not require it and therefore also occur under anaerobic conditions. The lipohydroperoxidase reaction includes as its first step the homolytic scission of the O–O bond of the hydroperoxy group (see Section VII.G). In contrast, the dioxygenation reaction is initiated by a homolytic cleavage of a C–H bond. Thus, the decisive feature of all lipoxygenase-catalyzed reactions is the homolytic cleavage of a sigma bond with the formation of intermediate radicals. In Reaction (5) (leukotriene A_4 synthase) both types of homolytic cleavage are probably involved (see Section VII.B).

A. POSITIONAL SPECIFICITY OF LIPOXYGENASES

The positional specificity of lipoxygenases and, therefore, the positional isomerism of lipoxygenase products is characterized by two factors: (a) positional specificity of the initial hydrogen removal and (b) the positional specificity of the subsequent oxygen insertion.

Depending on the numbers of double bonds h of the fatty acids there exist $(h - 1)$ possibilities for a hydrogen removal and therefore $2(h - 1)$ regioisomers of the primary oxygenation products. With linoleic acid $(h = 2)$ only one possibility for an initial hydrogen removal exists, whereas with arachidonic acid hydrogen atoms can be removed from three different double allylic methylene carbon atoms, C_m (C_7, C_{10}, C_{13}). From the mechanism of the lipoxygenase reaction (Fig. 2) it can be seen that two positional isomers of the oxygenation products can be formed if the initial hydrogen removal takes place from one methylene carbon atom C_m. If the mesomeric free radical which is formed after initial hydrogen removal reacts with the oxygen at the carbon nearer the carboxylic group of the fatty acid, the dioxygen is introduced at carbon C_{m-2}. Such a reaction should be designated as $(-)$. In contrast, a reaction involving a shift of the radical electron in the direction of the methyl end of the fatty acid should be designated as $(+)$.

The preference for the formation of one of the two possible isomers of the primary reaction products by most lipoxygenases may be explained in two different ways: (a) better stabilization of one of the mesomeric structures of the radical intermediate, which is formed via initial hydrogen removal at the enzyme and (b) orientation of the dioxygen introduced at the active site, so that only one of the possible mesomeric structures of the fatty acid radical is transformed to the hydroperoxy fatty acid radical. Such an oxygen orientation is known for hemoproteins [14].

From experiments with fatty acid congeners possessing positional isomerism of the double bonds it has been shown that, for various lipoxygenases showing a $[+2]$ rearrangement of the fatty acid radical, the position of hydrogen removal is determined by the distance from the methyl end of the fatty acid (15,16,17). In contrast, there is a lack of information concerning the signal recognition of the site of the hydrogen removal for lipoxygenases showing a $[-2]$ rearrangement of the fatty acid radical, for example, arachidonate 5-lipoxygenase from leukocytes and mast cells.

We could show in our laboratory that for the lipoxygenase from wheat showing a $[-2]$ shift of the radical electron (18) the place of the initial hydrogen removal is not related to the distance of the prochiral center from the methyl end, but rather to the distance from the carboxylic group (19). This conclusion was drawn from experiments using various fatty acids with positional isomerism of the double bonds as substrates. Linoleic acid, arachidonic acid, dihomo-γ-linoleic acid, and 5,8,11-all-*cis*-eicosatrienoic acid were good substrates, whereas 11,14,17-all-*cis*-eicosatrienoic acid and 11,14-all-*cis*-eicosadienoic acid were not oxygenated. These results as well as the pattern of the oxygenation products from arachidonic acid (5- and 8-HPETE) and linoleic acid (mainly 9-HPOD) indicate that the wheat lipoxygenase is able to remove hydrogen from either C_{10} or C_{11}, counted from the carboxylic group of the fatty acid, but not from the $(n - 8)$ double allylic carbon atom. It remains to be examined whether this principle is a general rule for $[-2]$ lipoxygenases. These regularities would imply an inverse topical arrangement of the substrate at the active site of $[+2]$- and $[-2]$ lipoxygenases (see Section V).

For fatty acids containing methylene interrupted double bonds, only the possibilities of $[+2]$ or a $[-2]$ rearrangement of the fatty acid radical during the lipoxygenase reaction exist. With substrates that contain a further double bond in conjugation with one of the dienes of a 1,4-pentadiene system the rearrangement of the fatty acid radical may take place as shown in Fig. 3.

A $[+2]$ and $[-4]$ rearrangement of the radical is possible if the additional double bond is situated between the carboxylic group of the fatty acid and the methylene carbon atom, C_m, from which the hydrogen is removed. If the additional double bond is situated between C_m and the methyl end of the fatty acid a $[-2]$ or $[+4]$ reaction is possible (Fig. 3). The $[+4]$ and $[-4]$ rearrangement of the fatty acid radical is of importance for the synthesis of leukotriene A_4 from 5_{DS}- or 15_{LS}-HPETE.

For a long time it has been assumed that lipoxygenases show an absolute positional specificity, that means that they remove hydrogen only from one double allylic carbon atom. Recently, however, experimental data were obtained suggesting that some lipoxygenases are able to remove hydrogen from different prochiral centers (20,21). For the purified lipoxygenase from reticulocytes this dual positional specificity has been shown unequivocally (22). The possible reason for such a dual activity may be a slight dislocation of the double allylic carbon atom from the enzyme-bound iron that participates in

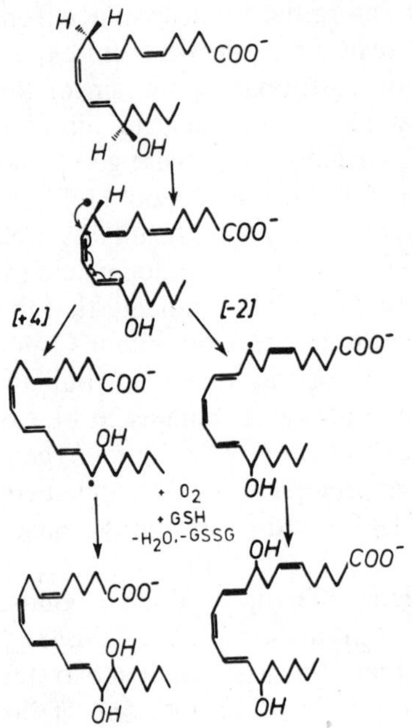

Figure 3. [+4] and [−2] rearrangement of the fatty acid radical during the lipoxygenase reaction with 15-HETE. The electron shift during the [+4] rearrangement is indicated by little arrows.

the hydrogen removal. As can be seen from Fig. 4, such a dislocation would bring a neighboring double allylic carbon atom closer to the iron, so that a hydrogen could be removed.

B. STEREOSPECIFICITY OF LIPOXYGENASES

The stereospecificity of lipoxygenases is characterized by the stereoselective removal of one of the two hydrogen atoms from the prochiral center, C_m, of the pentadiene system and by the stereospecific introduction of dioxygen at $C_{m+2(4)}$ or $C_{m-2(4)}$. The methylene carbon atom, C_m, of the pentadiene system is prochiral in the sense that the replacement of one of the two hydrogen atoms bound to it by a substituent with different priority results in a chiral assembly. If the hydrogen to be replaced is situated at the left side of the hydrocarbon chain according to the Fischer convention it is designated as L-hydrogen. Similarly, the D-hydrogen is that which leads to a D-configuration by application of the Fischer convention. If the replacement of one of the two hydrogen atoms by a substituent pos-

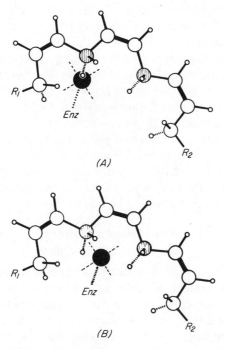

(A)

(B)

Figure 4.   Arrangement of the fatty acid at the active center of lipoxygenases with single (A) and dual (B) positional specificity. The black circles indicate the site of the enzyme involved in the hydrogen removal (probably the nonheme iron). A slight dislocation of one prochiral center from the iron brings another double allylic carbon atom closer to it, so that a hydrogen removal from both prochiral centers becomes possible.

sessing a higher priority than the other enantiotopic hydrogen but a lower one than that of the other two ligands (e.g., replacement by tritium) results in the R configuration according to ref. (23) this hydrogen atom is designated as pro-R. If the S configuration is obtained by application of the same rule the hydrogen atom is pro-S.

It has been shown for the lipoxygenases from soybeans (16,24), corn germs (24) and blood platelets (25) as well as for the leukotriene A_4-synthase (26), and cyclooxygenase from sheep vesicular glands (24) that they remove stereoselectively one of the two hydrogen atoms from the prochiral centers. Which hydrogen is removed depends on the stereospecificity of the enzymes (see Section V).

The steric analysis of the reaction products shows that most lipoxygenases exclusively introduce oxygen stereospecifically forming only one of the possible optical isomers. Racemic products, however, were found with the type-2 lipoxygenases from soybeans and green peas (28) and with the mercury-modified soybean lipoxygenase-1 (29). It is remarkable that racemic products are always accompanied by the occurrence of positional isomers. This behavior is assumed to be caused by a dissociation of the fatty acid radical from the active site of the enzymes and subsequent nonenzymatic rearrangement of the fatty acid radical and oxygen insertion forming different positional and optical isomers.

1. Specification of the Chiral and Prochiral Centers during Lipoxygenase Reaction

The configurations of the prochiral and chiral centers can be specified by the D/L system (Fischer convention) (30) or the S/R system (23). For the characterization of the stereochemistry of the lipoxygenase reaction the Fischer convention should be preferred, since the S/R system does not take into consideration genetic relations between lipoxygenase substrates and products. The application of the priority rules to the lipoxygenase reaction may give rise to assumptions of structural differences that do not correspond to reality. Two examples may serve to prove this contention.

1. It has been shown that the leukotriene A_4 synthase from human leukocytes removes stereoselectively the pro-$R(D_R)$ hydrogen from the prochiral center C_{10} of 5,8,11,14-all-*cis*-eicosatetraenoic acid (26). In similar experiments it has been shown that the leukotriene A_5 synthase from murine mastocytoma cells removes stereoselectively the pro-S (D_S) hydrogen from C_{10} of 5,8,11,14,17-all-*cis*-eicosapentaenoic acid (31). From Fig. 5 it is evident that one and the same hydrogen is removed in both cases despite its different designation as pro-R and pro-S. In contrast, the hydrogen atoms removed are D in both cases according to the Fischer convention. The reason for the different designation in the S/R system is the occurrence of the additional double bond in the methyl ligand in eicosapentaenoic acid, which gives that ligand a higher priority than the

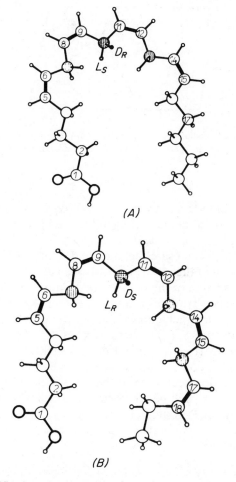

Figure 5.   Specification of the prochiral center C_{10} in arachidonic (A) and eicosapentaenoic (B) acid. The hydrogen that is removed during $LTA_{4(5)}$ synthesis is drawn in black.

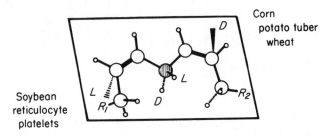

Figure 6. The antarafacial character of the lipoxygenase reaction (modified from (24). R_1, methyl ligand; R_2, carboxylic ligand. The thick arrows indicate the dioxygen introduction.

carboxylic ligand. In contrast, the carboxylic ligand possesses a higher priority than the methyl ligand in arachidonic acid.

2. Oxygenation of arachidonic acid to 15_{LS}-hydroperoxy-5,8,11,13(Z,Z,Z,E)-eicosatetraenoic acid (15_{LS}-HPETE) changes the configuration of the pro-S (L_S) hydrogen at C_{10} into a pro-R (L_R) since the conjugation of the double bonds in the methyl ligand from C_{10} makes the priority of this ligand higher than that of the carboxylic one. Therefore the arachidonate 12-lipoxygenase from blood platelets removes the pro-S (L_S) hydrogen from C_{n-11} of arachidonic acid (25) but the pro-R (L_R) hydrogen from 15_{LS}-HPETE (32). However, in both cases it is one and the same hydrogen atom as can be seen from its designation according to the Fischer convention.

However, it may still be convenient to specify the chiral centers according to the S/R system and add them as subscripts, since the priority rules are commonly used for the characterization of lipoxygenase products.

2. The Antarafacial Character of the Lipoxygenase Reaction

For soybean lipoxygenase, which stereoselectively removes the L_S-hydrogen from C_{n-8}, it has been shown that oxygen is introduced in the L_S-configuration (16,24). In contrast, corn germ lipoxygenase, which removes the D_R-hydrogen from linoleic acid, introduces oxygen in the D_S-position (24). Starting from the planar model of the unsaturated fatty acids (Fig. 1) hydrogen removal and dioxygen insertion take place from opposite sides of the pentadiene system (Fig. 6). This antarafacial character of the reaction was assumed to be a general principle for lipoxygenase-catalyzed reactions (24). Recently the antarafacial character has been shown for other lipoxygenases (blood platelets and leukocytes) as well as for leukotriene A_4 synthase and cyclooxygenase from sheep vesicular glands (see Section V). According to the antarafacial character, the lipoxygenases, which remove the D-hydrogen stereoselectively, introduce dioxygen in the D-position. Similarly, the L-lipoxygenases remove the L-hydrogen and introduce dioxygen in the L-position (see Table 1).

C.   CIS–TRANS ISOMERISM

From Fig. 2 it is evident that during the lipoxygenase reaction one of the double bonds is dislocated, whereas the other one remains uneffected. This double-bond shift is accompanied by a cis–trans isomerization so that a cis–trans structure of the dioxygenation product results. The cis–trans isomerization of the double bonds shifted has not only been shown for the primary dioxygenation products (hydroperoxy fatty acids) (4,16) but also for the more complex lipoxygenase products such as leukotriene A_4 (33), its hydrolysis products (34), and for the double dioxygenation products (35) containing a conjugated triene system. Since two double bonds are shifted during the formation of these products (Fig. 3) the conjugated triene geometry is cis, trans, trans. Some of the LTA_4 hydrolysis products possess an all-trans conjugated triene system. The complete triene system is shifted during their formation (see Section VII.E.2). Leukotriene B_4 is the only lipoxygenase metabolite apparently deviating from this rule, since one of the three double bonds shifted during its synthesis attains the cis configuration (see Section VII.C). However, one has to take into account that the conversion of LTA_4 to LTB_4 is a special type of epoxide hydrolysis rather than a lipoxygenase reaction.

IV.   Proposed Nomenclature for Lipoxygenases and Related Enzymes

Although knowledge of lipoxygenases and their reaction products has greatly increased in recent years, no general or differentiated nomenclature for these enzymes exists. In the enzyme nomenclature proposed by the Nomenclature Committee of the IUB, lipoxygenases are simply called linoleate: oxygen oxidoreductase

(EC1.13.11.12) (36). This designation is not very practicable, since it is restricted to linoleic acid, whereas the biologically most important substrate in animal cells is arachidonic acid. Moreover, it does not take into consideration stereochemical aspects. For this reason, in the current literature most animal lipoxygenases are designated according to the positional isomers of their reaction products with arachidonic acid. Thus, the lipoxygenase from blood platelets is called 12-lipoxygenase. Sometimes it is called $(n - 9)$-lipoxygenase. This nomenclature is more comprehensive since it is not only restricted to arachidonic acid and takes into consideration the positional specificity of the lipoxygenases. However, even this nomenclature does not completely reflect our knowledge of the positional and stereospecificity of the lipoxygenases. A comprehensive nomenclature of lipoxygenases should reflect the following properties of lipoxygenases: (a) positional specificity and stereospecificity of the initial hydrogen removal; (b) positional specificity and stereospecificity of dioxygen insertion; and (c) the type of reaction catalyzed, including the substrates.

The following formula is proposed to characterize lipoxygenases according to these points:

$$[C_{n-x}(C_x)] - L(D)_{pro-S(R)}:[\pm 2(4)] - L(D)_{S(R)}$$

polyenoic fatty acid/oxygen oxidoreductase

The terms before the colon characterize the specificity of the initial hydrogen removal and those following it characterize the specificity of the dioxygen introduction.

1. Positional specificity of the initial hydrogen removal: $[C_{n-x}(C_x)]$; $n - x$ or x indicates the position of the carbon atom where the hydrogen removal takes place; (n is the total number of carbon atoms).

The designation $[C_{n-x}]$ or $[C_x]$ depends on which terminus serves as signal for the recognition of the site of the hydrogen removal. $[C_{n-x}]$ indicates that the place of hydrogen removal is determined by the distance of the prochiral center from the methyl terminus. In contrast, $[C_x]$ indicates that for these lipoxygenases the place of hydrogen removal is determined by the distance from the terminus possessing the highest oxidation number, for example, the carboxylic group.

2. Stereospecificity of hydrogen removal: $D(L)_{\text{pro-}R(S)}$; steric configuration of the hydrogen abstracted. The pro characterizes the carbon atom as a prochiral center.

3. Positional specificity of oxygen insertion: $[\pm 2(4)]$; rearrangement of the fatty acid radical formed during the hydrogen removal. The free radical electron is shifted by 2 or 4 carbon atoms in the direction of the carboxylic ($-$) or the methyl ($+$) end of the fatty acid. There oxygen insertion takes place.

4. Stereospecificity of oxygen insertion: $D(L)_{R(S)}$; steric configuration of the chiral center after dioxygen insertion.

As already discussed, the specification of the chiral and prochiral centers according to the Fischer convention should be preferred. However, the specification according to the *S/R* system should be added as subscripts. In order to achieve uniformity according to this system it is proposed to use arachidonic acid, the biologically most important substrate, as model substrate.

V.   Specificity of Various Lipoxygenases

In 1967 Hamberg and Samuelsson were the first to clarify the complete stereochemistry of the dioxygenase reaction of soybean lipoxygenase (16). In experiments with stereospecifically labeled [3-^{14}C, $13D_R$-^{3}H]- and [3-^{14}C,$13L_S$-^{3}H]-8,11,14-all-*cis*-eicosatrienoic acid and with fatty acid homologues possessing positional isomerism of the double bonds the positional specificity and stereospecificity of both initial hydrogen removal and the subsequent dioxygen insertion of soybean lipoxygenase was studied (16). According to these results the soybean lipoxygenase should be designated as $[C_{n-8}]$-$L_{\text{pro-}S}$: $[+2]$-L_S lipoxygenase.

In contrast, the lipoxygenase from corn germs is a $[C_{11}]$-$D_{\text{pro-}R}$: $[-2]$-D_S lipoxygenase with linoleic acid, since the D_R-hydrogen from C_{11} is removed selectively and the hydroperoxy group is introduced at C_9 ($[-2]$ rearrangement of the fatty acid radical) in the D_S configuration (24). The positional specificity of the initial hydrogen removal $[C_x]$ has not been studied extensively. However, there is experimental evidence that the place of hydrogen removal does not relate to the distance of the prochiral center to the methyl end of the fatty acid (37).

The complete stereochemistry of the reaction of blood platelet lipoxygenase with arachidonic acid (15,25) and 5_{L_S}-HPETE (32) has been clarified. The results obtained indicate that blood platelets contain a $[C_{n-11}]$-L_{pro-S}:$[+2]$-L_S lipoxygenase and that the reaction shows antarafacial character. Moreover, the antarafacial character of the lipoxygenase reaction was shown to be true for the reaction of the $[C_{n-11}]$-L_{pro-S}:$[+2]$-L_S lipoxygenase from porcine leukocytes with arachidonic acid and 15_{L_S}-HPETE (32).

Recently, Hamberg (21) reported on an $(n-6)$ oxygenation of 6,9,12-all-cis-octadecatrienoic acid in addition to an $(n-9)$ oxygenation (ratio about 1:2) by human blood platelets . It was assumed that this product pattern is caused by a single lipoxygenase with dual positional specificity. In our opinion the different product pattern between arachidonic acid [exclusive $(n-9)$ oxygenation] and γ-linolenic acid may be caused by the different steric structure of the substrate (additional double bond in arachidonic acid and therefore a higher curvature of the horseshoe). This structural difference may cause a slight dislocation of the $(n-11)$ prochiral center from the place of the enzyme, where the hydrogen removal takes place (probably the enzyme-bound iron), bringing the $(n-8)$ prochiral center closer to it (Fig. 4). A similar fatty acid orientation might be the reason for the dual positional specificity of other lipoxygenases, for example, that from reticulocytes (17), wheat (19).

In Table I a number of lipoxygenases are summarized according to the proposed nomenclature. Unfortunately the stereospecificity of the initial hydrogen removal has not been shown for all lipoxygenases up till now. However, according to the antarafacial character one can predict which of the two hydrogen atoms from the prochiral center should be removed if the steric configuration of the dioxygenation product is established.

The reticulocyte lipoxygenase produces 15_{L_S}- and 12_{L_S}-HPETE from arachidonic acid in a ratio of about 15:1 (17). From copurification of both activities, from the homogeneity of the enzyme in the SDS gel electrophoresis, and from the inactivation behavior towards 5,8,11,14- and 5,8,11-eicosatetra(tri)ynoic acid it was evident that both products are formed by one enzyme (22). Therefore, the reticulocyte lipoxygenase is able to remove hydrogen from different double allylic methylene carbon atoms with arachidonic acid as substrate. From the steric structure of the oxygenation products and

TABLE I
Nomenclature of Various Lipoxygenases

Source	Nomenclature	Short form	References
Soybean 1	$[C_{n-8}]$-L_{pro-S}:$[+2]$-L_S	15-LOX	16, 24
Corn germ	$[C_{11}]$-D_{pro-R}:$[-2]$-D_S[a]	9-LOX[b]	24
Blood platelets	$[C_{n-11}]$-L_{pro-S}:$[+2]$-L_S	12-LOX	15, 25
Reticulocytes	$[C_{n-8}]$-L_{pro-S}:$[+2]$-L_S[c]	15-LOX	17, 22
	$[C_{n-11}]$-L_{pro-S}:$[+2]$-L_S[c]	12-LOX	17, 22
Leukocytes	$[C_7]$-D_{pro-S}:$[-2]$-D_S[a,c]	5-LOX	5
	$[C_{n-11}]$-L_{pro-S}:$[+2]$-L_S[c]	12-LOX	90
	$[C_{n-8}]$-L_{pro-S}:$[+2]$-L_S[c]	15-LOX	5, 90
Potato tuber	$[C_7]$-D_{pro-S}:$[-2]$-D_S[a,c]	5-LOX	20
Cyclooxygenase (first step)	$[C_{13}]$-L_{pro-S}:$[-2]$-L_R[a]	11-LOX	9, 22
Leukotriene A₄ synthase (5,6-LTA₄)	$[C_{10}]$-D_{pro-R}:$[-2]$[a] hydroperoxy fatty acid/epoxide synthase		26

[a] For lipoxygenases with a $[-]$ rearrangement of the fatty acid radical the positional specificity of the initial hydrogen removal has not been studied in detail. Therefore the position of the hydrogen removal with arachidonic acid is indicated.

[b] With linoleic acid as substrate.

[c] Conclusion from the stereochemistry of the dioxygenation product to the steroselective hydrogen removal according to the antarafacial character of the lipoxygenase reaction.

the antarafacial character of the lipoxygenase reaction, it is concluded that the L_S-hydrogen atoms from C_{n-8} and C_{n-11} are selectively removed during the reaction. Similarly one can assume that the arachidonate 5-lipoxygenase from leukocytes removes stereoselectively the D_S-hydrogen from C-7 of the arachidonic acid.

The cyclooxygenase, the initial enzyme of the prostaglandin formation, has been shown to be a lipoxygenase in the first reaction step (38–40). Hamberg and Samuelsson reported that the L_S-hydrogen from C_{13} of 8,11,14-all-*cis*-eicosatrienoic acid was removed stereoselectively during the formation of prostaglandin E_1 and $F_{1\alpha}$ (27). According to antarafacial character the dioxygen should be introduced in the L_R-configuration at C_{11}. It is impossible to experimentally check the stereospecificity of the oxygen introduction at C_{11} for the cyclooxygenase during its reaction with arachidonic acid, since the primary dioxygenation product undergoes rapid cyclization. However, from the stereochemical mechanism of prostaglandin formation it was concluded that both dioxygens at C_{11} (L_R) and C_{15} (L_S) are introduced in the L-configuration (9,41).

The leukotriene A_4 synthase was shown to remove the D_R-hydrogen from C_{10} of $5D_S$-HPETE. This hydrogen is situated at the opposite side to the hydroperoxy group related to the plane determined by the conjugated diene. Therefore even this reaction follows the antarafacial character (26).

From Table I it is evident that lipoxygenases showing a [+2] rearrangement of the fatty acid radical (reticulocytes, soybean, and blood platelets) stereoselectively remove the L-hydrogen, whereas the [−2] lipoxygenases (arachidonate 5-lipoxygenase from leukocytes, corn germs, and LTA$_4$ synthase) remove the D-hydrogen. This behavior may also reflect a different orientation of the fatty acids at the active site of the two classes of lipoxygenases and support the conclusion drawn before (see Section III.A).

VI.   The Possible Biological Importance of Primary Lipoxygenase Products

The physiological importance of the hydroperoxy fatty acids as primary products of the lipoxygenase reaction is not well understood. From *in vitro* experiments it is known that lipoxygenases as well as cyclooxygenase require a certain hydroperoxide level as activator for the dioxygenase reaction (42,43). However, for the activation only the occurrence of a hydroperoxy group is required, whereas the steric structure does not appear to be of importance. On the other hand, a high cellular level of hydroperoxy fatty acids may inactivate both lipoxygenase and cyclooxygenase (44,45). A steady state of hydroperoxide-generating and hydroperoxide-consuming processes seems to be important for the autoregulation of the lipoxygenase and cyclooxygenase pathways.

Lipoxygenase products, especially 5-, 12-, and 15-H(P)ETE, were proposed to have regulatory functions on the lipoxygenase pathway in different cells (46). The formation of $12L_S$-HETE in blood platelets as well as the formation of $5D_S$-HETE in rabbit peritoneal polymorphonuclear leukocytes are inhibited by $15L_S$-HETE (47,48). On the other hand, the formation of 5-HETE, 5,12-diHETE, and other leukotrienes by mast/basophil cells is stimulated by $15L_S$-HETE

(49,50). The primary product of platelet lipoxygenase, 12$_{L_S}$-HPETE, as well as 15$_{L_S}$-HPETE (51,52) inhibit platelet aggregation via inhibition of thromboxane synthesis. 5$_{D_S}$-HPETE and 5$_{D_S}$-HETE were found to be mediators of the IgE-induced histamine release from human basophils (53). There seems to be a nearly absolute requirement for the spatial structure of these lipoxygenase metabolites, since 11-HPETE, 15-HPETE, and 13-HPOD were 100-fold less active than 5-HPETE.

12$_{L_S}$-HETE was shown to be an attractant for aortic smooth muscle cells (54), whereas 15$_{L_S}$-HETE acts proangiogen (60). Thus, they may be of importance as mediators in vasculary diseases. Recently, it has been reported that 12$_{L_S}$-HETE and 15$_{L_S}$-H(P)ETE inhibit the glucose-induced insulin secretion by isolated pancreatic islets *in vitro*. In contrast, 5$_{D_S}$-HETE and 5epi-HETE strongly induce insulin secretion in a low glucose medium (55). Moreover, epoxyeicosanoids were found to be mediators of the insulin and glucagon release *in vitro* (56). Bronchoconstrictory and chemotactic activities of the HETE's are only weak as compared with those of the more complex lipoxygenase metabolites such as leukotrienes B$_4$, C$_4$ and histamine (57).

VII.   Stereochemistry of Complex Lipoxygenase Products

A.   5$_{D_S}$-TRANS-5,6,-OXIDO-7,9,11,14($E,E,Z,Z,$)-EICOSATETRAENOIC ACID (5,6-LTA$_4$)

Leukocytes from different species as well as mastocytoma cells (58,59) convert exogenous or endogenously produced 5$_{D_S}$-HPETE to LTA$_4$. During the LTA$_4$ synthesis by human leukocytes the D$_R$-hydrogen from C$_{10}$ of the arachidonic acid is stereoselectively removed (26). With 5,8,11,14,17-eicosapentaenoic acid as substrate the leukotriene A$_5$ synthase of mastocytoma cells removes the D$_S$-hydrogen from C$_{10}$ (31). As discussed before, it is one and the same hydrogen atom (Fig. 5).

Recently Shimizu et al. reported that the purified lipoxygenase from potato tubers, an enzyme that is able to remove hydrogen from C$_7$ and C$_{10}$ of arachidonic acid and dihomo-γ-linolenic acid, converts 5$_{D_S}$-HPETE to 5,6,-LTA$_4$ (20). These experiments as well as the results obtained with the pure reticulocyte lipoxygenase (see Section

VII.B) indicate that lipoxygenases are able to form epoxyleuko-trienes if they remove hydrogen from C_{10} of 5- or 15-HPETE.

In contrast to the formation of 5,6,-LTA$_4$ and 14,15-LTA$_4$ by the lipoxygenases from potato tubers and reticulocytes, respectively, in which the oxygenase and leukotriene synthase activities reside in the same protein, the 5,6,-LTA$_4$ synthesis in guinea pig leukocytes appears to require two distinct enzymes; the lipoxygenase forming 5D$_S$-HPETE and the LTA$_4$-synthase. Both activities can be distin-guished by their sensitivity towards inactivation by 5,8,11,14-eico-satetraynoic acid (ETYA). LTA$_4$-synthase is strongly inhibited by this acetylenic fatty acid, whereas the arachidonate-5 lipoxygenase is not affected under comparable conditions (61). However, it should be mentioned that there is a strong species and tissue dependence of the inhibitory effect of ETYA on the 5-HETE formation (107).

The similarities of the reaction mechanism, the antarafacial char-acter, and the inhibition by ETYA indicate that the LTA$_4$-synthase activity is a special type of a lipoxygenase reaction. According to the nomenclature proposed it can be designated as $[C_{10}]$-D$_{\text{pro-}R}$:$[-4]$ hydroperoxy fatty acid: epoxide synthase. The $[C_{n-11}]$-L$_{\text{pro-}S}$: $[+2]$-L$_S$ lipoxygenase from blood platelets (intact cells) is not able to synthesize LTA$_4$ from 5D$_S$-HPETE (32).

B.   15L$_S$-TRANS-14,15-OXIDO-5,8,10,12(Z,Z,E,E)-
EICOSATETRAENOIC ACID (14,15-LTA$_4$)

15L$_s$-HPETE is converted by blood platelets, porcine (32) and human leukocytes (62,63) to an unstable intermediate, which was demonstrated to be 14,15-LTA$_4$ by its hydrolysis products. This transformation is initiated by a stereoselective removal of the L$_R$-hydrogen from C_{10} of the substrate (it is the L$_{\text{pro-}S}$-hydrogen in arachidonic acid) (32). It is assumed that the $[C_{n-11}]$-L$_{\text{pro-}S}$:$[+2(4)]$-L$_S$ lipoxygenases from these cells are responsible for the formation of 14,15-LT A$_4$. As seen from Fig. 7, the stereoselective hydrogen removal takes place antarafacially to the configuration of the hy-droperoxide group.

The pure lipoxygenase from reticulocytes converts 15L$_S$-HPETE to 14, 15-LTA$_4$ probably by its $[C_{n-11}]$-L$_{\text{pro-}S}$: $[+2(4)]$-L$_S$ activity (64). It is proposed (Fig. 8) that the formation of 14,15-LTA$_4$ involves a lipoxygenase reaction (hydrogen removal and $[+4]$ shift of the rad-

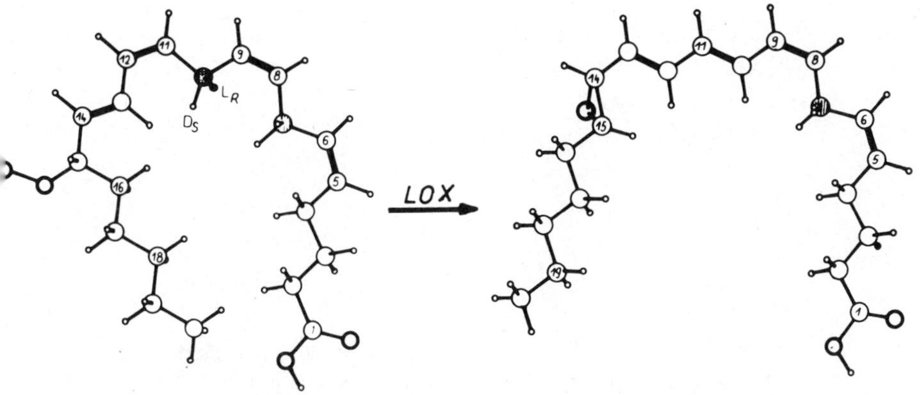

Figure 7. Stereochemistry of 14,15-LTA$_4$ formation from 15L$_S$-HPETE. The L$_R$-hydrogen at C$_{10}$ of 15L$_S$-HPETE that is removed during the 14,15-LTA$_4$ synthesis is drawn in black. It is situated at the opposite side of the plane determined by the double bonds as referred to the hydroperoxy group (antarfacial character).

ical electron) and a hydroperoxidase reaction (formation of the alkoxy radical). The biradical so formed is subsequently stabilized by epoxide formation. Soybean lipoxygenase, which does not possess sizable $[n - 11]:[+2(4)]$ activity with arachidonic acid, does not form 14,15-LTA$_4$ under comparable conditions (64). As expected with 15L$_S$-HETE as substrate, no LTA$_4$ is formed by porcine and human leukocytes, blood platelets, and purified reticulocyte lipoxygenase, since a hydroperoxidase reaction is impossible.

Recently it has been shown that 14,15-LTA$_4$ can be transformed by rat basophilic leukemia cells to a 14L$_R$-S-glutathionyl-15L$_S$-hydroxy-(threo)5,8,10,12(Z,Z,E,E,)-eicosatetraenoic acid (14,15-LTC$_4$) (65). Human leukocytes, however, are not able to do so. The possible biological importance of these metabolites is not clarified yet. Their activity to act as bronchoconstrictors is very low as compared with the 5D$_S$-hydroxy-6-D$_R$-S-glutathionyl-7,9,11,14(E,E,Z,Z,)-eicosatetraenoic acid (LTC$_4$). During the reaction of hemoglobin with 15L$_S$-HPETE (66), 14,15-LTA$_4$ was also described as a minor product.

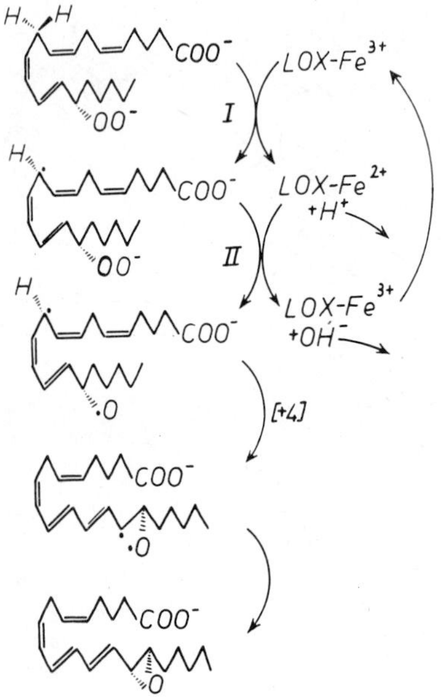

Figure 8. Proposed mechanism of the 14,15-LTA_4 synthesis by the pure reticulocyte lipoxygenase

I Lipoxygenase reaction-Hydrogen removal from C_{10} of 15_{LS}-HPETE and [+ 4] rearrangement of the fatty acid radical.

II   Lipohydroperoxidase   reaction-Homolytic scission of the hydroperoxy group.

C. $5_{DS},12_{DR}$-DIHYDROXY-6,8,10,14(Z,E,E,Z) EICOSATETRAENOIC ACID (LTB_4)

LTB_4 is formed by an enzymatic hydrolysis of LTA_4. After ionic cleavage of the epoxide the complete triene system is rearranged and the second hydroxyl group is introduced at C_{12}. Experiments with $H_2^{18}O$ indicated that the second oxygen introduced came from water which disproves the double dioxygenation origin of the product (67,68). The enzyme responsible for this reaction has been purified and characterized from human leukocytes (69). It is remarkable that despite the rearrangement of the whole conjugated triene system one of the double bonds remains in the cis structure in contrast to the nonenzymatic hydrolysis products of LTA_4 (70,71). The occurrence of the cis double bond between C_6 and C_7 in the conjugated triene system is absolutely necessary for the biological activity of LTB_4 in the lung. The trans–cis–trans, trans–trans–cis, and all-trans isomers are without any effects on the contraction of parenchymal strips and prostaglandin release from the lungs (72). LTB_4 as well as some other leukotrienes are further metabolized via ω-oxidation

(73). ω-oxidized LTB_4 causes contraction of lung parenchymal strips with a dose dependence comparable to LTB_4 (74). On the other hand, ω-hydroxylated LTB_4 is a less potent chemotactic factor than LTB_4 (75).

D.   5-$_{DS}$-HYDROXY,6$_{DR}$-S-GLUTATHIONYL(THREO)7,9,11,14(*E,E,Z,Z,*)-EICOSATETRAENOIC ACID (LTC$_4$)

The leukotrienes of the C group are formed by a glutathione transfer to the LTA_4 (76,77) catalyzed by special glutathione transferases (78). The cleavage of the epoxide is achieved by a nucleophilic attack of the glutathione at the allylic C_6 position. Glutathione is introduced in the D_R-position, whereas the oxygen of the epoxide remains at C_5 in the D_S-position. The structure of the biosynthesized LTC_4 was compared with LTC_4 obtained by chemical total synthesis (33) and found to be the same. From LTC_4 the LTD_4 (removal of glutamate by γ-glutamyltranspeptidase) (79) and LTE_4 (additional removal of glycine by a dipeptidase) are formed. The analogous leukotrienes arising from eicosapentaenoic acid (LTC_5, LTD_5, and LTE_5) and from eicosatrienoic acid (LTC_3, LTD_3, and LTE_3) are synthesized and metabolized in the same manner (80).

The steric configuration of the SRS-A leukotrienes described here is absolutely necessary for their biological function to act as potent bronchoconstrictors. Structural changes of the configuration significantly decrease the activity. Derivatization of LTC_4 and LTD_4 leading to positional and optical isomers of the *S*-peptidyl moiety as well as a change of the peptidyl moiety leading to an altered relationship between the eicosanoid and the peptidyl domains (substitution of D-cysteine for L-cysteine) drastically decrease the biological activity. Changes of the double bond geometry are also accompanied by a loss of biological activity (87).

The formation of 11-trans leukotrienes was observed in several biological systems (82). It is proposed that the trans isomerization of the double bond between C_{11} and C_{12} is caused by a nonenzymatic reversible addition of a thiyl radical (83).

E.   STEREOCHEMISTRY OF Di-HETEs

From animal sources a variety of dihydroxyeicosatetraenoic acids have been isolated that differ from each other with respect to their positional isomerism and stereoisomerism of the hydroxyl groups as well as to the cis–trans isomerism of the double bonds. Generally these compounds can be divided into two groups: (a) double oxygenation products and (b) 5,6,-LTA$_4$ and 14,15-LTA$_4$ hydrolysis products (enzymatic hydrolysis with the formation of LTB$_4$, non-enzymatic hydrolysis).

1. Double Dioxygenation Products

a. 14,15-Dihydroxy-5,8,10,12(Z,Z,E,E,)-eicosatetraenoic Acid

Human blood platelets and porcine leukocytes convert 15$_{LS}$-H(P)ETE to 14,15-dihydroxy-5,8,10,12($Z,Z,E,E,$)-eicosatetraenoic acid probably by their $(C_{n-11}]_{Lpro-S}$:$[+2]$ L$_S$ lipoxygenase activity (32). The reaction is initiated by a stereoselective removal of the C_{10} L$_R$-hydrogen from 15$_{LS}$-HPETE (it is the L$_S$-hydrogen in arachidonic acid) that is followed by a $[+4]$ rearrangement of the radical electron and a stereospecific (L$_R$) dioxygen insertion at C_{14}. Based on ^{18}O experiments and comparison of the relative amounts of erythro- and threo-14,15-diols it was determined that up to 90% of the erythro-14,15-diols originate from double dioxygenation and only 10% from hydrolysis of 14,15-LTA$_4$. After oxidative ozonolysis the absolute configuration of the native erythro-14,15-DiHETE was shown to be 98% 14$_{LR}$,15$_{LS}$ and 2% 14$_{DS}$,15$_{DR}$. These results show the possibility of the formation of 14,15-DiHETE via a lipoxygenase reaction shunting previous synthesis of 14,15-LTA$_4$. The participation of an activated oxygen species such as the superoxide anion (84) seems to be unlikely during the formation of this product. The steric characteristics of this reaction confirm the antarafacial character of initial hydrogen removal and subsequent oxygen insertion assuming a planar structure of the substrate (32).

With human leukocytes, however, Rådmark et al. (84) observed the formation of two isomers of 14,15-DiHETE. Two ^{18}O atoms are introduced in both isomers if the experiments are performed under an ^{18}O$_2$ atmosphere. The occurrence of two isomers argues against their origin as double dioxygenation products formed by a lipoxygenase. A possible participation of activated oxygen species, which are produced in leukocytes in large amounts (85), has been proposed. The superoxide anion, for instance, may attack the 14,15-LTA$_4$ as a nucleophilic to form isomers of 14,15-DiHETE.

The reticulocyte lipoxygenase with its $[C_{n-11}]$-$_{Lpro-S}$:$[+2]$-L$_S$ activity, the steric characteristics of which are comparable with those of blood platelet lipoxygenase, also converts 15L$_S$-HPETE to 14,15-DiHETE to a small extent (64); its origin as a 14,15-LTA$_4$ hydrolysis product or double dioxygenation product remains to be established via analysis of stereoisomers.

b.   8,15-Dihydroxy-5,9,11,13(Z,E,Z,E,)-eicosatetraenoic Acid

Human blood platelets convert 8,11,14-all-*cis*-eicosatrienoic acid to 8L$_R$, 15L$_S$-dihydroxy-9,11,13(E,Z,E,)-eicosatrienoic acid (32) without formation of sizable amounts of 14,15-LTA$_3$ hydrolysis products. With arachidonic acid the corresponding 8L$_R$, 15L$_S$-dihydroxy-5,9,11,13(Z,E,Z,E,)-eicosatetraenoic acid and its 8-epimer (107) was formed in addition to the 14,15-LTA$_4$- derived hydrolysis products.[18]O experiments and the double geometry of the conjugated triene system indicate the double oxygenation reaction and exclude a 14,15 = LTA$_4$ hydrolysis product (108).

The purified soybean lipoxygenase-1 converts 15L$_S$-H(P)ETE to 8D$_S$, 15L$_S$-DiH(P)ETE (35,86). The dioxygenation at C$_8$ must be initiated by hydrogen removal from C$_{10}$ with a subsequent [− 2] rearrangement of the fatty acid radical. These results seem to contradict the specificity of soybean lipoxygenase, since no hydrogen removal from C$_{10}$ as well as no [− 2] rearrangement of the radical has been observed with arachidonic acid as substrate. However, it is by no means certain that the hydroperoxy compounds behave in an identical manner as the fatty acids with respect to the positional specificity and stereospecificity of the initial hydrogen removal and the subsequent radical rearrangement (see Section VII.E.1.d.)

It is of particular interest that the oxygenation of 15L$_S$-H(P)ETE to 8,15-DiH(P)ETE by blood platelets and soybean lipoxygenase show different steric characteristics. The formation of 8L$_R$, 15L$_S$-DiH(P)ETE by the blood platelet lipoxygenase should be initiated by a removal of the L$_R$-hydrogen from C$_{10}$ of 15-H(P)ETE (L$_S$-hydrogen in arachidonic acid). This is the same hydrogen atom that is removed during the oxygenation of arachidonic acid. Therefore it is concluded that 15-H(P)ETE that still contains the (n − 11) double allylic methylene carbon atom shows a similar steric arrangement at the enzyme as arachidonic acid. In contrast, the oxygenation of 15L$_S$-H(P)ETE to 8D$_S$, 15L$_S$-DiH(P)ETE by the soybean lipoxygenase requires a removal of the D$_S$-hydrogen from C$_{10}$ of 15-H(P)ETE (antarafacial character). As shown in Section VII.E.1.d, there are experimental evidences which indicate that the lack of the $(n - 8)$ prochiral center and the presence of the hydro(pero)xy group in 15-H(P)ETE favor an inverse orientation of this substrate at the enzyme as compared with arachidonic acid. 8_{DS}, 15_{LS}-DiHETE also has been isolated as a minor product from mixed human leukocytes (87).

The purified reticulocyte lipoxygenase, the specificity of which in many aspects resembles the soybean enzyme, also converts arachidonic acid and 15_{LS}-HPETE to 8,15-DiHPETE. The complete stereochemistry has not been clarified. Starting from arachidonic acid, two molecules of atmospheric ^{18}O were incorporated, whereas anaerobiosis completely abolished the 8,15-DiHPETE formation.

c. 5,12-Dihydroxy-6,8,10,14($E,Z,E,Z,$)-eicosatetraenoic Acid

5_{DS}-H(P)ETE is converted by human blood platelets to 5_{DS}, 12_{LS}-dihydroxy-6,8,10,14($E,Z,E,Z,$)-eicosatetraenoic acid. Moreover, this double dioxygenation product containing the characteristic alternating cis, trans geometry of the conjugated triene system was isolated from human leukocytes after incubation with arachidonic acid and 12_{LS}-HETE (88,89) as well as from porcine leukocytes after incubation with 5_{DS}-HETE (90).

On the basis of the antarafacial character of the lipoxygenase reaction it is proposed that during the dioxygenation of 5_{DS}-H(P)ETE by human blood platelets the L_S-hydrogen is removed from C_{10} of the substrate. It is likely that the same enzyme that catalyzes the 12_{LS}-HPETE formation (platelet$[C_{n-11}]$-L_{pro-S}:$[+2]$-L_S-lipoxygenase) is responsible for the conversion of 5_{DS}-H(P)ETE to 5,12-DiHETE.

15_{LS}-HPETE is converted by this enzyme to 14,15-LTA$_4$ (see Section VII.B). In contrast 5_{DS}-HPETE is not converted to the corresponding 5,6-LTA$_4$ for two reasons: (a) the platelet lipoxygenase is unable to remove the D_R-hydrogen from the C_{10} of the 5_{DS}-HPETE, that would be necessary for 5,6,-LTA$_4$ synthesis and (b)this enzyme catalyzes only a $[+4]$ rearrangement of the fatty acid radical; for 5,6-LTA$_4$ synthesis the $[-4]$ rearrangement is necessary.

5_{DS},12_{LS}-dihydroxy-6,8,10,14($E,Z,E,Z,$)-eicosatetraenoic acid as a double dioxygenation product must be distinguished from its 8-trans isomer, which arises from hydrolysis of 5,6-LTA$_4$ (see Section VII.E.2).

d. 5,15-Dihydroxy-6,8,11,13(E,Z,Z,E)-eicosatetraenoic Acid

Porcine and rat leukocytes (91,92) convert arachidonic acid to 5,15-DiHETE, the stereochemistry of which has not been established. From ^{18}O experiments it may be concluded that this product is formed from arachidonic acid by successive dioxygenation by $[C_7]:[-2]$-(arachidonate 5-lipoxygenase) and $[C_{n-8}]:[+2]$ (arachidonate 15-lipoxygenase) lipoxygenase.

The purified lipoxygenases from soybeans and reticulocytes are able to dioxygenate arachidonic acid to 5,15-DiHPETE (35,64). For the soybean enzyme the steric structure of the product was determined as 5_{DS}, 15_{LS}-dihydroperoxy-6,8,11,13(E,Z,Z,E,)-eicosatetraenoic acid.

It should be mentioned that high enzyme concentrations are necessary for this reaction; k_{cat} is one order of magnitude lower than for the reaction with arachidonic acid. Moreover, the K_m value for 15_{LS}-HPETE is 50-fold higher than that for arachidonic acid (35).

Recently we found that both the reticulocyte and soybean lipoxygenase convert 15-HETE in the presence of traces of hydroperoxylinoleic acid to 5,15- and 8,15-DiHETE (after triphenylphosphine reduction of the primary products). For the reticulocyte enzyme the maximal rate of this reaction was determined to be about one order of magnitude lower than that for arachidonic acid dioxygenation, whereas the K_m values for 15-HETE and arachidonic acid correspond to each other (93).

As discussed for 8,15-DiHETE, the formation of 5,15-DiHETE seems to contradict the specificity of both soybean and reticulocyte lipoxygenase. However, it must be taken into consideration that the steric structure of 15-H(P)ETE differs from that of arachidonic acid. The lack of the ($n - 8$) prochiral center which is necessary for arachidonic acid dioxygenation and the presence of the hydrophilic hydro(pero)xy group at C_{15} in the region, which had served as signal for the recognition of the initial hydrogen removal during arachidonate dioxygenation, may cause an inverse orientation of the substrate at the enzyme so that the carboxylic group would constitute the signal and hydrogen removal from C_7 would be plausible.

Indeed, with the reticulocyte lipoxygenase we obtained several types of experimental evidence that strongly suggest that 15-HETE shows an inverse orientation at the active site of the enzyme as compared with arachidonic acid.

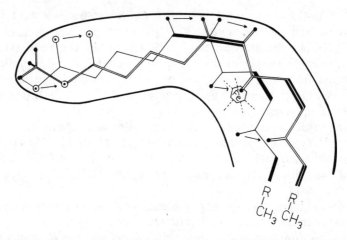

Figure 9. Influence of the topical arrangement of 15-HETE and 15-HETE methylester on the positional specificity of their dioxygenation. ——, 15-HETE; ==, 15-HETE methyl ester; ●, hydrogen atoms; ⊙, oxygen atoms; thick lines indicate the double bonds; Fe is the enzyme bound nonheme iron responsible for the hydrogen removal. The arrows indicate the movement of the molecule after methylation.

(a) derivatives of 15-HETE with increased hydrophobicity of the carboxylic ligand such as the methyl ester or the corresponding alcohol exhibit a much higher oxygenation rate than the free hydroxy acid (93).

(b) methylation of the carboxylic group increases the regioselectivity of the 15-HETE dioxygenation. 15-HETE is converted to 5,15- and 8,15-DiHETE (after borohydride reduction of the primary products) in a ratio of 60:40. With 15-HETE methyl ester this ratio is about 95:5. The dual oxygenase activity with 15-HETE may be due to the similar distances of the prochiral carbon atoms C_7 and C_{10} from the hydrogen abstracting center of the enzyme (nonheme iron). Methylation of the carboxylic group would lengthen this ligand by one C–C bond. Therefore C_7 approaches the iron, whereas C_{10} is more distant (Fig. 9). This topical arrangement must favour both an increased rate of the reaction and an increased regioselectivity. Experiments with other 15-HETE derivatives are in line with the assumption of an inverse orientation (93).

Further experimental evidence for the inverse orientation of the

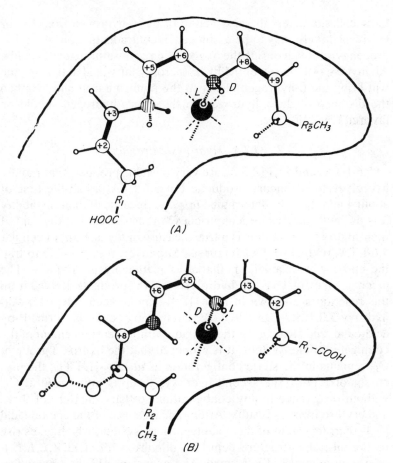

Figure 10.   Inverse spatial arrangement of polyunsaturated fatty acids (A) and hydroperoxy fatty acids (B) at the active site of lipoxygenases. The inverse orientation of both lipoxygenase substrates shown here explains the positional and stereospecificity of the formation of double oxygenation products by lipoxygenases (see text).

hydro(pero)xy fatty acids at the enzyme as compared with the polyunsaturated fatty acids is furnished by the stereochemistry of the double oxygenation products of soybean lipoxygenase. As already discussed, in both the 8- and 5-position of 8,15- and 5,15-DiHETE, the oxygen is introduced in the D_S-position. According to the antarafacial character, the abstraction of the D-hydrogen from C_{10} or C_7, respectively, has to be assumed. In contradistinction, the L-hydrogen is removed during the oxygenation of unsaturated fatty acids. An inverse orientation of the hydro(pero)xy fatty acid at the enzyme will bring the D-hydrogen closer to the nonheme iron and lengthen the distance of the L-hydrogen so that the D-hydrogen should be favored (Fig. 10).

2. LTA₄-Hydrolysis Products

5,6-LTA₄ and 14,15-LTA₄ are very unstable products that rapidly hydrolyze to secondary products. For 5,6-LTA₄ a half-life time of about 4 min (68) was determined *in vitro*. Because of their instability it is difficult to measure leukotrienes A themselves, but the sum of their hydrolysis products is a true measure of the amount of cellular 5,6-LTA₄ or 14,15-LTA₄ synthesis. Moreover, it is possible to trap the epoxide with acidic methanol as methanolysis products. The mechanism of 14,15-LTA₄ hydrolysis and the products derived from this reaction are shown in Fig. 11. As can be seen, 14epi,15L_S-dihydroxy-5,8,10,12($Z,Z,E,E,$)-eicosatetraenoic acid is formed by acid-catalyzed cleavage of the epoxide without rearrangement of the conjugated triene system. It is proposed that the hydroxyl group at C_{15} remains in the steric configuration as in 15L_S-HPETE, the precursor of LTA₄. The second hydroxyl group is introduced at C_{14} without preference to any configuration so that a mixture of threo and erythro isomers results. After ionic cleavage of the epoxide and [+ 1] rearrangement of the complete triene system, which gives rise to the all-trans structure 8epi,15L_S-dihydroxy5,9,11,13($Z,E,E,E,$)-eicosatetraenoic acid is formed. The occurrence of the C_8-epimers and the all-trans structure of the conjugated triene system indicate the origin as 14,15-LTA₄ hydrolysis products and exclude a possible double dioxygenation nature.

5,6-LTA₄ hydrolyzes in a similar manner and the products formed show the same characteristics described for the hydrolysis products of 14,15-LTA₄ with the exception of the positional isomerism of the hydroxy groups (34,68). The products formed are 5D_S,6epi-dihydroxy-7,9,11,14($E,E,Z,Z,$)-eicosatetraenoic acid and 5D_S,12epi-dihydroxy-6,8,10,14($E,E,E,Z,$)-eicosatetraenoic acid (12epi-6-*trans*-LTB₄).

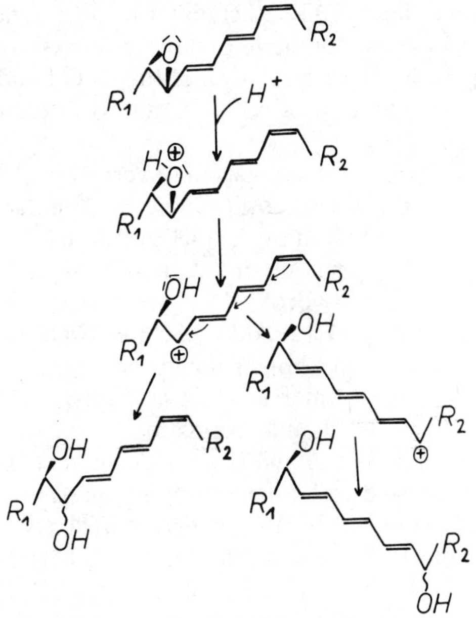

Figure 11.   Nonenzymatic hydrolysis of epoxy-leukotrienes.

F. EPOXYHYDROXY DERIVATIVES AND TRIHYDROXY FATTY ACIDS

It has been reported (see Section III), that lipoxygenases may act as lipohydroperoxidases under certain conditions. During this reaction lipoxygenases catalyze the breakdown of hydroperoxypolyenoic fatty acids forming a variety of secondary products.(1).

The purified lipoxygenase from soybeans converts 13Ls-hydroperoxy-9Z,11E-octadecadienoic acid, 13Ls-HPOD, into threo-11-hydroxy-12,13-oxido-9c-octadecenoic acid (94). The reaction seems to be a special hydroperoxidase-type reaction, since both oxygen atoms present in the product were derived from the hydroperoxy group of the substrate. Similarly, during the hemin-catalyzed conversion of $^{18}O_2$-hydroperoxy fatty acids into epoxyhydroxy derivatives both ^{18}O atoms are retained in the reaction products (95). It is proposed that both the hemin- and lipoxygenase-catalyzed reaction are initiated by a homolytic cleavage of the hydroperoxy group forming an alkoxy radical that gives rise to the formation of an epoxy-monoene radical. This radical intermediate is rapidly recombined with the enzyme- or hemin-bound hydroxyl radical that is formed during the homolytic scission of the O–O bond. This mechanism implies the retention of both ^{18}O atoms in the epoxhydroxy derivatives.

Purified reticulocyte lipoxygenase converts 15$_{LS}$-HPETE to 13-hydroxy-14,15-oxido-5,8,11-eicosatrienoic acid under aerobic and anaerobic conditions in addition to 14,15-LTA$_4$(64).

The high-speed supernatant of rat lung homogenates converts 12$_{LS}$-HPETE to 8epi-hydroxy-11,12-oxido-5,9,14-eicosatrienoic acid and 10epi-hydroxy-11,12-oxido-5,8,14-eicosatrienoic acid (96). Although the precise double bond geometry has not been established it is proposed that the former product contains a (Z,E,Z)-system, whereas in the latter all double bonds are in the cis form. These products called hepoxilins A and B were also detected in pancreatic islets and are proposed to be mediators of glucose induced insulin secretion (106). Hydrolysis of the epoxy-hydroxy-eicosatrienoic acids leads to the formation of 10,11,12- and 8,11,12-trihydroxy-eicosatrienoic acids.

Epoxyhydroxy and trihydroxy derivatives are also found after the reaction of hemoglobin with hydroperoxylinoleic acid (97), 12-HPETE (98), 15-HPETE (99) and as arachidonic acid metabolites from rat blood platelets (100).

G. TRIHYDROXYEICOSANOIDS CONTAINING A CONJUGATED TETRAENE SYSTEM (LIPOXINS)

Human polymorphonuclear leukocytes (101,102) convert 15$_{LS}$-HPETE to 5,6,15-trihydroxy-7,9,11,13-eicosatetraenoic acid (lipoxin A) and 5,14,15-trihydroxy-6,8,10,12-eicosatetraenoic acid (lipoxin B). The chemical total synthesis of lipoxin isomers has been described recently (103).

Porcine leukocytes convert 15-hydroperoxy-5,8,11,13,17-eicosapentaenoic acid to 5,6,15-trihydroxy-7,9,11,13,17-eicosapentaenoic acid, called lipoxene A, and 5,14,15-trihydroxy-6,8,10,12,17-eicosapentaenoic acid, called lipoxene B. As shown by SP-HPLC analysis comparable amounts of both trihydroxyeicosanoid isomers are formed (104).

Recently, it has been shown in our laboratory that the purified reticulocyte lipoxygenase converts 5,15-DiHETE during a lipoxygenase reaction to a product possessing a UV spectrum very similar to that shown for 5,6,15-trihydroxy-7,9,11,13-eicosatetraenoic acid (105). It is proposed that this reaction is initiated by hydrogen removal from C_{10} of the 5,15-DiHETE and a subsequent [+4] or [−4] rearrangement of the radical electron forming hydroperoxylipoxins. From data obtained by reversed phase high pressure liquid chromatography and gas chromatography/mass spectroscopy it may be concluded that 6-hydroperoxy,5,15-hydroxy,7,9,11,13-eicosatetraenoic acid is the main product of this reaction. Enzymatic reduction of this oxygenation product by gluthathione peroxidase or glutathione-S-transferase leads to the formation of lipoxin A.

As proposed in (105), lipoxenes are synthezised from 15-hydroperoxyeicosapentaenoic acid via a hydrolysis of either 5,6-oxido, 15-hydroxyeicosapentaenoic acid or 5-hydroxy,14,15-oxidoeicosapentaenoic acid by porcine leukocytes. The formation of lipoxins during the oxygenation of 5,15-DiHETE by the pure reticulocyte lipoxygenase indicate, that a sequential triple dioxygenation of arachidonic acid may lead to the formation of lipoxins. However, both types of mechanism show great similarities since the sequential action of three hydrogen removing enzymes is necessary for the synthesis starting from the polyenoic fatty acid.

VIII.   GENERAL CONCLUSIONS

1. Lipoxygenases, which dioxygenate poly-cis unsaturated fatty acids to the corresponding hydroperoxy fatty acids, catalyze a variety of other reactions under certain conditions: (a) dioxygenation of hydro(pero)xy fatty acids forming double dioxygenation products, (b) formation of triple dioxygenation products via sequential triple dioxygenation of arachidonic acid, sequential double dioxygenation of 15-HETE, and dioxygenation 5,15-DiHETE, (c) breakdown of hydroperoxy fatty acids, (d) formation of epoxy leukotrienes. The first step of PGG_2 synthesis is also a lipoxygenase reaction.

2. According to their stereospecificity lipoxygenases generally can be divided into two classes: (a) D-lipoxygenases (potato tubers, corn germs, wheat, and 5-lipoxygenase from leukocytes), which initially remove the D-hydrogen from the prochiral center. These lipoxygenases introduce dioxygen in the D-position (antarafacial character). (b) L-lipoxygenases (soybeans, platelets, and reticulo-

cytes); hydrogen removal and oxygen insertion in the L-position. D-lipoxygenases shown a [−2] rearrangement of the fatty acid radical, whereas L-lipoxygenases show a [+2] rearrangement. The cyclooxygenase is the only exception from this rule.

3. For the L-lipoxygenases ([+2]) the place of the hydrogen removal is determined by the distance of the prochiral center from the methyl end of the fatty acids. For the D-lipoxygenases ([−2]) preliminary results suggest that the place of hydrogen removal relates to the distance from the carboxylic group. For both types of lipoxygenases the radical rearrangement takes place in the direction of the fatty acid terminus, which serves as a signal for the hydrogen removal.

4. The specificity of lipoxygenases may differ for the reaction with polyunsaturated fatty acids and hydro(pero)xy unsaturated fatty acids. This behavior may be due to a different orientation of the substrate at the active site of the enzyme, changing the positional specificity and stereospecificity of the hydrogen removal and oxygen insertion as well as the radical rearrangement.

5. A comprehensive nomenclature for lipoxygenases and related enzymes reflecting the stereospecificity and positional specificity of the initial hydrogen removal and the subsequent dioxygen insertion is proposed.

6. For the characterization of the stereochemistry of the lipoxygenase reaction the Fischer convention should be preferred. Specification of the chiral and prochiral centers according to the S/R system for the reaction with arachidonic acid as model substrate should be added as subscripts.

7. The antarafacial character seems to be a general feature not only for the dioxygenation of polyunsaturated fatty acids but also for the formation of more complex lipoxygenase products such as epoxy leukotrienes or double dioxygenation products.

8. The biological activity of lipoxygenase products is closely related to their steric structure. Changes of the positional and optical isomerism of the functional groups as well as changes of the geometric properties of the double bonds strongly influence the physiological activity.

ABBREVIATIONS

15_{L_S}-H(P)ETE 15_{L_S}-hydro(pero)xy-5,8,11,13(Z,Z,Z,E)-eicosatetraenoic acid;

12L$_S$-H(P)ETE 12L$_S$-hydro(pero)xy-5,8,10,14(Z,Z,E,Z)-eicosa-tetraenoic acid;

5D$_S$-H(P)ETE 5D$_S$-hydro(pero)xy-6,8,11,14(E,Z,Z,Z)-eicosa-tetraenoic acid;

LT leukotriene;
PG prostaglandin;
13L$_S$-HPOD 13L$_S$-hydroperoxy-9Z,11E-octadecadienoic acid;
LOX lipoxygenase.

References

1. Veldink, G.A., Vliegenthart, J.F.G., and Boldingh, J., *Prog. Chem. Fats Other Lipids*, **15**, 131–166 (1977).

2. Schewe, T., Rapoport, S.M., and Kühn, H., this volume.

3. Ho, P.P.K., Walters, P.C., and Sullivan, H.R., *Biochem. Biophys. Res. Commun.*, **76**, 398–405 (1977).

4. Borgeat, P., Hamberg, M., and Samuelsson, B., *J. Biol. Chem.*, **251**, 7816–7820 (1976).

5. Borgeat, P. and Samuelsson, B. *Proc. Natl. Acad. Sci. USA*, **76**, 2148–2152 (1979).

6. Goetzl, E.J. *Biochem. Biophys. Res. Commun.*, **101**, 344–350 (1981).

7. Rapoport, S.M., Schewe, T., Wiesner, R., Halangk, W., Ludwig, P., Höhne, M., Tannert, C., Hiebsch, C., and Klatt, D., *Eur. J. Biochem.*, **96**, 545–561 (1979).

8. Kühn, H., Pönicke, K., Halle, W., Wiesner, R., Schewe, T., and Förster, W., *Prostaglandins, Leukotrienes and Medicine*, **17**, 291–303 (1985).

9. Gunt, P. and Shen, T.Y., *J. Med. Chem.*, **20**, 1146–1152 (1977).

10. Salvetti, F., Buttinoni, A., and Ceserani, R., *Eur. J. Med. Chem. Chim. Ther.*, **16**, 81–90 (1981).

11. Schewe, T., Halangk, W., Hiebsch, C., and Rapoport, S.M., *FEBS Lett.*, **60**, 149–152 (1974).

12. Kühn, H., Pliquett, F., Wunderlich, S., Schewe, T., and Krause, W. *Biochim. Biophys. Acta*, **735**, 283–290 (1983).

13. de Groot, J., Veldink, G.A., Vliegenthart, J.F.G., Boldingh, J., Wever, R., and van Gelder, B.F., *Biochim. Biophys. Acta*, **377**, 71–79 (1975).

14. Shaanan, B., *Nature*, **296**, 683–684 (1982).

15. Nugteren, D.H. *Biochim. Biophys. Acta*, **380**, 299–307 (1975).

16. Hamberg, M. and Samuelsson, B., *J. Biol. Chem.*, **424**, 5329–5335 (1967).

17. Bryant, R.W., Bailey, J.M., Schewe, T., and Rapoport, S.M., *J. Biol. Chem.*, **257**, 6050–6055 (1982).

18. Graveland, A., *Lipids*, **8**, 606–611 (1973).

19. Kühn, H., Heydeck, D., Wiesner, R., and Schewe T., *Biochem. Biophys. Acta*, **830**, 25–29 (1985).

20. Shimizu, T., Rådmark, O., and Samuelsson, B. *Proc. Natl. Acad. Sci. USA*, **81**, 689–693 (1984).

21. Hamberg, M., *Biochem. Biophys. Res. Commun.*, **117**, 593–600 (1983).

22. Kühn, H., Wiesner, R., Schewe, T., and Rapoport, S.M., *FEBS Lett.*, **153**, 353–356 (1983).

23. Cahn, R.S., Ingold, C.K., and Prelog, V., *Experientia*, **12**, 81–94 (1956).

24. Egmond, M.R., Vliegenthart, J.F.G., and Boldingh, J., *Biochem. Biophys. Res. Commun.*, **48**, 1055–1060 (1972).

25. Hamberg, M. and Hamberg, G., *Biochem. Biophys. Res. Commun.*, **95**, 1090–1097 (1980).

26. Maas, R.L., Ingram, C.D., Taber, D.F., Oates, J.A., and Brash, A.R., *J. Biol. Chem.*, **257**, 13515–13519 (1982).

27. Hamberg, M. and Samuelsson, B., *J. Biol. Chem.*, **424**, 5336–5343 (1967).

28. van Os, C.P.A., Rijke-Schilder, G.P.M., and Vliegenthart, J.F.G., *Biochim. Biophys. Acta*, **575**, 479–484 (1979).

29. Spaapen, L.J.M., Verhagen, J., Veldink, G.A., and Vliegenthart, J.F.G., *Biochim. Biophys. Acta*, **618**, 153–162 (1980).

30. IUPAC Commission on the nomenclature of Organic Chemistry *Pure Appl. Chem.*, **45**, 13–30 (1976).

31. Hammarström, S., *J. Biol. Chem.*, **258**, 1427–1430 (1983).

32. Maas, R.L. and Brash, A.L., *Proc. Natl. Acad. Sci. USA*, **80**, 2884–2888 (1983).

33. Corey, E.J., Clark, D.A., Goto, G., Marfat, A., Mioskowski, C., Samuelsson, B., and Hammarström, S., *J. Am. Chem. Soc.*, **102**, 1436–1439 (1980).

34. Borgeat, P. and Samuelsson, B., *J. Biol. Chem.*, **254**, 7865–7869 (1979).

35. van Os, C.P.A., Rijke-Schilder, G.P.M., van Halleek, H., Verhagen, J., and Vliegenthart, J.F.G., *Biochim. Biophys. Acta*, **663**, 177–193 (1981).

36. *Enzyme Nomenclature*, Recommendations of the Nomenclature Committee of IUB, Academic, New York, (1979), pp. 770–777.

37. Hamberg, M., *Anal. Biochem.*, **43**, 515–526 (1971).

38. Hamberg, M. and Samuelsson, B., *J. Biol. Chem.*, **242**, 5344–5354 (1967).

39. Nugteren, D.H., Beerthuis, P.K., and van Dorp, D.A., *Prostaglandins*, S. Bergström and B. Samuelsson, Eds., Proceedings of the 2nd Nobel Symposium Stockholm, Almquist & Wiksell, Stockholm 1966.

40. Hemler, M.E., Crawford, C.G., and Lands, W.E.M., *Biochemistry*, **17**, 1777–1779 (1978).

41. Samuelsson, B., *J. Am. Chem. Soc.*, **87**, 3011–3012 (1965).

42. Hemler, M.E., Graff, G., and Lands, W.E.M., *Biochem. Biophys. Res. Commun.*, **85**, 1325–1331 (1978).

43. Smith, W.L. and Lands, W.E.M., *J. Biol. Chem.*, **247**, 1038–1047 (1972).

44. Egan, R.W., Paseton, J., and Kuehl, F.A., *J. Biol. Chem.*, **251**, 7329–7335 (1976).

45. Härtel, B., Ludwig, P., Schewe, T., and Rapoport, S.M., *Eur. J. Biochem.*, **126**, 353–357 (1982).

46. Vanderhoek, J.Y., Bryant, R.W., and Bailey, J.M., *Biochem. Pharmacol.*, **31**, 3463–3467 (1982).

47. Vanderhoek, J.Y., Bryant, R.W., and Bailey, J.M., *J. Biol. Chem.*, **255**, 5996–5998 (1980).

48. Vanderhoek, J.Y., Bryant, R.W., and Bailey, J.M., *J. Biol. Chem.*, **255**, 10064–10066 (1980).

49. Maclouf, J., Fruteau de Laclos, B., and Borgeat, P., *Proc. Natl. Acad. Sci. USA*, **79**, 6042–6046 (1982).

50. Vanderhoek, J.Y., Tase, N.S., Bailey, J.M., Goldstein, A.L., and Pluznik, D.H., *J. Biol. Chem.*, **257**, 12191–12195 (1982).

51. Siegel, M.I., McConnel, R.T., Abrahams, S.L., Porter, N.A., and Cuatrecasas, P., *Biochem. Biophys. Res. Commun.*, **89**, 1273–1280 (1979).

52. Vericel, E. and Lagarde, M. *Lipids*, **15**, 472–474 (1980).

53. Peters, S.P., Schleimer, R.P., Marone, G., Kagey-Sobotka, A., Siegel, M.I., and Lichtenstein, L.M., in *Leukotrienes and Other Lipoxygenase Products*, B. Samuelsson and R. Paoletti, Eds., Raven, New York, 1982, pp. 315–324.

54. Nakao, J., Ooyama, T., Chang, W.C., Murota, S., and Orimo, H., *Atherosclerosis*, **43**, 143–148 (1982).

55. Yamamoto, S., Ishii, M., Nakadate, T., Nakaki, T., and Kato, R., *J. Biol. Chem.*, **258**, 12149–12152 (1983).

56. Falk, J.R., Manna, S., Moltz, J., Chacos, N. and Capdevila, J., *Biochem. Biophys. Res. Commun.*, **114**, 743–749 (1983).

57. Copas, J.L., Borgeat, P., and Gardiner, P.L., *Prostaglandins Leukotrienes and Medicine*, **8**, 105–114 (1982).

58. Borgeat, P. and Samuelsson, B., *Proc. Natl. Acad. Sci. USA*, **76**, 3213–3217 (1979).

59. Rådmark, O., Malmsten, C., Samuelsson, B., Goto, G., Marfat, A., and Corey, E.J., *J. Biol. Chem.*, **255**, 11828–11831 (1980).

60. Yamaja, B.N., Ganley, C., and Stuart, M.J., Pediatrics **75**, 911–915 (1985).

61. Bokoch, G.M. and Reed, W. *J. Biol. Chem.*, **256**, 4156–4159 (1981).

62. Lundberg, U., Rådmark, O., Malmsten, C., and Samuelsson, B., *FEBS Lett.*, **126**, 127–132 (1981).

63. Jubiz, W., Rådmark, O., Lindgren, J.A., Mälmsten, C., and Samuelsson, B., *Biochem. Biophys. Res. Commun.*, **99**, 976–986 (1981).

64. Bryant, R.W., Schewe, T., Rapoport, S.M. and Bailey, J.M., *J. Biol. Chem.*, **260**, 3548–3555 (1985).

65. Sok, D.-E., Han, C.-O., Shiek, W.-R., Zouk, B.-N., and Sik, C.J., *Biochem. Biophys. Res. Commun.*, **104**, 1363–1370 (1982).

66. Sok, D.-E., Chung, T., and Sik, C.J. *Biochem. Biophys. Res. Commun.*, **110**, 273–279 (1983).

67. Borgeat, P. and Samuelsson, B., *J. Biol. Chem.*, **254**, 2643–2646 (1979).

68. Samuelsson, B., in *Leukotrienes and Other Lipoxygenase Products*, B. Samuelsson and R. Paoletti, Eds., Raven, New York, 1982, pp. 1–17.

69. Rådmark, O., Shimizu, T., Jörnwall, H., and Samuelsson, B., *J. Biol. Chem.*, **259**, 12339–12345 (1984).

70. Corey, E.J., Marfat, A., Goto, G., and Brion, F., *J. Am. Chem. Soc.*, **102**, 7984–7985 (1980).

71. Corey, E.J., Hopkins, P.B., Barton, A.E., Bagenter, B., and Borgeat, P., *Tetrahedron*, **38**, 2653–2657 (1982).

72. Sirois, P., Roy, S., Borgeat, P., Picard, S., and Corey, E.J., *Biochem. Biophys. Res. Commun.*, **99**, 385–390 (1981)

73. Lindgren, A., Hansson, G., Claesson, H.-E., and Samuelsson, B., in *Leukotrienes and Other Lipoxygenase Products*, B. Samuelsson and R. Paoletti, Eds., Raven, New York, 1982, pp. 53–60.

74. Hansson, G., Lindgren, J.A., Dahlen, S.E., Hedquist, P., and Samuelsson, B., *FEBS Lett.*, **130**, 107–112 (1981).

75. Palmblad, J.W., Lindgren, A.M., Rådmark, J.O., Hansson, G., and Malmsten, C.L., *FEBS Lett.*, **144**, 81–84 (1982).

76. Hammarström, S. and Samuelsson, B., *FEBS Lett.*, **122**, 83–86 (1980).

77. Hammarström, S., Samuelsson, B., Clark, D.A., Goto, G., Marfat, A., Mioskowski, C., and Corey, E.J., *Biochem. Biophys. Res. Commun.*, **92**, 946–953 (1980).

78. Back, M.K., Brashler, J.R., Morton, D.R., Steel, L.K., Kaliner, M.A., and Hugli, T.E., in *Leukotrienes and Other Lipoxygenase Products*, B. Samuelsson and R. Paoletti, Eds., Raven, New York, 1982, pp. 103–114.

79. Hammarström, S., *J. Biol. Chem.*, **256**, 9573–9578 (1981).

80. Örning, L., Bernström, K., and Hammarström, S., *Eur. J. Biochem.*, **120**, 41–45 (1981).

81. Lewis, R.A., Austen, K.F., Drazen, J.M., Soter, N.A., Figueirodo, J.C., and Corey, E.J., in *Leukotrienes and Other Lipoxygenase Products*, B. Samuelsson and R. Paoletti, Eds., Raven, New York, 1982, pp.137–151.

82. Morris, H.R., Taylor, G.W., Piper, P.J., Samhoun, M.N., and Tippins, J.R., *Prostaglandins*, **19**, 185–201 (1980).

83. Attrache, V., Sok, D.-E., Pai, J.-K., and Sik, C.J., *Proc. Natl. Acad. Sci. USA*, **78**, 1523–1526 (1981).

84. Rådmark, O., Lundberg, U., Jubiz, W., Malmsten, C., and Samuelsson, B., in *Leukotrienes and Other Lipoxygenase Products*, B. Samuelsson and R. Paoletti, Eds., Raven, New York, 1982, pp. 61–70.

85. Badwey, J.A. and Karnowsky, M.L., *Ann. Rev. Biochem.*, **49**, 695–726 (1980).

86. Bild, G.S., Ramadoss, C.S., Linn, S., and Axelrod, B., *Biochem. Biophys. Res. Commun.*, **74**, 949–954 (1977).

87. Shak, S., Perez, D.H., and Goldstein, I.M. *J. Biol. Chem.*, **258**, 14948–14953 (1982).

88. Borgeat, P., Fruteau de Laclos, B., Picard, S., Drapeau, J., Vallerand, P., and Corey, E.J., *Prostaglandins*, **23**, 713–724 (1982).

89. Lindgren, J.A., Hansson, G., and Samuelsson, B., *FEBS Lett.*, **128**, 329–335 (1981).

90. Yoshimoto, T., Miyamoto, Y., Ochi, K., and Yamamoto, S., *Biochim. Biophys. Acta*, **713**, 638–646 (1982).

91. Borgeat, P., Fruteau de Laclos, B., Picard, S., Vallerand, P., and Sirois, P., in *Leukotrienes and Other Lipoxygenase Products*, B. Samuelsson and R. Paoletti, Eds., Raven, New York, 1982, pp. 45–51.

92. Maas, R.L., Brash, A.R., and Oates, J.A., in *Leukotrienes and Other Lipoxygenase Products*, B. Samuelsson and R. Paoletti, Eds., Raven, New York, 1982, pp. 29–44.

93. Kühn, H., Wiesner, R., Schewe, T., Lankin, V., Alder, L., and Nekrasov, S., submitted for publication (1985).

94. Garssen, G.J., Veldink, G.A., Vliegenthart, J.F.G., and Boldingh, J., *Eur. J. Biochem.*, **62**, 33–36 (1976).

95. Dix, A.T. and Lawrence, L.J., *J. Am. Chem. Soc.*, **105**, 7001–7002 (1983).

96. Pace-Asciak, C.R., Granström, E., and Samuelsson, B., *J. Biol. Chem.*, **258**, 6835–6840 (1982).

97. Hamberg, M., *Lipids*, **10**, 87–92 (1975).

98. Pace-Asciak, C.R., *Biochim. Biophys. Acta*, **793**, 485–488, 1984.

99. Bryant, R.W. and Bailey, J.M., *Prog. Lipid Res.*, **20**, 279–281 (1981).

100. Bryant, R.W. and Bailey, J.M., *Prostaglandins*, **17**, 9–18 (1979).

101. Serhan, C.N., Hamberg, M., and Samuelsson, B., *Proc. Natl. Acad. Sci. USA*, **81**, 5335–5339 (1984).

102. Serhan, C.N., Hamberg, M., and Samuelsson, B., *Biochem. Biophys. Res. Commun.*, **118**, 943–949 (1984).

103. Adams, J., Fitzsimmons, B.J., Girard, Y., Leblanc, Y., Evans, J.F., and Rokach, J., *J. Am. Chem. Soc.* **107**, 464–469 (1985).

104. Wong, P.Y.K., Hughes, R., and Lam, B., *Biochem. Biophys. Res. Commun.*, **126**, 763–772 (1985).

105. Kühn, H., Wiesner, R., and Stender, H., *FEBS Lett.*, **177**, 255–259 (1984).

106. Pace-Asciak, C.R. and Martin, J.M., Prostaglandins, Leukotrienes and Medicine **16**, 173–180 (1984).

107. Hansson, G., Malmsten, C. and Rådmark, O., in *Prostaglandins and Related Substances* (Pace Asciak, C. and Granström, E. eds.) Elsevier, Amsterdam, pp. 127–167, 1983.

108. Brash, A.R., Maas, R.L., Oates, J.A., *J. Allergy and Clinical Immunology* **74**, 316–323 (1985).

AUTHOR INDEX

Numbers in parentheses are reference numbers and indicate that the author's work is referred to although his name is not mentioned in the text. Numbers in *italics* indicate pages on which the complete reference appears.

SUBJECT INDEX

335

CUMULATIVE INDEX, VOLUMES 1–58

A. Author Index

CUMULATIVE INDEX, VOLUMES 1–58

B. Subject Index

VOL.   PAGE